T0178342

Animal Movement

STATISTICAL MODELS FOR TELEMETRY DATA

Animal Movement

STATISTICAL MODELS FOR TELEMETRY DATA

Mevin B. Hooten
U.S. Geological Survey
Colorado Cooperative Fish and Wildlife Research Unit
Department of Fish, Wildlife, and Conservation Biology
Department of Statistics
Colorado State University

Devin S. Johnson
National Oceanic and Atmospheric Administration
National Marine Fisheries Service

Brett T. McClintock
National Oceanic and Atmospheric Administration
National Marine Fisheries Service

Juan M. Morales
Grupo de Ecología Cuantitativa
INIBIOMA-CRUB, CONICET
Universidad Nacional del Comahue

CRC Press
Taylor & Francis Group
Boca Raton London New York

CRC Press is an imprint of the
Taylor & Francis Group, an **informa** business

Cover photo credits, left to right: Elk (*Cervus canadensis*; Célie Intering), spotted seal (*Phoca largha*; Dave Withrow), and mountain lion (*Puma concolor*; Jacob Ivan, Colorado Parks and Wildlife).

CRC Press
Taylor & Francis Group
6000 Broken Sound Parkway NW, Suite 300
Boca Raton, FL 33487-2742

First issued in paperback 2021

© 2017 by Taylor & Francis Group, LLC
CRC Press is an imprint of Taylor & Francis Group, an Informa business

No claim to original U.S. Government works

Version Date: 20160908

ISBN 13: 978-1-03-209718-3 (pbk)
ISBN 13: 978-1-4665-8214-9 (hbk)

This book contains information obtained from authentic and highly regarded sources. Reasonable efforts have been made to publish reliable data and information, but the author and publisher cannot assume responsibility for the validity of all materials or the consequences of their use. The authors and publishers have attempted to trace the copyright holders of all material reproduced in this publication and apologize to copyright holders if permission to publish in this form has not been obtained. If any copyright material has not been acknowledged please write and let us know so we may rectify in any future reprint.

Except as permitted under U.S. Copyright Law, no part of this book may be reprinted, reproduced, transmitted, or utilized in any form by any electronic, mechanical, or other means, now known or hereafter invented, including photocopying, microfilming, and recording, or in any information storage or retrieval system, without written permission from the publishers.

For permission to photocopy or use material electronically from this work, please access www.copyright.com (http://www.copyright.com/) or contact the Copyright Clearance Center, Inc. (CCC), 222 Rosewood Drive, Danvers, MA 01923, 978-750-8400. CCC is a not-for-profit organization that provides licenses and registration for a variety of users. For organizations that have been granted a photocopy license by the CCC, a separate system of payment has been arranged.

Trademark Notice: Product or corporate names may be trademarks or registered trademarks, and are used only for identification and explanation without intent to infringe.

Publisher's Note
The publisher has gone to great lengths to ensure the quality of this reprint but points out that some imperfections in the original copies may be apparent.

Library of Congress Cataloging-in-Publication Data

Names: Hooten, Mevin B., 1976-
Title: Animal movement : statistical models for telemetry data / Mevin B. Hooten [and three others].
Description: Boca Raton : CRC Press, 2017. | Includes bibliographical references and indexes.
Identifiers: LCCN 2016034976 | ISBN 9781466582149 (hardback : alk. paper)
Subjects: LCSH: Animal behavior--Mathematical models. | Home range (Animal geography)--Mathematical models. | Biotelemetry.
Classification: LCC QL751.65.M3 A55 2017 | DDC 591.501/5118--dc23
LC record available at https://lccn.loc.gov/2016034976

Visit the Taylor & Francis Web site at
http://www.taylorandfrancis.com

and the CRC Press Web site at
http://www.crcpress.com

Contents

Preface

With the field of animal movement modeling evolving so rapidly, navigating the expanding literature is challenging. It may be impossible to provide an exhaustive summary of animal movement concepts, biological underpinnings, and behavioral theory; thus, we view this book as a starting place to learn about the fundamental suite of statistical modeling tools available for providing inference concerning individual-based animal movement.

Notice that the title is focused on "*statistical models* for telemetry data." The set of existing literature related to animal movement is massive, with thousands of individual papers related to the general topic. All of this information cannot be synthesized in a single volume; thus, we focus on the subset of literature mainly concerned with parametric statistical modeling (i.e., statistical approaches for inverse modeling based on data and known probability distributions, mainly using likelihood and Bayesian methods). There are many other approaches for simulating animal movement and visualizing telemetry data; we leave most of those for another volume.

Our intention is that this book reads more like a reference than a cookbook. It provides insight about the statistical aspects of animal movement modeling. We expect two types of readers: (1) a portion of readers will use this book as a companion reference for obtaining the background necessary to read scientific papers about animal movement, and (2) the other portion of readers will use the book as a foundation for creating and implementing their own statistical animal movement models.

We designed this book such that it opens with an overview of animal movement data and a summary of the progression of the field over the years. Then we provide a series of chapters as a review of important statistical concepts that are relevant for the more advanced animal movement models that follow. Chapter 4 covers point process models for learning about animal movement; many of these rely on uncorrelated telemetry data, but Section 4.7 addresses spatio-temporal point processes. Chapters 5 through 6 are concerned with dynamic animal movement models of both the discrete- and continuous-time flavors. Finally, Chapter 7 describes approaches to use models in sequence, properly accommodating the uncertainty from first-stage models in second-stage inference.

We devote a great deal of space to spatial and temporal statistics in general because this is an area that many animal ecologists have received no formal training in. These subjects are critical for animal movement modeling and we recommend at least a light reading of Chapters 2 and 3 for everyone. However, we recognize that readers already familiar with the basics of telemetry data, as well as spatial and temporal statistics, may be tempted to skip ahead to Chapter 4, only referring back to Chapters 2 and 3 for reference.

Finally, despite the rapid evolution of animal movement modeling approaches, no single method has risen to the top as a gold standard. This lack of a universally accepted framework for analyzing all types of telemetry data is somewhat unique in the field of quantitative animal ecology and can be daunting for new researchers just wanting to do the right thing. On the other hand, it is an exciting time in animal ecology because we can ask and answer new questions that are fundamental to the biology, ecology, and conservation of wildlife. Each new statistical approach for analyzing telemetry data brings potential for new inference into the scientific understanding of critical processes inherent to living systems.

Acknowledgments

The authors acknowledge the following funding sources: NSF DMS 1614392, CPW T01304, NOAA AKC188000, PICT 2011-0790, and PIP 112-201101-58. The authors are grateful to (in alphabetical order) Mat Alldredge, Chuck Anderson, David Anderson, Ali Arab, Randy Boone, Mike Bower, Randy Brehm, Brian Brost, Franny Buderman, Paul Conn, Noel Cressie, Kevin Crooks, Marìa del Mar Delgado, Bob Dorazio, Tom Edwards, Gabriele Engler, John Fieberg, James Forester, Daniel Fortin, Marti Garlick, Brian Gerber, Eli Gurarie, Ephraim Hanks, Dan Haydon, Trevor Hefley, Tom Hobbs, Jennifer Hoeting, Gina Hooten, Jake Ivan, Shea Johnson, Gwen Johnson, Layla Johnson, Matt Kaufman, Bill Kendall, Carey Kuhn, Josh London, John Lowry, Jason Matthiopoulos, Joc Margraf, Leslie McFarlane, Josh Millspaugh, Ryan Neilson, Joe Northrup, Otso Ovaskainen, Jim Powell, Andy Royle, Henry Scharf, Tanya Shenk, John Shivik, Bob Small, Jeremy Sterling, David Theobald, Len Thomas, Jay Ver Hoef, Lance Waller, David Warton, Gary White, Chris Wikle, Perry Williams, Ken Wilson, Ryan Wilson, Dana Winkelman, George Wittemyer, Jamie Womble, Jun Zhu, and Jim Zidek for various engaging discussions about animal movement, assistance, collaboration, and support during this project. The findings and conclusions in this book by the NOAA authors do not necessarily represent the views of the National Marine Fisheries Service, NOAA. Any use of trade, firm, or product names is for descriptive purposes only and does not imply endorsement by the U.S. Government.

Authors

Mevin B. Hooten is an associate professor in the Departments of Fish, Wildlife, and Conservation Biology, and Statistics at Colorado State University. He is also assistant unit leader in the U.S. Geological Survey, Colorado Cooperative Fish and Wildlife Research Unit. Dr. Hooten earned a PhD in statistics at the University of Missouri. His research focuses on the development of statistical methodology for spatial and spatio-temporal ecological processes.

Devin S. Johnson is a statistician at the National Oceanic and Atmospheric Administration, National Marine Fisheries Service. Dr. Johnson earned a PhD in statistics at Colorado State University. His research focuses on the development and application of statistical models for ecological data, with special focus on marine mammals. He is also the creator and maintainer of the "crawl" R package.

Brett T. McClintock is a statistician at the National Oceanic and Atmospheric Administration, National Marine Fisheries Service. Dr. McClintock earned a PhD in wildlife biology and MS in statistics at Colorado State University. His research focuses on the development and application of statistical models for ecological data with a primary focus on marine mammals.

Juan M. Morales is a researcher from CONICET (Consejo Nacional de Investigaciones Científicas y Técnicas–National Scientific and Technical Research Council) and a professor at Universidad Nacional del Comahue in Bariloche, Argentina. Dr. Morales earned a PhD in ecology at the University of Connecticut and his research focus is on animal movement and spatial ecology.

1 Introduction

The movement of organisms is a fundamentally important ecological process. Voluntary movement is a critical aspect of animal biology and ecology. Humans have been keenly interested in the movement of individual animals and populations for millennia. Over 2000 years ago, Aristotle wrote about the motion of animals, and the associated philosophical and mathematical concepts, in his book, *De Motu Animalium* (Nussbaum 1978). Historically, it was critical to understand how and where wild food sources could be obtained. Thus, early humans were natural animal movement modelers. In modern times, we are interested in the movement of animals for scientific reasons and for making decisions regarding the management and conservation of natural resources (Cagnacci et al. 2010).

The study of wild animals can be challenging. Animals are often elusive and reside in remote or challenging terrain. Many animals have learned to minimize exposure to perceived threats, which, unfortunately for us, include the well-intentioned biologist approaching them with binoculars or a capture net. Therefore, it is no surprise that the development of animal-borne telemetry devices has revolutionized our ability to study animals in the wild (Cagnacci et al. 2010; Kays et al. 2015). Animal telemetry has helped us overcome many of the practical, logistical, and financial challenges of direct field observation. Telemetry data have opened windows that allow us to address some of the most fundamental ecological hypotheses about space use ("Where is the animal?"), movement ("How did the animal get there?," "Where could it go?"), resource selection ("Where does the animal like to be?"), and behavior ("What is the animal doing?") (Figure 1.1).

1.1 BACKGROUND ON ANIMAL MOVEMENT

Animal movement plays important roles in the fitness and evolution of species (e.g., Nathan et al. 2008), the structuring of populations and communities (e.g., Turchin 1998), ecosystem function (Lundberg and Moberg 2003), and responses to environmental change (e.g., Thomas et al. 2004; Trakhtenbrot et al. 2005; Jønsson et al. 2016). The scientific study of animal movement has a deep history, and we are unable to explore all of the ecological implications and methodological developments in a single volume. Instead, we focus on several specific inferential methods that can provide valuable ecological insights about animal movement and behavior from telemetry data.

The importance of animal movement in larger-scale ecosystem function probably inspired the Craighead brothers to develop and deploy the first radio collars on grizzly bears (*Ursus arctos*) from Yellowstone National Park in the 1960s (Craighead and Craighead 1972). Satellite tracking devices are now capable of pinpointing animal locations at any moment, remote sensing provides ever refined environmental

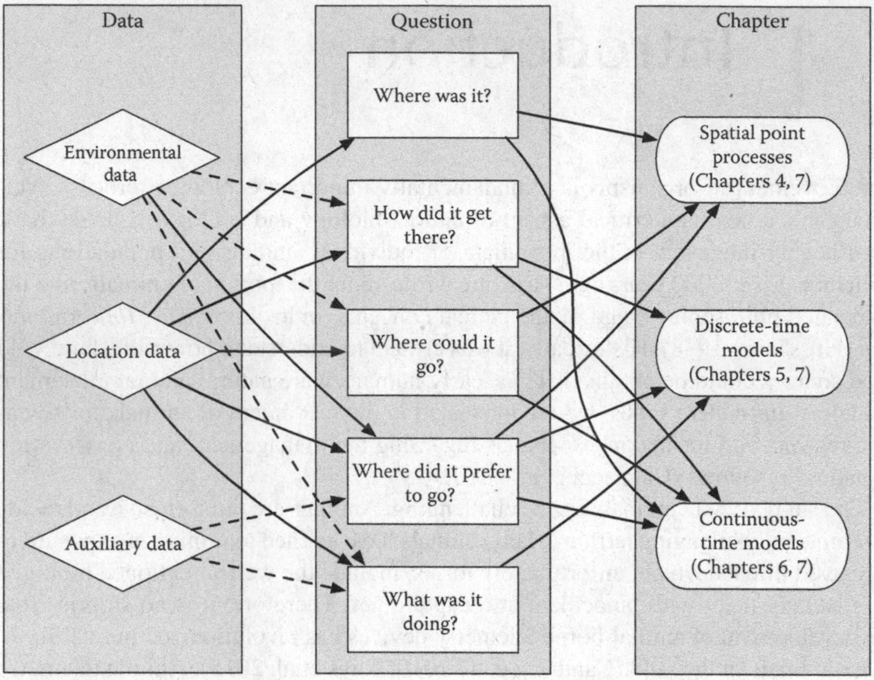

FIGURE 1.1 Relationships among data types, analytical methods, and some fundamental questions of movement ecology. Location data are the cornerstone of all of the analysis methods described in this book. Environmental data, such as those acquired from remote sensing, are useful in drawing connections between animals and their surroundings. Auxiliary biotelemetry data, such as accelerometer or dive profile data, can help address questions about animal behavior. Dashed lines indicate where data can be helpful for addressing particular questions, but are not essential.

data, and biotelemetry tags allow for the simultaneous collection of important physiological and behavioral information from wild animals. These technological advances will lead to a better understanding of how individual decisions affect demographic parameters and ultimately translate into population dynamics. In this sense, animal movement can provide the long-sought bridge between behavior, landscape ecology, and population dynamics (Lima and Zollner 1996; Wiens 1997; Morales et al. 2010; Kays et al. 2015).

In what follows, we provide a brief summary of research findings, existing knowledge, and analytic approaches for important aspects of animal movement ecology. We organized these topics into 10 sections:

1. Population dynamics
2. Spatial redistribution
3. Home ranges, territories, and groups
4. Group movement and dynamics
5. Informed dispersal and prospecting

6. Memory
7. Individual condition
8. Energy balance
9. Food provision
10. Encounter rates and patterns

1.1.1 Population Dynamics

In classical models of population dynamics, predators and prey encounter each other in proportion to their overall abundance over space and reproductive rates decrease as the global population density increases. This is because traditional models of population and community dynamics assume we are dealing with many individuals that are well mixed (Turchin 2003). Such "mean field" representations of population dynamics can provide good approximations when the physical environment is relatively homogeneous and organisms are highly mobile, or when organisms interact over large distances. However, when the external environment or the limited mobility of organisms results in lack of mixing, the conditions experienced by a particular member of a population or community can be quite different from the mean (Lloyd 1967; Ovaskainen et al. 2014; Matthiopoulos et al. 2015). That is, when *per capita* vital rates are affected by varying local conditions, the observed population and community dynamics can differ markedly from mean field predictions.

Population dynamics involve births, deaths, immigration, and emigration; modern tracking technology, together with new statistical models, can greatly improve our understanding of these processes. The individuals that comprise a population can vary in several traits and individual behavior can change in response to internal and external stimuli. Individual traits and behavior determine the way they interact with the environment and other organisms while the conditions that individuals experience ultimately translate to their performance (i.e., growth, survival, and reproduction).

Survival analysis can be used to model changes in hazard with time and in relation to covariates such as location, age, body condition, and habitat type. Detailed tracking through satellite telemetry enables spatial information and survival data to be combined at small temporal scales, leading to an increasingly sophisticated understanding of the determinants of survival (Murray 2006; Haydon et al. 2008; Schick et al. 2013). Likewise, changes in movement behavior can be used to infer reproductive events in some species (Long et al. 2009). However, to take full advantage of these data, new analytic techniques should take into account the sequential nature of individual survival and reproduction. For example, the chance of an animal dying of starvation depends on its history of encounters with food items and foraging decisions.

Coupling demographic data with movement models is an area of active research, but is still somewhat nascent. Spatial capture–recapture (SCR) models provide a way to formally connect animal encounter data with movement processes; we refer the interested reader to Royle et al. (2013) and references therein for additional details. The methods presented in this book will be critical for formally integrating location data and demographic data in future SCR modeling efforts.

1.1.2 Spatial Redistribution

Classical reaction–diffusion models, such as those used by Fisher (1937) to describe the spread of an advantageous mutation within a population assume that mortality and recruitment rates depend linearly on local population density and that individuals move at random over a large and homogeneous area. Early implementations of these models were also used to describe the dynamics of population invasion and range expansion (e.g., Skellam 1951; Andow et al. 1990; Shigesada and Kawasaki 1997), and later, were embedded in a hierarchical statistical modeling framework (e.g., Wikle 2003; Hooten and Wikle 2008; Hooten et al. 2013a) to provide inference about spreading populations.

Diffusion equations have been justified as a good approximation to the displacement of individuals performing a "random walk."* Although we know that animals do not move at random, the diffusion approximation can still be sufficient at certain (usually large) scales and also serves as a null model to compare with more complex models (Turchin 1998).

More general forms of movement can be taken into account by formulating spatial population models as integral equations. These have commonly been formulated in discrete time, yielding integro-difference equations where local population growth is combined with a "redistribution kernel" that describes the probability that an individual moves from its current location to another one in a given time-step.† The temporal scale of these models is usually set to match reproductive events so that the redistribution kernel represents successful dispersal rather than regular movement. A great deal of theoretical and empirical work has explored the consequences of kernel shape, particularly in the tail of the distribution, on invasion speed (Kot et al. 1996; Powell and Zimmermann 2004).

There are many ways to make spatial population models more realistic and appropriate for particular species, places, and scales of interest. A good starting point is to consider the spatial structure of the population, which is generally accepted as an important prerequisite for more accurate ecological predictions (Durrett and Levin 1994; Hanski and Gaggiotti 2004).‡ The spatial structure of populations can range from classical closed populations to a set of subpopulations with different degrees of interaction (Thomas and Kunin 1999). As different degrees of connectivity among subpopulations can have important dynamical consequences, researchers are increasingly interested in understanding how connectivity arises from the interaction among individual phenotypes, behaviors, and the structure of landscapes.

One particular feature of the models described thus far is that every individual is assumed to move according to the same kernel (whether Gaussian or otherwise). However, detailed tracking of individual movements consistently reveals differences among individuals. Theoretical and empirical studies have shown how the characteristics of redistribution kernels can depend on differences among individuals (Skalski and Gilliam 2000; Fraser et al. 2001; Morales and Ellner 2002; Delgado

---

* We describe random walks in discrete and continuous time in Chapters 5 and 6.

† We describe redistribution kernels and integral equation models for movement in Chapters 4 and 6.

‡ See Chapter 2 for a brief primer on spatial statistics.

and Penteriani 2008), and on the interplay between individual behavior and features of the underlying landscape (Johnson et al. 1992; McIntyre and Wiens 1999; Fahrig 2001; Ricketts 2001; Morales et al. 2004; Mueller and Fagan 2008), including reactions to habitat boundaries (Schultz and Crone 2001; Morales 2002; Schtickzelle and Baguette 2003; Ovaskainen 2004; Haynes and Cronin 2006). In particular, population heterogeneity produces leptokurtic (i.e., heavy tailed) redistribution kernels when a subset of individuals consistently moves longer distances than others (Skalski and Gilliam 2000; Fraser et al. 2001).

Several factors can explain why two individuals belonging to the same population move differently. They may be experiencing different environments of heterogeneous landscapes; they can also have different phenotypes or condition, different past experiences (e.g., Frair et al. 2007), or even different "personalities" (Fraser et al. 2001; Dall et al. 2004). In a theoretical study, Skalski and Gilliam (2003) modeled animals switching between fast and slow random walk movement states and found that the resulting redistribution kernel depended on the total time spent in each of the states and not on the particular sequence of changes. This theoretical result highlights the importance of animals' time budgets for scaling movement processes (Figure 1.2).

It is common to consider that individuals have a small set of movement strategies (Blackwell 1997; Nathan et al. 2008), and the time allocation to these different behaviors (or "activity budgets") can depend on the interaction between their motivation and the structure of the landscape they occupy (Morales et al. 2004, 2005). The results

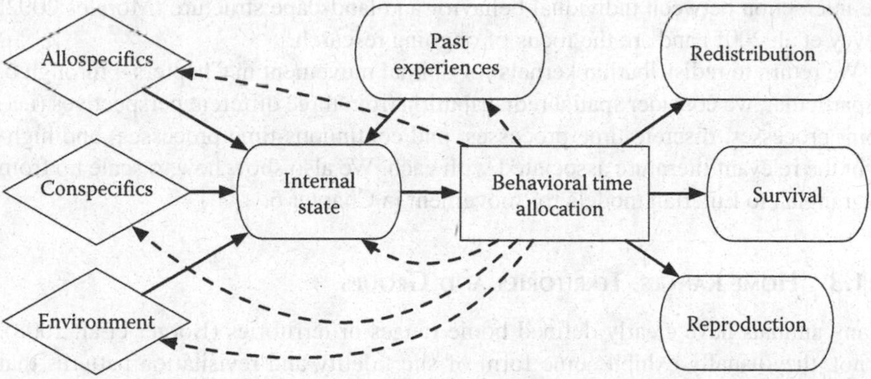

FIGURE 1.2 Mechanistic links between animal movement and population dynamics adapted from Morales et al. (2010). We consider an unobserved individual internal state that integrates body condition (e.g., energy reserves, reproductive status). Several factors affect the dynamics of this internal state, including social interactions with conspecifics, trophic or other interaction with allospecifics (other species), and abiotic environmental effects and dynamics. Internal state dynamics determine the organism's time allocation to different behaviors (e.g., food acquisition, predator avoidance, homing, and landscape exploration) but is also modulated by past experiences and phenotypic trails such as behavioral predispositions. As different behaviors imply different movement strategies, the time budget determines the properties of the spatial redistribution that describes space use. Time allocation to different behaviors also affects individual survival and reproduction, and hence, overall population dynamics.

of Skalski and Gilliam (2003) imply that knowing the fraction of time allocated to each behavior makes it possible to derive suitable redistribution kernels.

A common reaction in the visual inspection of movement data is to intuit that individuals are moving differently at different times. As a result, several techniques (including many *ad hoc* procedures) have been developed to identify and model changes in movement behavior from trajectory data (reviewed in Patterson et al. 2008; Schick et al. 2008; Gurarie et al. 2016). Clustering models, such as those we describe in Chapter 5, can be difficult to reliably implement because biologically different movement behaviors can lead to very similar trajectories. For example, it may be difficult to distinguish relative inactivity (e.g., resting) from intense foraging, within a small patch, based on horizontal trajectory alone. However, as physiological and other information becomes available through biotelemetry devices, we may gain greater insight into how animals allocate time to different tasks and how this allocation changes in different environments (McClintock et al. 2013), thus providing a mechanistic way to model redistribution kernels conditional on individual state.

Another result from Skalski and Gilliam (2003) is that a mixture of movement states converges to simple diffusion if given enough time. The sum of n independent and identically distributed random variables with finite variance will be Gaussian distributed as n increases. Thus, if all individuals in a population move according to the same stochastic process, we would expect that, at some time after the initiation of movement, the distribution of distance moved becomes Gaussian because the distance traveled is the sum of movement vectors. However, this depends on the rate of convergence and independence assumption. Still, similar results may relate to the interaction between individual behavior and landscape structure (Morales 2002; Levey et al. 2005) and are the focus of ongoing research.

We return to redistribution kernels for animal movement in Chapters 4 through 6. In particular, we consider spatial redistribution from three different perspectives (i.e., point processes, discrete-time processes, and continuous-time processes) and highlight the relevant literature associated with each. We also show how to scale up from Lagrangian to Eulerian models for movement in Chapter 6.

1.1.3 Home Ranges, Territories, and Groups

Many animals have clearly defined home ranges or territories (Borger et al. 2008). If not, they usually exhibit some form of site fidelity and revisitation patterns that are not captured by simple random walks. Most likely, these animals will spend their reproductive life in a region that is small compared to their movement capabilities. Substantial progress has been made in developing mechanistic models of animal movement with territorial behavior (e.g., Moorcroft et al. 1999; Smouse et al. 2010; Moorcroft and Lewis 2013; Giuggioli and Kenkre 2014). However, territoriality models typically describe the space use by particular individuals (or members of a wolf pack, for example) rather than an entire population. As a result, they have not yet been linked to models of population demography.

For territorial animals, the carrying capacity of a particular region or landscape can be determined by competition for space. When the environment provides a limited number of essential items, such as nest cavities, the maximum number of breeders is bounded and surplus individuals form a population of nonbreeders referred to as "floaters" (Brown 1969; Penteriani and Delgado 2009). When dispersal or mortality create vacancies in previously occupied territories, floaters may become a crucial population reserve for filling these empty territories. Floaters can also have a negative effect on population growth through interference, conflict, or disturbance. Furthermore, the aggressive behavior of breeders can also decrease the carrying capacity of the population.

We describe basic methods for estimating home ranges and core areas in Chapter 4, and discuss methods for modeling interactions among individuals at the end of Chapter 5. However, the formal statistical modeling of floaters, together with individual-level behavior and territoriality is still developing and an open area of research.

1.1.4 Group Movement and Dynamics

Understanding the distribution of social animals over landscapes requires scaling up from individual movement patterns to groups of individuals and populations (Okubo et al. 2001). Most models of group dynamics focus on relatively short temporal scales (Couzin et al. 2005; Eftimie et al. 2007; Strandburg-Peshkin et al. 2015). However, the interaction between the group structure of a population and the movement of individuals is also relevant at longer time scales (e.g., Fryxell et al. 2007). Long time scales in group dynamics are particularly relevant for reintroduced species, where a balance of spread and coalescence processes will determine how individuals distribute themselves over the landscape. Often, individual survival and fecundity are higher in groups, so that the successful persistence of the introduced population may depend on coalescence dominating and limiting the spreading process, thereby enabling the establishment of a natural group structure within the release area. Haydon et al. (2008) developed movement models for North American elk (*Cervus canadensis*), reintroduced to Ontario, where, as in Morales et al. (2004), animals can switch between exploratory (large daily displacements and small turning angles) and encamped behavior (small daily displacements and frequent reversals in direction). The rate of switching among these movement modes depended on whether individuals were part of a group or not. Haydon et al. (2008) combined their movement models with analysis of mortality and fecundity to build a spatially explicit, individual-based model for the dynamics of the reintroduced elk population. Their analysis showed that elk moved farther when they were solitary than when they were in a group, and that mortality risk increased for individuals that moved progressively away from the release location. The simulation model showed how the population rate of increase and the spatial distribution of individuals depended on the balance of fission and fusion processes governing group structure.

New approaches for studying the interaction among individuals in groups are appearing regularly in the literature. For example, Scharf et al. (In Press) developed a discrete-time model that captures the alignment and attraction of killer whales (*Orcinus orca*) in Antarctica, and Russell et al. (2016b) used point processes to model interactions among individual guppies (*Poecilia reticulata*). Using high temporal resolution telemetry data from a group of baboons (*Papio anubis*) in Kenya, Strandburg-Peshkin et al. (2015) analyzed individual movement in relation to one another. They found that, rather than following dominant individuals, baboons are more likely to follow others when multiple initiators of movement agree, suggesting a democratic collective action emerging from simple rules. In a study of fission–fusion dynamics of spider monkeys (*Ateles geoffroyi*), Ramos-Fernández and Morales (2014) found that group composition and cohesion affected the chance that a particular individual will leave or join a group. As another example, Delgado et al. (2014) found that dispersing juveniles of eagle owl (*Bubo bubo*) were generally attracted to conspecifics, but the strength of attraction decreased with decreasing proximity to other individuals. However, despite this progress, models for animals that decide to leave their territory or abandon a group, and how they explore and choose where to establish new territories or home ranges, have yet to appear in the literature.

1.1.5 Informed Dispersal and Prospecting

Dispersal involves the attempt to move from a natal or breeding site to another breeding site (Clobert 2000), and is essential for species to persist in changing environments (Ronce 2007). The redistribution modeling ideas we introduced in the previous sections represent dispersal as a random process that may be sensitive to the spatial structure of the landscape or the presence of conspecifics. However, there is a great deal of evidence indicating that individuals are capable of sophisticated and informed decision-making when choosing a new place to live (Bowler and Benton 2005; Stamps et al. 2005, 2009). Clobert et al. (2009) proposed the concept of "informed dispersal" to convey the idea that individuals gather and exchange information at all three stages of dispersal (i.e., departure, transience, and settlement). Thus, movement involves not only the exchange of individuals among habitat patches but also information transfer across the landscape. Animals can acquire information about the environment by "looking" at others' morphology, behavior, or reproductive success (Danchin et al. 2004; Dall et al. 2005). For example, in an experiment with the common lizard (*Lacerta vivipara*), Cote and Clobert (2007) quantified emigration rate from artificial enclosures that received immigrants. They found that when local populations received immigrants that were reared under low population density, the emigration rate of the local population increased, providing evidence that immigrants supplied information about the density of surrounding populations, probably via their phenotype.

We only have a rudimentary understanding of how individuals integrate different sources of information to make movement and dispersal decisions. Long-term tracking is needed to study how animals adjust to the changing characteristics of their home ranges or territories, and under what conditions they are likely to search for a new home. Detailed tracking of juveniles may shed light on the processes of exploration (i.e., transience) and settlement. In particular, movement data can be used to test ideas about search strategies, landscape exploration, and the importance of past experience in biasing where animals decide to attempt breeding.

1.1.6 Memory

The importance of previous experiences and memory is increasingly being recognized and explicitly considered in the analysis of telemetry data (e.g., Dalziel et al. 2008; McClintock et al. 2012; Avgar et al. 2013; Fagan et al. 2013; Merkle et al.

2014). Smouse et al. (2010) provide a summary of the approaches used to include memory in movement models. Formulating memory models has largely been a theoretical exercise but the formal connection with data is possible. For example, the approach used to model the effect of scent marking in mechanistic home range models (Moorcroft and Lewis 2013) could be easily adapted to model memory processes. Avgar et al. (2015) fit a movement model that included perceived quality of visited areas and memory decays to telemetry data from migrating Caribou. It is less clear what role memory plays in population dynamics.

Forester et al. (2007) describe how certain discrete-time movement models can be reformulated to provide inference about memory. We explain these ideas in Chapter 5. In continuous-time models, Hooten and Johnson (2016) show how to utilize basis function specifications for smooth stochastic processes to represent different types of memory and perception processes. We discuss these functional movement modeling approaches in Chapter 6.

1.1.7 Individual Condition

Recognizing that the contribution of a particular individual to the population is a function of its fitness has historically promoted the development of physiological-, age-, and stage-structured population models (Caswell 2001; Ellner and Rees 2006; Metz and Diekmann 2014). Body condition integrates nutritional intake and demands, affecting both survival and reproduction. For example, studies of ungulates living in seasonal environments have found that percent body fat in early winter is a very good predictor for whether animals die, live without reproducing, or live and reproduce (Coulson et al. 2001; Parker et al. 2009). Also, many populations show "carryover effects" where conditions experienced during a time period influence vital rates in future periods, which has the potential to generate many different population responses (Ratikainena et al. 2008; Harrison et al. 2011). Movement decisions and habitat use affect energy balance and body condition in animals. Linking individual condition to movement and space use is challenging because we usually need to recapture individuals to assess percent body fat, for example. However, some marine mammals perform "drift dives," using their buoyancy to change depth without active propulsion and with their rate of drift determined largely by their lipid-to-lean-mass ratio Biuw et al. (2003). Working with Southern elephant seals (*Mirounga leonina*), Schick et al. (2013) modeled changes in individual condition as a function of travel distance and foraging events. They also linked changes in behavior due to human disturbances to population-level effects.

The animal movement models we describe in Chapters 4 through 6 are mostly focused on modeling individuals. However, when scaling up inference to the population level (using random effects for parameters or other hierarchical modeling approaches), it may be important to account for variation in body condition among individuals to help describe differences in movement parameters. See Sections 4.5 and 5.2 for examples of accounting for individual-level differences when obtaining inference at the population level.

1.1.8 Energy Balance

Many aspects of life history evolution, behavioral ecology, and population dynamics depend on how individuals consume resources and on how they allocate energy to growth and reproduction. Food acquisition is an important driver of animal movement to the point that relationships between scaling of space use and daily distance traveled in relation to body mass and trophic requirements has been hypothesized (Jetz et al. 2004; Carbone et al. 2005).

Technological developments in biotelemetry allows the possibility of observing a suite of relevant physiological data such as heart rate and core temperature, in addition to individual location (Cooke et al. 2004; Rutz and Hays 2009). Furthermore, accelerometers can be used for detailed movement path reconstruction and for recording energy expenditure, activity budgets (i.e., ethograms), and rare behavioral events such as prey captures (Wilson et al. 2007, 2008; Williams et al. 2014; Bidder et al. 2015). Combined with detailed environmental maps, these data could lead to empirically based models of animal performance in the wild, linking behavioral decisions with space use, survival, and reproduction (Figure 1.2).

The formal integration of energy balance information into dynamic statistical animal movement models is still in early development stages (Shepard et al. 2013). However, many approaches we describe in Chapters 4 through 6 allow for the use of auxiliary data pertaining to energy-intensive behavior. For example, Section 5.2.5 describes how to integrate dive data for marine mammals into discrete-time movement models.

1.1.9 Food Provision

Food acquisition in poor habitats (or in good habitats that have been depleted) demands more searching time and energy, which is reflected in their movement patterns (e.g., Powell 1994). These effects are best documented in central place foragers such as nesting birds or pinnipeds that forage at sea but breed on land. Many of these animals forage at particular oceanographic features (Boersma and Rebstock 2009) that change in location and quality from year to year. Magellanic penguins (*Spheniscus magellanicus*) breeding at Punta Tombo, Argentina showed a decrease in reproductive success with increasing average foraging trip duration (Boersma and Rebstock 2009). Also, penguins stayed longer at feeding sites in more distant foraging areas, presumably to feed themselves and recover from the increased cost of swimming (Boersma and Rebstock 2009). Thus, satellite telemetry technology has allowed a better understanding of the interplay between landscape or seascape variability and breeding success.

In Chapter 5, we show how to use discrete-time movement models to cluster animal paths into different behavioral types, which can help identify food acquisition modes based on telemetry data. We also demonstrate how to account for food-related aspects of movement in the continuous-time setting discussed in Chapter 6.

1.1.10 Encounter Rates and Patterns

The "functional response" is a key component of population models that include trophic interactions; it describes the rate of prey consumption by individual predators as a function of prey density (Holling 1959a,b). The dynamics and persistence of interacting populations usually depend on the shape and dimensionality of functional responses (Turchin 2003). Mechanistically, the functional response depends on encounter rates. Thus, a useful null model for encounter rates is one where individuals move randomly and independently of each other. More than 150 years ago, Maxwell (1860) calculated the expected rates of molecular collisions of an ideal gas as a function of density, particle size, and speed.* The ideal gas model has been used and rediscovered in many ways, including Lotka's justification of predator-prey encounters being proportional to predator speed and size and to predator and prey densities. As a recent example, the scaling of home ranges with body size derived by Jetz et al. (2004) assumes that the proportion of resources lost to neighbors is related to encounter rates as calculated from the ideal gas model for known scaling relationships of speed, population density, and detection distance.

The thorough review by Hutchinson and Waser (2007) shows many more examples of the application of Maxwell's model plus several refinements, including different assumptions about detection, speed, and density. Recently, Gurarie and Ovaskainen (2013) presented analytical results and a taxonomy for a broad class of encounter processes in ecology. The movement of animals almost certainly deviates from the assumptions of Maxwell's model and we can use information about the characteristics of movement paths from real animals to derive better predictions of encounter rates, or in the case of carnivores, kill rates (e.g., Merrill et al. 2010).

Environmental heterogeneity can also be an important determinant in encounter rates and group dynamics. For example, Flierl et al. (1999) used individual-based models of fish groups to study the interplay among the forces acting on the individuals and the transport induced by water motion. They found that flows often enhanced grouping by increasing the encounter rate among groups and thereby promoting merger into larger groups.† In general, habitat structure will affect encounter rates among individuals of the same species but also among predators and prey.

Encounter rates and population dynamics are also altered when predators or prey form social groups. Fryxell et al. (2007) developed simple models of group-dependent functional responses and applied them to the Serengeti ecosystem. They found that grouping strongly stabilizes interactions between lions and wildebeest, suggesting that social groups, rather than individuals, were the basic building blocks for these predator–prey systems.

As satellite tracking devices become more affordable, and larger numbers of individuals can be tracked in the same study areas, we can expect to learn more about interactions among individuals. Furthermore, the use of additional telemetry technologies can make this more feasible. For example, Prange et al. (2006) used proximity detectors in collars fitted to free-living raccoons and were able to obtain accurate information in terms of detection range, and duration of contact. Animal-borne video systems also may help identify social interactions and foraging events for a focal individual (Hooker et al. 2008; Moll et al. 2009). Hence, the study of encounters offers great opportunities for marrying theory with data and to greatly improve our understanding of spatial dynamics.

---

* Assuming independent movements in any direction and with normally distributed velocities.
† Although the grouping effect breaks down for strong flows.

As animals face similar constraints and environmental heterogeneity, it is expected that they will exhibit similar movement rules and patterns. Early enthusiasm surrounding Levy flights and walks is now being taken with a bit more caution (e.g., Pyke 2015), but it is valuable to identify common movement rules based on individual animal's morphology, physiology, and cognitive capacity. There is also much theoretical and empirical work needed to better understand the costs and benefits of different movement strategies. Scharf et al. (In Press) described a method for inferring time-varying social networks in animals based on telemetry data. Using data from killer whales, Scharf et al. (In Press) developed a model that was motivated by encounter rate approaches that clustered similarities in movement patterns to learn about underlying binary networks that identified groups of individuals and how they change over time. We discuss these ideas more at the end of Chapter 5.

1.2 TELEMETRY DATA

Animal telemetry data are varied. This variation is an advantage because different field studies often have very different objectives and logistical (or financial) constraints. At a minimum, most animal-borne telemetry devices provide information about animal location. The earliest devices were very high frequency (VHF) radio tags designed for large carnivores and ungulates.* VHF tags emit a regular radio wave signal (or pulse) at a specific frequency. A beeping sound (or ping) is heard whenever the signal is picked up by a nearby receiver that is tuned to this frequency, and the pings get louder as the receiver approaches the tag. As one hones in on the pings, the location of an animal with a VHF tag can be either closely approximated or confirmed by visual sighting. Accurate radio telemetry data acquisition requires practice and, often, triangulation. Radio tracking can sometimes be very challenging from the ground; thus, radio relocation surveys are often performed from small aircraft. Many VHF tags include a sensor that triggers a faster pulse rate after a prespecified length of inactivity that is believed to be indicative of mortality or other events (e.g., hibernation). The analysis of radio telemetry data has historically been limited to descriptive statistical models of space use, home range delineation, survival, and abundance (e.g., White and Garrott 1990; Millspaugh and Marzluff 2001; Manly et al. 2007), but more sophisticated movement models have also been applied to radio telemetry data (e.g., Dunn and Gipson 1977; Moorcroft et al. 1999). Early VHF tags were too large for many smaller species, but improvements in battery technology now permit tags that are small enough for birds and even insects. The primary limitations of VHF tags are the limited range of radio signals and the cost and effort required to reliably locate animals via radio tracking. Radio tracking technology may seem archaic in the age of smart phones, but it still offers a relatively inexpensive and long battery-lived alternative to modern telemetry devices.

Since the mid-1990s, modern telemetry devices have been capable of storing and transmitting information about an individual animal's location as well as internal and

* We refer to "tags" generically here; for most terrestrial mammals, the telemetry devices are attached to neck collars and fitted to the individual animals. Telemetry devices have been fitted to animals in a variety of other ways.

external characteristics (e.g., heart rate, temperature, depth/altitude). Because modern telemetry devices can include additional sensors unrelated to location acquisition, the terms "biotelemetry" and "biologging" are increasingly used for describing modern animal telemetry techniques and devices (e.g., Cooke et al. 2004). There are two main types of modern (non-VHF) animal telemetry tags. These are often called storing (or "archival") and sending (or "transmitting") tags. Archival tags can be smaller than transmitting tags and store vast amounts of biotelemetry information, such as high-resolution accelerometer data, but they possess no mechanism for data transmission. Therefore, archival tags must be recovered from the animal before any data can be accessed. Transmitting tags send data in the form of electromagnetic waves to nearby receivers (similar to VHF tags) or to orbiting communications satellites. Satellite transmitting tags allow researchers to retrieve biotelemetry data without needing to recover or be close to the tag. Similar to archival tags, transmitting tags can store vast amounts of data. However, satellite tags require line of sight for transmission, and this limitation often necessitates careful consideration when designing and programming satellite tags. For example, marine animals do not surface long or frequently enough to transmit large quantities of biotelemetry data, so researchers must often make difficult trade-offs between data quality and quantity based on the specific objectives of their study (e.g., Breed et al. 2011).

Whether of the archival or transmitting type, most modern biotelemetry tags rely on satellites for determining an animal's location. Tags that are equipped with an internal global positioning system (GPS) usually provide the most accurate locations currently available. GPS location errors (i.e., the distance between the observed and true location of the individual) tend to be less than 50 m, but GPS tags need to transmit larger data payloads and tend to be larger in size. Therefore, GPS tags are ideal for larger, terrestrial species in open habitat, but they are typically unsuitable for aquatic species such as marine mammals and fish.

Although not as accurate as GPS, Argos tags are a popular option for marine and small terrestrial species. Argos tags rely on a system of polar-orbiting satellites to decode the animal's location from a relatively tiny packet of transmitted information. Argos tags can quickly transmit data to satellites within the brief intervals that marine mammals surface to breathe because the transmission packets are small. The main drawback of Argos tags is the limited size and duration of transmissions; this limits the quantity and quality of onboard biotelemetry data that can be recovered. Argos tags tend to perform best at higher latitudes (due to the polar orbits of the satellites), but location errors can typically range from hundreds to thousands of meters (e.g., Costa et al. 2010; Brost et al. 2015).[*]

As a compromise between GPS and Argos, Fastloc-GPS (Wildtrack Telemetry System Limited, Leeds, UK) tags compress a snapshot of GPS data and quickly transmit via the Argos satellite system. With location errors typically between 50 and 1000 m, Fastloc-GPS is considerably more accurate than Argos overall.

Biotelemetry technology is rapidly improving,[†] and there are many tag designs and data collection capabilities that we have not covered in this brief introduction. These

---

[*] We describe specific aspects of Argos data and potential remedies in Chapters 4 and 5.

[†] See Kays et al. (2015) for a recent overview of tag technology.

include light-sensing "geologgers" for smaller species (e.g., Bridge et al. 2011), archival "pop-up" tags popular in fisheries (e.g., Patterson et al. 2008), proximity detectors (e.g., Ji et al. 2005), acoustic tags (e.g., McMichael et al. 2010), "life history" tags (Horning and Hill 2005), accelerometer tags (e.g., Lapanche et al. 2015), and automatic trajectory representation from video recordings (Pérez-Escudero et al. 2014). In what follows, we primarily focus on the analysis of location data such as those obtained from VHF, GPS, and Argos tags. However, many of the methods we present can utilize location information arising from other sources, as well as incorporate auxiliary information about the individual animal's internal and external environment that is now regularly being collected from modern biotelemetry tags. Winship et al. (2012) provide a comparison of the fitted movement of several different marine animals when using GPS, Argos, and light-based geolocation tags.

1.3 NOTATION

A wide variety of notation has been used in the literature on animal movement data and modeling. This variation in statistical notation used makes it challenging to maintain consistency in a comprehensive text on the subject. We provided this section, along with Table 1.1, in an attempt to keep expressions as straightforward as possible. We recommend bookmarking this section on your first reading so that you may return to it quickly if the notation becomes confusing.

Conventional telemetry data consist of a finite set of spatially referenced geographic locations ($\mathbf{S} \equiv \{\mathbf{s}_1, \ldots, \mathbf{s}_i, \ldots, \mathbf{s}_n\}$) representing the individual's observed location at a set of times spanning some temporal extent of interest (e.g., a season or year). We use the notation, $\{\boldsymbol{\mu}_1, \ldots, \boldsymbol{\mu}_n\}$ to represent the corresponding true positions of the animal. Sometimes, the observed telemetry data are assumed to be the true positions (i.e., no observation error); however, in most situations, they will be different. The times at which locations are observed can be thought of as fixed and part of the "design," or as observed random variables. In either case, a statistical notation with proper time indexing becomes somewhat tricky. To remain consistent with the broader literature on point processes (and with Chapter 2), we assume that there are n telemetry observations collected at times $\mathbf{t} \equiv (t_1, \ldots, t_i, \ldots, t_n)'$ such that $t_i \in \mathcal{T}$ and $\mathbf{t} \subset \mathcal{T}$. The seemingly redundant time indexing accounts for the possibility of irregularly spaced data in time. If the differences ($\Delta_i = t_i - t_{i-1}$) between two time points at which we have telemetry observations are all equal, we could just as easily use the direct time indexing where the data are $\mathbf{s}_t$ for $t = 1, \ldots, T$. In that case, we have $T = n$. From a model-building perspective, it is sometimes less cumbersome to index telemetry observations in time (i.e., $\mathbf{s}_t$) and deal with temporal irregularity during the implementation. However, there are some situations, for example, when the points are serially dependent, where we need the Δ_i notation. A further perspective on notation arises when considering that the true animal location process is a continuous process in time. To formally recognize this, we often index the observed location vectors as $\mathbf{s}(t_i)$ (or $\boldsymbol{\mu}(t_i)$, in the case of the true positions). The parenthetical notation at least admits that we are often modeling animal locations as a continuous function. Thus, prepare yourself to see all types of indexing, both in this text and in the vast animal movement literature.

TABLE 1.1
Statistical Notation

Notation	Definition
i	Observation index for $i = 1, \ldots, n$ total observations.
t	Time point at which the data or process occurs (in the units of interest).
$\mathcal{T}$	The set of times at which the process exists; typically compact interval in continuous time such that $t \in \mathcal{T}$.
t_i	Time associated with observation i.
T	Either largest time in observations or process, or upper temporal endpoint in study, depending on context.
$\mathbf{s}_i$	Observed telemetry observation for $i = 1, \ldots, n$. $\mathbf{s}_i$ is a 2×1 vector unless otherwise stated. Also written as: $\mathbf{s}(t_i)$ in continuous-time context.
$\mathcal{S}$	The spatial support for the observed telemetry observations (i.e., $\mathbf{s} \in \mathcal{S}$).
$\boldsymbol{\mu}_i$	True individual location (i.e., position) for $i = 1, \ldots, n$. $\boldsymbol{\mu}_i$ is a 2×1 vector unless otherwise stated. Also written as: $\boldsymbol{\mu}(t_i)$ in continuous-time context.
$\mathcal{M}$	The spatial support for the true individual locations (i.e., $\boldsymbol{\mu}(t) \in \mathcal{M}$). Typically, the support for the true locations $\mathcal{M}$ is a subset of the support for the observed locations $\mathcal{S}$ (i.e., $\mathcal{M} \subset \mathcal{S}$).
$\mathbf{X}$	A "design" matrix of covariates, which will often be decomposed into rows $\mathbf{x}_i$ for row i, depending on the context in which it is used.
$\boldsymbol{\beta}$	Vector of regression coefficients (i.e., $\boldsymbol{\beta} \equiv (\beta_1, \beta_2, \ldots, \beta_p)'$), where p is the number of columns in $\mathbf{X}$.
$\boldsymbol{\beta}'$	The "prime" symbol (′) denotes a vector or matrix transpose (e.g., converts a row vector to a column).
σ^2	Variance component associated with the observed telemetry data, true position process, or a model parameter.
$\boldsymbol{\Sigma}$	Covariance matrix for either a parameter vector such as $\boldsymbol{\beta}$ (if subscripted) or the data or process models.
$f(\cdot)$, $[\cdot]$	Probability density or mass function. $p()$, $P()$, and $\pi()$ are used in other literature. The $[\cdot]$ has become a Bayesian convention for probability distributions.
$E(y)$	Expectation of random variable y; an integral if y is continuous and sum if y is discrete.
$\propto$	Proportional symbol. Often used to say that one probability distribution is proportional to another (i.e., only differs by a scalar multiplier).

1.4 STATISTICAL CONCEPTS

We focus mostly on parametric statistical models* in this book; thus, we rely on both Bayesian and non-Bayesian models using maximum likelihood. Occasionally, for example, in Chapters 2 through 4, we present statistical methods that are nonparametric or involve implementation methods that do not involve Bayesian or maximum likelihood approaches. A generic data model statement will appear as $y_i \sim [y_i|\boldsymbol{\theta}]$,

* Parametric statistical models involve the specification of known probability distributions with parameters that are unknown but estimated in the model fitting procedure.

where y_i are the observations (we use $\mathbf{s}_i$ for telemetry observations instead of y_i) for $i = 1, \ldots, n$, $\boldsymbol{\theta}$ are the data model parameters, and the bracket notation "$[\cdot]$" represents a probability distribution. The data model is often referred to as the "likelihood" by Bayesians, but the likelihood used in maximum likelihood estimation (MLE) is proportional to the joint distribution of the data conditioned on the parameters. When the observations are conditionally independent, the likelihood is often written as $[\mathbf{y}|\boldsymbol{\theta}] = \prod_{i=1}^{n}[y_i|\boldsymbol{\theta}]$, where individual data distributions can be multiplied to obtain the joint distribution because of independence. To fit the model using MLE, the likelihood is usually maximized numerically to find the optimal parameter values $\hat{\boldsymbol{\theta}}$.

The Bayesian approach involves the specification of a probability model for the parameters, $\boldsymbol{\theta} \sim [\boldsymbol{\theta}]$, that depend on fixed hyperparameters assumed to be known. The prior probability distribution should contain information about the parameters that is known before the data are collected, except for cases where regularization-based model selection is desired (Hooten and Hobbs 2015), in which case, the prior can be tuned based on a cross-validation procedure. Rather than maximizing the likelihood, the Bayesian approach seeks to find the conditional distribution of the parameters given the data (i.e., the posterior distribution)

$$[\boldsymbol{\theta}|\mathbf{y}] = \frac{[\mathbf{y}|\boldsymbol{\theta}][\boldsymbol{\theta}]}{\int [\mathbf{y}|\boldsymbol{\theta}][\boldsymbol{\theta}]\,d\boldsymbol{\theta}}, \tag{1.1}$$

where $\mathbf{y}$ is a vector notation for all the observations and the denominator in Equation 1.1 equates to a scalar constant after the data have been observed. For complicated models, the multidimensional integral in the denominator of Equation 1.1 cannot be obtained analytically (i.e., exactly by pencil and paper) and must be either numerically calculated or avoided using a stochastic simulation procedure. Markov chain Monte Carlo (MCMC; Gelfand and Smith 1990) allows us to obtain samples from the posterior distribution while avoiding the calculation of the normalizing constant in the denominator of Equation 1.1. MCMC algorithms have many advantages (e.g., easy to develop), but also limitations (e.g., can be time consuming to run).

Hierarchical models are composed of a sequence of nested probability distributions for the data, the process, and the parameters (Berliner 1996). For example, a basic Bayesian hierarchical model is

$$y_{i,j} \sim [y_{i,j}|z_i, \boldsymbol{\theta}], \tag{1.2}$$

$$z_i \sim [z_i|\boldsymbol{\beta}], \tag{1.3}$$

$$\boldsymbol{\theta} \sim [\boldsymbol{\theta}], \tag{1.4}$$

$$\boldsymbol{\beta} \sim [\boldsymbol{\beta}], \tag{1.5}$$

where z_i is an underlying process for individual i and $y_{i,j}$ are repeated measurements for each individual ($j = 1, \ldots, J$). Notice that the process model parameters $\boldsymbol{\beta}$ also require a prior distribution if the model is Bayesian. The posterior for this model is a generalized version of Equation 1.1 such that

$$[\mathbf{z}, \boldsymbol{\theta}, \boldsymbol{\beta} | \mathbf{y}] = \frac{[\mathbf{y}|\mathbf{z}, \boldsymbol{\theta}][\mathbf{z}|\boldsymbol{\beta}][\boldsymbol{\theta}][\boldsymbol{\beta}]}{\iiint [\mathbf{y}|\mathbf{z}, \boldsymbol{\theta}][\mathbf{z}|\boldsymbol{\beta}][\boldsymbol{\theta}][\boldsymbol{\beta}]\, d\mathbf{z}\, d\boldsymbol{\theta}\, d\boldsymbol{\beta}}. \tag{1.6}$$

Throughout the remainder of this book, we use both Bayesian and non-Bayesian models for statistical inference in the settings where they are appropriate. Many complicated hierarchical models are easier to implement from a Bayesian perspective, but may not always be necessary. Hobbs and Hooten (2015) provide an accessible description of both Bayesian and non-Bayesian methods and model-building strategies as well as an overview of basic probability and fundamental approaches for fitting models. Hereafter, we remind the reader of changes in notation and modeling strategies as necessary without dwelling on the details of a full implementation because those can be found in the referenced literature.

1.5 ADDITIONAL READING

The timeless reference describing the mathematics of animal movement processes is Turchin (1998), and while newer references exist, Turchin (1998) is still the default for many scientists. For a newer synthesis, the special issue in the *Philosophical Transactions of the Royal Society of London B* provided a cross section of contemporary ideas for modeling animal movement and analyzing telemetry data (see Cagnacci et al. 2010 for an overview). Schick et al. (2008) proposed a general hierarchical modeling structure to modeling telemetry data that many contemporary efforts now follow.

Historical, but still very relevant, references describing approaches for collecting and analyzing telemetry data include White and Garrott (1990), Kenward (2000), Millspaugh and Marzluff (2001), and Manly et al. (2007), although they focused more on vital rates (e.g., survival), resource selection, and home range estimation from radio telemetry data because that technology preceded current satellite telemetry devices.

Connecting telemetry data with population demographic data is still nascent. However, the field of SCR models is advancing rapidly and a few developments of SCR models have formally incorporated telemetry data to better characterize space use and resource selection. Also, individual-based movement models, in general, provide us with a better understanding about how animals are interacting with each other and their environment and the learning that is gained from fitting them can be used to develop smart demographic models that best account for features of population and community dynamics that depend on movement.

2 Statistics for Spatial Data

Spatial statisticians are often asked how conventional spatial statistics are relevant for animal ecology. In fact, there is an apparent gap between spatial statistics research and animal ecology research. To clarify what we mean by "spatial statistics," any statistical procedure—estimation, prediction, or modeling—that explicitly uses spatial information in data could be referred to as spatial statistics. In many cases, "spatial statistics" conventionally implies that second-order (i.e., covariance) estimation or modeling is employed to characterize dependence in the data or process, perhaps in addition to the first-order estimation (i.e., mean). Even though it fits into the generally accepted definition, a linear model with spatially indexed covariates is not typically thought of as spatial statistics. Furthermore, point process models belong in the realm of spatial statistics,* even though they are often only considered from a first-order perspective (though not always). Thus, given the inherent ambiguity with the terminology, we describe the classical models for each of the three main spatial processes:

1. Spatial point processes
2. Continuous spatial processes
3. Discrete spatial processes

Each of these processes and associated statistical methods is relevant for analyzing telemetry data. In Chapters 4 and 6, we show how spatial statistical concepts can be employed to analyze telemetry data. We do not intend this chapter to be comprehensive, but rather to serve as a reference for the important spatial processes in the formulation of animal movement models in the following chapters. See Cressie (1993) and Cressie and Wikle (2011) for additional material and references concerning spatial and spatio-temporal statistical modeling.

2.1 POINT PROCESSES

Point processes appear in many different settings, including geographical and temporal settings, but they can generally arise in any multidimensional real space. The basic concept that separates spatial point processes (SPPs) from the other spatial processes described in this chapter is that the *locations* associated with an observed SPP are the random quantities of interest. Continuous spatial processes (CSPs) also involve locations that are points in space, but the points are assumed to be fixed and known, instead of random. In CSPs (see the next section), we are often interested in other characteristics associated with the locations (e.g., soil moisture measurements taken at a set

* When the points fall in some geographic space.

of spatial locations), and it is those characteristics that are the random quantities of interest. For SPPs, we may also be interested in other characteristics associated with the points such as size, condition, or another variable associated with the point, but the point location is of primary interest. An SPP containing auxiliary information is referred to as a marked SPP.

Many types of SPPs have been studied and models have been formulated to provide inference using observed SPP data. In the two-dimensional (2-D) spatial setting, where the size (n) of the SPP is known, we can formulate a basic model for an SPP with data represented by location vectors $\mathbf{s}_i$ (containing the coordinates in some geographic space) such that $\mathbf{s}_i \sim f(\mathbf{s})$ for $i = 1, \ldots, n$ and with support $\mathbf{s}_i \in \mathcal{S}$. The probability density function (PDF) f stochastically controls the placement of the points $\mathbf{s}_i$, as it would for any other random variable. In the situation where n is unknown before observing the SPP, the size of the SPP is also random, and thus, a component of the overall random process that arises.

Consider a set of observed telemetry data for an individual bobcat (*Lynx rufus*; Figure 2.1). In this case, the positions of the individual are measured at an irregular set of times and presented in geographic space. Bobcat occur throughout much of North America and have been the subject of several scientific studies involving telemetry data. The data presented in Figure 2.1 were collected at the Welder Wildlife Foundation Refuge in southern Texas, USA (Wilson et al. 2010) using VHF telemetry techniques. We return to these data in what follows to demonstrate spatial statistical methods. Telemetry data, such as those shown in Figure 2.1, are often treated as SPP data, but doing so relies on several assumptions that we discuss in more detail as they arise.

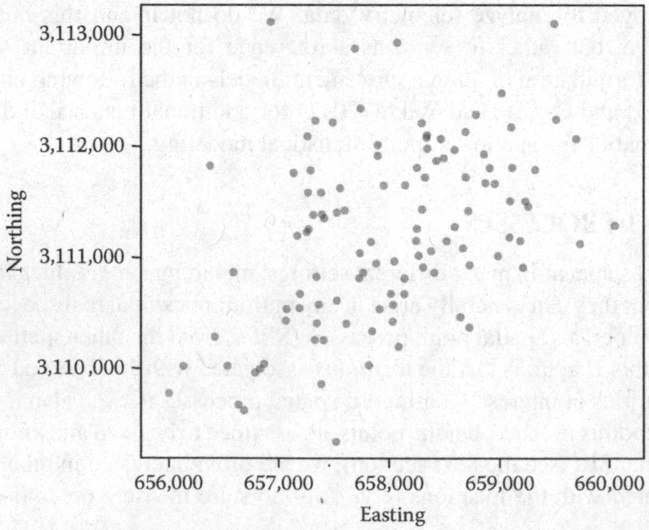

FIGURE 2.1 Measured positions ($\mathbf{s}_i$, for $i = 1, \ldots, 110$ in UTM) of an individual bobcat in the Welder Wildlife Foundation Refuge.

2.1.1 Homogeneous SPPs

If an SPP arises from a uniform probability distribution over $\mathcal{S}$, then it is called homogeneous, implying that the density giving rise to the points does not vary over the support* $\mathcal{S}$. These SPPs are commonly referred to as "complete spatial random" (CSR) processes in the spatial statistics literature and they are often used as a null model for testing whether the observed SPP arises from a probability distribution with spatially varying density.

Readers often find that the formal point process literature is very technical and difficult to understand. This is partially due to the mathematical rigor needed to derive theoretical results pertaining to point processes. It is often simpler to describe point processes in terms of how they are simulated, rather than how they may be formally specified in the statistical literature. For example, consider the homogeneous Poisson SPP, where the size of the SPP is an unknown random variable as well as the actual locations of the points. To simulate the homogeneous Poisson SPP in a statistical software is a trivial two-step procedure:

1. Sample $n \sim \text{Pois}(\lambda)$,
2. Sample $\mathbf{s}_i \sim \text{Unif}(\mathcal{S})$ for $i = 1, \ldots, n$,

where the intensity parameter λ is set *a priori* and equal to the expected size of the point process (i.e., $E(n) = \lambda$) and "Unif" is a multivariate uniform distribution (usually 2-D) for $\mathbf{s}_i \equiv (s_{1,i}, \ldots, s_{d,i})'$ in d dimensions. The resulting set of points $\mathbf{S} \equiv (\mathbf{s}_1, \ldots, \mathbf{s}_n)'$ is a realization from a homogeneous Poisson SPP (they will also be CSR).

Using an intensity of $\lambda = 100$, we simulated two independent realizations (i.e., random sets of points) from a 2-D CSR Poisson SPP (Figure 2.2). In this case, the first

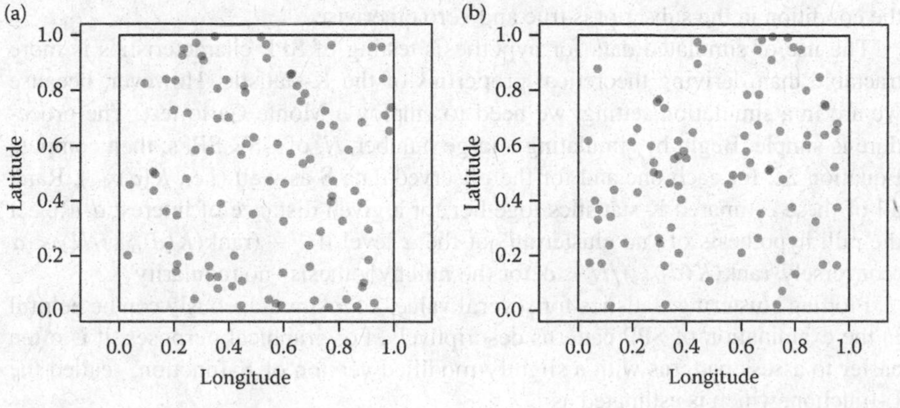

FIGURE 2.2 Simulated positions ($\mathbf{s}_{1,i}$, for $i = 1, \ldots, 94$ observations in panel (a) and $\mathbf{s}_{2,i}$, for $i = 1, \ldots, 84$ observations in panel (b)) on a unit square outlined in black.

---

* "Support" is the term used in statistics to describe the space where the random variable lives, or in other words, the values that $\mathbf{s}_i$ can take on.

simulated point process $\mathbf{S}_1$ contained $n_1 = 94$ observations (Figure 2.2a), whereas the second ($\mathbf{S}_2$) contained only $n_1 = 84$ observations (Figure 2.2b). In Figure 2.2, we see that both simulated point processes are different and show a somewhat "random" organization of spatial positions. They tend to occur throughout the spatial support (i.e., the unit square), but are neither regular (i.e., perfectly spaced out) nor clustered (i.e., grouped tightly together).

Few real SPPs are actually thought to be homogeneous in space. Rather, as previously mentioned, we merely leverage our ability to easily simulate CSR processes so that we can compare their behavior with observed SPPs. We need a formal way to compare SPPs (i.e., summary statistic) that has an intuitive interpretation and can be easily computed for both the real data and the numerous simulated data sets. The Ripley's K-statistic describes the degree of clustering or regularity in a point process. The K-statistic is often specified as an expectation:

$$K(d) = \lambda^{-1}\text{E}(\#\text{ of points within } d \text{ of any point}), \tag{2.1}$$

where d represents the distance for which we desire inference (e.g., is the process clustered or regular within distance d?). The K-statistic can be affected by edges of the spatial domain when points are close to it; thus, an edge-corrected estimator for the K-statistic was proposed by Ripley (1976):

$$\hat{K}(d) = \hat{\lambda}^{-1} \sum_{i \neq j} w(\mathbf{s}_i, \mathbf{s}_j)^{-1} 1_{\{d_{i,j} \leq d\}} / n, \tag{2.2}$$

where $d_{i,j}$ is the distance between point i and point j, $\hat{\lambda} = n/\text{area}(\mathcal{S})$, and $w(\mathbf{s}_i, \mathbf{s}_j)$ is the proportion of the circumference of a circle, centered at $\mathbf{s}_i$, that is inside the study region $\mathcal{S}$. The term $1_{\{d_{i,j} \leq d\}}$ is an indicator variable that is equal to one when the condition in the subscript is true and zero otherwise.

The use of simulated data for hypothesis testing of SPP characteristics is more tractable than deriving theoretical properties of the K-statistic. However, because we are in a simulation setting, we need to employ a Monte Carlo test. The procedure is simple. Begin by simulating a large number, N, of CSR SPPs; then compute Equation 2.2 for each one and for the observed data $\mathbf{S}$ as well (i.e., $\hat{K}(d)_{\text{obs}}$). Rank all of these estimated K-statistics together for a given distance of interest d. Reject the null hypothesis of "no clustering" at the α level if $1 - (\text{rank}(\hat{K}(d)_{\text{obs}})/N) < \alpha$ (conversely, $\text{rank}(\hat{K}(d)_{\text{obs}})/N < \alpha$ for the null hypothesis "no regularity").

Plotting clustering statistics for several values of d simultaneously can be helpful in the examination of SPP patterns descriptively. For graphical purposes, it is often easier to assess patterns with a slightly modified version of K-function,* called the L-function, which is estimated as

$$\hat{L}(d) = \sqrt{\hat{K}(d)/\pi} - d. \tag{2.3}$$

* We use the term "function" here because $K(d)$ and $L(d)$ are considered for a range of values of d.

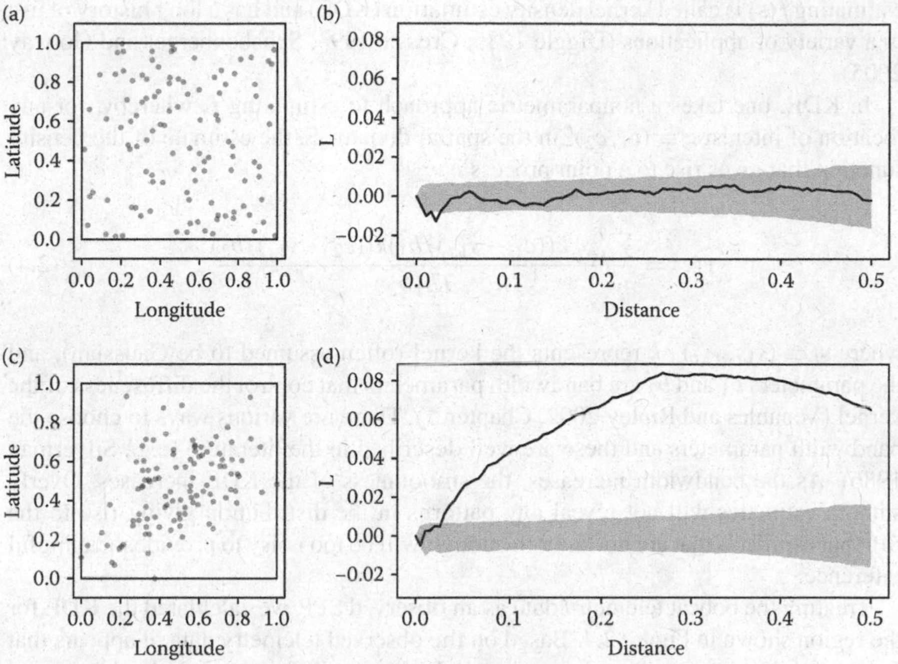

FIGURE 2.3 Simulated CSR SPP (a) and scaled bobcat telemetry SPP (c). Associated $\hat{L}(d)$ functions appear in the right panels (b: CSR and d: bobcat). The gray regions in panels (b) and (d) represent 95% intervals based on Monte Carlo simulation from 1000 CSR processes in the same spatial domain.

Significantly positive values of $\hat{L}(d)$ indicate clustering at distance d, whereas significantly negative values indicate regularity.

For example, we compared the bobcat data shown earlier (i.e., Figure 2.1) with a simulated CSR SPP with the same number of observations ($n = 110$) in Figure 2.3 over a range of distances d. Notice that we have rescaled the bobcat telemetry data (Figure 2.3c) so that they fit within the same bounding box as the CSR process (Figure 2.3a). While the CSR process stays largely within the simulation interval (region shown in gray), the bobcat telemetry SPP shows evidence of clustering beyond a distance of approximately 0.03 (in the scaled domain).*

2.1.2 Density Estimation

An estimate for the PDF $f(\mathbf{s})$, based on an observed SPP, is a common form of desired inference. In such cases, there are a variety of parametric and nonparametric approaches to estimate the density of an SPP and they depend on the desired form of inference and utility. One of the most commonly used nonparametric methods for

* Approximately 120 m in the original untransformed domain.

estimating $f(\mathbf{s})$ is called kernel density estimation (KDE) and has a long history of use in a variety of applications (Diggle 1985; Cressie 1993; Schabenberger and Gotway 2005).

In KDE, one takes a nonparametric approach to estimating f, whereby, for any location of interest $\mathbf{c} \equiv (c_1, c_2)'$ in the spatial domain $\mathcal{S}$, the estimate of the density function that gives rise to a point process is

$$\hat{f}(\mathbf{c}) = \frac{\sum_{t=1}^{T} k((c_1 - s_{1,t})/b_1)k((c_2 - s_{2,t})/b_2)}{Tb_1b_2}, \tag{2.4}$$

where $\mathbf{s}_t \equiv (s_{1,t}, s_{2,t})'$, k represents the kernel (often assumed to be Gaussian), and the parameters b_1 and b_2 are bandwidth parameters that control the diffuseness of the kernel (Venables and Ripley 2002, Chapter 5). There are various ways to choose the bandwidth parameters and these are well described in the literature (e.g., Silverman 1986). As the bandwidth increases, the smoothness of the KDE increases. Overly smooth estimates will not reveal any patterns in the distribution giving rise to the SPP but estimates that are not smooth enough will be too noisy to provide meaningful inference.*

Treating the bobcat telemetry data as an observed SPP, we calculated the KDE for the region shown in Figure 2.4. Based on the observed telemetry data, it appears that the spatial density function giving rise to the bobcat SPP is irregularly shaped and nonuniform. These results agree with the estimated L-function based on the same data (Figure 2.3c and d).

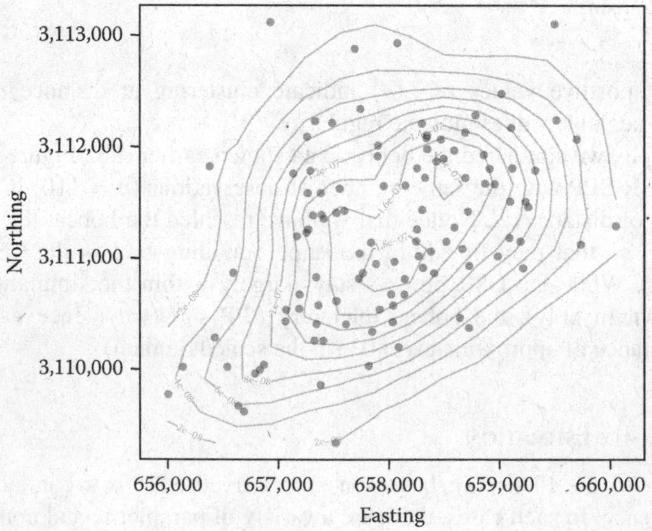

FIGURE 2.4 Kernel density estimate (shown as contours) for the bobcat telemetry data.

* See Fieberg (2007) for a review of KDE methods for telemetry data.

2.1.3 Parametric Models

The various descriptive and nonparametric approaches are excellent for illuminating patterns in SPPs; however, we are often interested in model-based inference concerning potential drivers of the patterns we observe. The heterogeneous Poisson SPP model can provide inference for predictors associated with the intensity function. Instead of a single value, λ, governing the number of points in the homogeneous SPP, we now consider an intensity function, $\lambda(\mathbf{s}|\boldsymbol{\beta})$, that varies over the domain $\mathcal{S}$ and is controlled by the parameters $\boldsymbol{\beta}$. Important properties of the heterogeneous Poisson SPP are (Illian et al. 2008):

1. For any subregion $\mathcal{B} \subseteq \mathcal{S}$, the number of events occurring within $\mathcal{B}$, $n(\mathcal{B})$ is a Poisson random variable with intensity

$$\tilde{\lambda}(\mathcal{B}|\boldsymbol{\beta}) = \int_{\mathcal{B}} \lambda(\mathbf{u}|\boldsymbol{\beta}) d\mathbf{u}. \tag{2.5}$$

   The expression (2.5) implies that the expected total number of points in $\mathcal{S}$ is $E(n(\mathcal{S})) = \tilde{\lambda}(\mathcal{S}|\boldsymbol{\beta})$.
2. For any J regions, $\mathcal{B}_1, \ldots, \mathcal{B}_J \subseteq \mathcal{S}$, that do not overlap, the number of points in each subregion, $n(\mathcal{B}_1), \ldots, n(\mathcal{B}_J)$, are independent Poisson random variables.

Independence in a Poisson SPP can be interpreted as a lack of interaction among the points; that is, point locations are not a function of each other directly.

For statistical inference, we wish to estimate $\boldsymbol{\beta}$ given an observed SPP. Estimation can be accomplished with maximum likelihood using the Poisson SPP likelihood function. The likelihood function can be constructed by conditioning on n observed points $\mathbf{s}_1, \ldots, \mathbf{s}_n$ distributed as

$$\begin{aligned} f(\mathbf{s}_1, \ldots, \mathbf{s}_n) &= \prod_{i=1}^{n} \frac{\lambda(\mathbf{s}_i|\boldsymbol{\beta})}{\tilde{\lambda}(\mathcal{S}|\boldsymbol{\beta})} \\ &= \frac{\prod_{i=1}^{n} \lambda(\mathbf{s}_i|\boldsymbol{\beta})}{\tilde{\lambda}(\mathcal{S}|\boldsymbol{\beta})^n}. \end{aligned}$$

From the first property above, $n \sim$ Poisson $(\tilde{\lambda}(\mathcal{S}|\boldsymbol{\beta}))$ implies that the full likelihood is the joint PDF of $(\mathbf{s}_1, \ldots, \mathbf{s}_n, n)$ as a function of $\boldsymbol{\beta}$:

$$L(\boldsymbol{\beta}) = \frac{n! \prod_{i=1}^{n} \lambda(\mathbf{s}_i|\boldsymbol{\beta})}{\tilde{\lambda}(\mathcal{S}|\boldsymbol{\beta})^n} \times \frac{\tilde{\lambda}(\mathcal{S}|\boldsymbol{\beta})^n e^{-\tilde{\lambda}(\mathcal{S}|\boldsymbol{\beta})}}{n!} \tag{2.6}$$

$$= \left(\prod_{i=1}^{n} \lambda(\mathbf{s}_i|\boldsymbol{\beta})\right) \exp(-\tilde{\lambda}(\mathcal{S}|\boldsymbol{\beta})), \tag{2.7}$$

The $n!$ term in Equation 2.6 arises because it does not matter in which order the points are observed. Thus, the indices can be permuted $n!$ different ways. The log of Equation 2.7 yields the classic form of the log likelihood for the Poisson SPP:

$$l(\boldsymbol{\beta}) = \sum_{i=1}^{n} \log \lambda(\mathbf{s}_i|\boldsymbol{\beta}) - \int_{\mathcal{S}} \lambda(\mathbf{s}|\boldsymbol{\beta}) d\mathbf{s}. \tag{2.8}$$

Typically, $\boldsymbol{\beta}$ controls the relationship between $\lambda(\mathbf{s}|\boldsymbol{\beta})$ and a vector of spatially referenced covariates, $\mathbf{x}(\mathbf{s})$. A commonly assumed relationship between the intensity and the covariates is $\lambda(\mathbf{s}|\boldsymbol{\beta}) = \exp(\mathbf{x}'(\mathbf{s})\boldsymbol{\beta})$ such that the regression coefficients $\boldsymbol{\beta}$ imply the strength of a relationship. Substituting this intensity function into Equation 2.8 results in

$$l(\boldsymbol{\beta}) = \sum_{i=1}^{n} \mathbf{x}'(\mathbf{s}_i)\boldsymbol{\beta} - \int_{\mathcal{S}} \exp(\mathbf{x}'(\mathbf{s})\boldsymbol{\beta}) d\mathbf{s}. \tag{2.9}$$

One can now proceed with standard statistical model fitting, from a likelihood or Bayesian perspective, by either maximizing Equation 2.9 or assigning a prior distribution for $\boldsymbol{\beta}$ (e.g., $\boldsymbol{\beta} \sim \text{N}(\mathbf{0}, \boldsymbol{\Sigma}_\beta)$) and finding the posterior distribution of $\boldsymbol{\beta}|\mathbf{s}_1, \ldots, \mathbf{s}_n$.

The main challenge in fitting the Poisson SPP model is that the integral on the right-hand side of Equations 2.8 and 2.9 must be computed at every step in an optimization or sampling algorithm because it contains the parameter vector $\boldsymbol{\beta}$. The added computational cost of the required numerical integration can lead to cumbersome algorithms for direct maximization of the log likelihood and this is compounded if the model is fit using MCMC. However, recent findings have shown that inference using the inhomogeneous Poisson SPP model can be achieved in a wide variety of ways, often with readily available statistical software. See Berman and Turner (1992), Baddeley and Turner (2000), and Illian et al. (2013) for more detailed descriptions. We provide a very brief introduction to the general approaches used in model fitting in what follows.

There are two basic methods for approximating the log likelihood in Equation 2.9. First, if $\mathbf{x}(\mathbf{s})$ is defined on a grid of cells over $\mathcal{S}$, then we can use the first property of Poisson SPP and sum all of the events occurring in each cell, y_j. Each y_j is an independent Poisson random variable with rate $\lambda_j = \exp(a_j + \mathbf{x}_j'\boldsymbol{\beta})$, where a_j is the area of cell j. One can use any statistical software to fit a Poisson regression with offset equal to a_j. This does not really numerically approximate the likelihood, but rather uses a summary of the raw data that retain much of the original information and has a more usable likelihood. The second technique approximates the likelihood function itself. The likelihood approximation is known as the Berman–Turner device (Berman and Turner 1992), which can be described as

1. Partition $\mathcal{S}$ into J regions (e.g., grid cells) and take the centroids, $\mathbf{c}_1, \ldots, \mathbf{c}_J$, as quadrature points. The integral is then approximated with

$$\int_{\mathcal{S}} \exp(\mathbf{x}'(\mathbf{s})\boldsymbol{\beta}) d\mathbf{s} \approx \sum_{j=1}^{J} w_j \exp(\mathbf{x}'(\mathbf{c}_j)\boldsymbol{\beta}),$$

where w_j is the area of the jth region.

2. After substituting the approximation into Equation 2.9 and combining the quadrature points and data together, we obtain

$$l(\boldsymbol{\beta}) \approx \sum_{j=1}^{n+J} w_j \left(y_j \mathbf{x}'(\mathbf{s}_j)\boldsymbol{\beta} - \exp(\mathbf{x}'(\mathbf{s}_j)\boldsymbol{\beta})\right), \tag{2.10}$$

where y_j equals $1/w_j$ if $\mathbf{s}_j$ is an observed location and 0 if $\mathbf{s}_j$ is a quadrature point. Notice that Equation 2.10 is in the form of a weighted Poisson log likelihood with rate $\exp(\mathbf{x}'(\mathbf{s}_j)\boldsymbol{\beta})$; thus, any generalized linear model (GLM) software can be used to fit this approximation.

When the number of points is known *a priori*, the SPP likelihood in Equation 2.6 simplifies to

$$L(\boldsymbol{\beta}) = \frac{n! \prod_{i=1}^{n} \lambda(\mathbf{s}_i|\boldsymbol{\beta})}{\tilde{\lambda}(\mathcal{S}|\boldsymbol{\beta})^n}, \tag{2.11}$$

resulting in the log likelihood

$$l(\boldsymbol{\beta}) = \log(n!) + \sum_{i=1}^{n} \log \lambda(\mathbf{s}_i|\boldsymbol{\beta}) - n \cdot \log \int_{\mathcal{S}} \lambda(\mathbf{s}|\boldsymbol{\beta}) d\mathbf{s}. \tag{2.12}$$

The integral in Equation 2.12 must still be calculated to maximize the likelihood with respect to $\boldsymbol{\beta}$. Similar methods can be used to fit the SPP model when n is known and we describe several of these in Chapter 4. The general form of PDF,

$$\frac{\lambda(\mathbf{s}_i|\boldsymbol{\beta})}{\tilde{\lambda}(\mathcal{S}|\boldsymbol{\beta})}, \tag{2.13}$$

for a point ($\mathbf{s}_i$) arising from a point process distribution has been referred to as a "weighted distribution" in the statistical literature (e.g., Patil and Rao 1976, 1977, 1978).

There are three other useful classes of point process models for analyzing animal telemetry data that we briefly mention here and expand upon in Chapter 4 where they are directly discussed in reference to telemetry data models. The first class of models is log Gaussian Cox process (LGCP) model. The LGCP model is a simple extension to the Poisson SPP in Equation 2.9 with intensity function modeled as

$$\lambda(\mathbf{s}|\boldsymbol{\beta}) = \exp(\mathbf{x}'(\mathbf{s})\boldsymbol{\beta} + \eta(\mathbf{s})), \tag{2.14}$$

where $\eta(\mathbf{s})$ is a random spatial process as described in the following sections. Estimation for the LGCP model follows by using one of the previously described approximations with either a random effect component in the model (i.e., a generalized linear mixed model [GLMM]) or, alternatively, a Poisson generalized additive model (GAM), where $\eta(\mathbf{s})$ is a spatial spline or basis function. The LGCP model is useful for modeling clustering of points that is not fully explained by the covariates.

The second class of models that extends the Poisson SPP is Gibbs spatial point processes (GSPPs). The GSPP extends the Poisson SPP by allowing interactions among points. GSPPs are a very broad class of models, but a very useful and flexible subset is the pairwise interacting processes. In a GSPP, one usually conditions on the observed number of points. For a pairwise interacting GSPP, conditioning on the observed number of points results in a likelihood of the form

$$L(\boldsymbol{\beta}) = z_{\boldsymbol{\beta}} \exp\left\{ -\sum_{i=1}^{n} \alpha(\mathbf{s}_i|\boldsymbol{\beta}) + \sum_{i=1}^{n-1}\sum_{l=i+1}^{n} \phi(\delta_{il}|\boldsymbol{\beta}) \right\}, \tag{2.15}$$

where $\alpha(\mathbf{s}_i, \boldsymbol{\beta})$ is a spatial effect (e.g., $\mathbf{x}'(\mathbf{s}_i)\boldsymbol{\beta}$), ϕ is a potential function that decreases with increasing distance between points ($\delta_{ij} \equiv ||\mathbf{s}_i - \mathbf{s}_j||$) and controls the interaction among points, and $z_{\boldsymbol{\beta}}$ is a normalizing term that ensures the likelihood is a PDF with respect to $\mathbf{s}_1, \ldots, \mathbf{s}_n$. While the likelihood in Equation 2.15 appears relatively benign, the $z_{\boldsymbol{\beta}}$ needed is usually analytically intractable and cannot be easily evaluated. However, Baddeley and Turner (2000) and Illian et al. (2013) have examined methods similar to the previously described approximations for fitting GSPPs. In Chapter 4, we illustrate a method similar to Illian et al. (2013) for developing and fitting a GSPP model specifically for animal telemetry data.

2.2 CONTINUOUS SPATIAL PROCESSES

In our review of conventional spatial statistics, we now introduce the most widely known class of spatial models. Often called "geostatistical" models for their roots in the geological sciences and mining industry (Cressie 1990), these models are relevant for CSPs. Unlike with SPPs where the random quantity of interest is the location of the data, with CSPs, the locations are known, but the random quantity of interest (i.e., response variable) is a measured characteristic (or set of characteristics) at the known locations.

To formalize this, consider a variable of interest $y(\mathbf{s}_i)$ measured at a set of spatial locations $\mathbf{s}_1, \ldots, \mathbf{s}_n$ in the spatial region $\mathcal{S}$ of interest (note that $\mathcal{S}$ was referred to as the "support" in the preceding section). For CSPs, we often seek to (1) characterize their relationship with other spatially varying covariates and/or (2) predict at a set of unobserved locations.

An example of data arising from a CSP are average maximum temperatures collected at locations throughout the Midwestern United States (Figure 2.5; from the U.S. historical climate network; see Wikle 2010b for details). In viewing these data, we see that the temperatures generally increase from north to south and have a nonlinear pattern from east to west. A goal in analyzing these data might be to use the

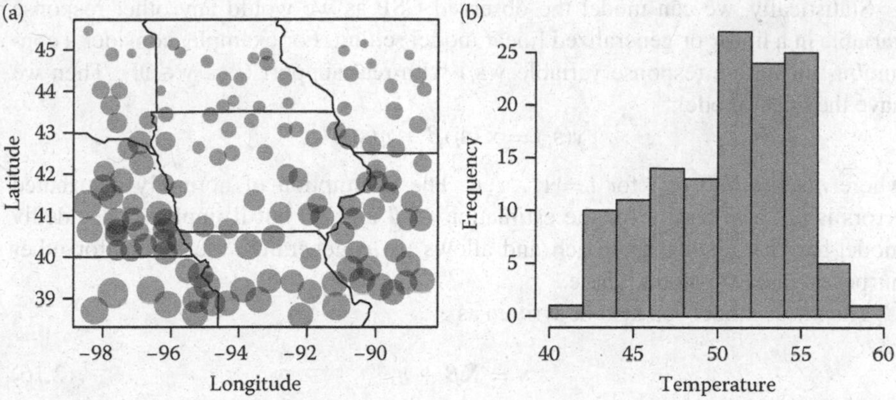

FIGURE 2.5 (a) Temperature data in February 1941, from portions of eight Midwestern states (states outlined in black). Relative temperature values indicated by circle size. (b) Frequency histogram for the average maximum temperatures in degrees Fahrenheit.

information about the measurement locations to predict temperature throughout the region (i.e., the states of South Dakota, Nebraska, Kansas, Minnesota, Iowa, Missouri, Michigan, and Illinois).

2.2.1 Modeling and Parameter Estimation

We begin by describing a model-based procedure for goal 1 first (inference) and then move to goal 2 (prediction). For example, suppose we wish to use the geographic position to explain temperature in the Midwestern United States. To help explain the general trend in temperature across the study region, we can use longitude, latitude, or various transformations of both (Figure 2.6).

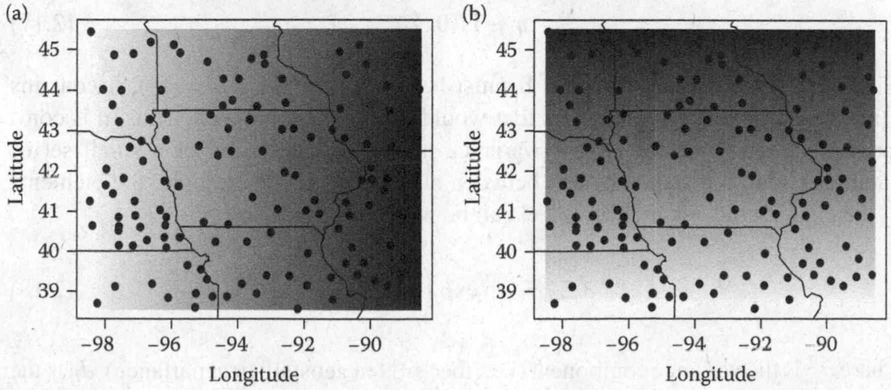

FIGURE 2.6 Covariates (i.e., predictor variables) for temperature: (a) longitude and (b) latitude, shown as spatial maps (larger values shown in darker shade). U.S. state boundaries overlaid in black. Points represent measurement locations.

Statistically, we can model the observed CSP as we would any other response variable in a linear or generalized linear model setting. For example, consider a continuous univariate response variable $y(\mathbf{s}_i)$ with real support (i.e., $y \in \Re$). Then we have the linear model:

$$y(\mathbf{s}_i) = \mathbf{x}'(\mathbf{s}_i)\boldsymbol{\beta} + \eta(\mathbf{s}_i),$$

where $\eta(\mathbf{s}_i) \sim N(0, \sigma^2)$ for $i = 1, \ldots, n$. The assumption of normally distributed errors is not a necessity for the estimation of $\boldsymbol{\beta}$ and σ^2 but it implies a decidedly model-based statistical approach and allows us to generalize the model for other purposes; thus, we retain it here.

The linear model can also be written as

$$\mathbf{y} = \mathbf{X}\boldsymbol{\beta} + \boldsymbol{\eta}, \tag{2.16}$$

where $\mathbf{y}$ is an $n \times 1$ response vector, $\mathbf{X}$ is the $n \times q$ "design matrix" containing a vector of ones and $q - 1$ covariates, $\boldsymbol{\beta}$ is the $q \times 1$ vector of regression coefficients, $\boldsymbol{\eta} \sim \text{N}(\mathbf{0}, \sigma^2\mathbf{I})$ is the $n \times 1$ vector of errors that is multivariate normal, and $\mathbf{I}$ is the $n \times n$ identity matrix (i.e., a matrix with ones on the diagonal and zeros elsewhere). The matrix notation allows for much more compact model specifications and for easier analytical and computational calculations relevant to the model. We use a similar notation (2.16) for regression specifications throughout this book.

In model-based geostatistics,* we assume that broad spatial patterns in the data can be explained by the covariates through the first-order (i.e., mean or "trend") term $\mathbf{X}\boldsymbol{\beta}$. Then, any remaining structure in the data has to be absorbed by the error $\boldsymbol{\eta}$. If the errors are not independent and identically distributed (iid), then the conventional regression model (2.16) is not appropriate because the assumptions are not met and the model cannot be relied upon for correct inference. In the special case where the errors may be spatially correlated, we could add a structured second-order (i.e., covariance) process to the model. That is, we let the errors be dependent such that $\text{cov}(\eta(\mathbf{s}_i), \eta(\mathbf{s}_j)) \neq 0$; or in matrix notation, we have the model for $\boldsymbol{\eta}$:

$$\boldsymbol{\eta} \sim \text{N}(\mathbf{0}, \boldsymbol{\Sigma}). \tag{2.17}$$

Given that the covariance matrix $\boldsymbol{\Sigma}$ must be symmetric (i.e., $\boldsymbol{\Sigma} = \boldsymbol{\Sigma}'$), it contains $n(n-1)/2$ covariance parameters that would need to be estimated. Thus, it is common to parameterize the error covariance matrix as a function of a small set of parameters and a distance matrix between all locations. For example, the elements of the exponential covariance model can be written as

$$\Sigma_{ij} = \sigma^2 \exp\left(-\frac{d_{ij}}{\phi}\right), \tag{2.18}$$

where σ^2 is the variance component (i.e., the "sill" in geostatistical parlance), d_{ij} is the distance between locations $\mathbf{s}_i$ and $\mathbf{s}_j$ (often written as $d_{ij} \equiv ||\mathbf{s}_i - \mathbf{s}_j||$, where the double bar notation implies a "norm"), and ϕ is a spatial range parameter. As ϕ increases,

* The phrase "model-based geostatistics" was coined by Peter Diggle (Diggle et al. 1998).

the range of spatial structure in the second-order process η also increases. Thus, in fitting this model, there is only one additional parameter (ϕ) to estimate beyond the $q+1$ parameters in the conventional regression model (2.16). Also note that the covariance matrix (2.17) can be written as $\mathbf{\Sigma} \equiv \sigma^2 \mathbf{R}(\phi)$, where $\mathbf{R}(\phi) \equiv \exp(-\mathbf{D}/\phi)$ for pairwise distance matrix $\mathbf{D}$.*

Numerous covariance models have been used to capture different types of spatial dependence in the errors (e.g., Matern, Gaussian, spherical), and some are more general than others. There are many excellent spatial statistics references, but Banerjee et al. (2014) (p. 21) provided a particularly useful succinct summary of covariance models.

Geostatistical models can be fit using generalized least squares (GLS), maximum likelihood, or Bayesian methods. In the nonparametric setting, the residuals $e(\mathbf{s}_i) = y(\mathbf{s}_i) - \hat{y}(\mathbf{s}_i)$, arising from a model fit based on Equation 2.16, are used to empirically characterize the covariance using either a covariogram or a variogram. The covariogram ($c(\mathbf{s}_i, \mathbf{s}_j)$) and variogram ($2\gamma(\mathbf{s}_i, \mathbf{s}_j)$) are directly related to each other under certain conditions by $c(\mathbf{s}_i, \mathbf{s}_j) = c(0) - \gamma(\mathbf{s}_i, \mathbf{s}_j)$. Under the assumption of stationarity, the variogram for the errors can be expressed as

$$\begin{aligned} 2\gamma(\mathbf{s}_i, \mathbf{s}_j) &= \mathrm{Var}(\eta(\mathbf{s}_i) - \eta(\mathbf{s}_j)) \\ &= \mathrm{E}\left((\eta(\mathbf{s}_i) - \eta(\mathbf{s}_j))^2\right), \end{aligned} \tag{2.19}$$

where the last equality arises because $\eta(\mathbf{s}_i)$ and $\eta(\mathbf{s}_j)$ are assumed to have a constant mean. The variogram is then estimated with the "empirical variogram"

$$2\hat{\gamma}(\mathbf{s}_i, \mathbf{s}_j) = \frac{1}{n_b} \sum_{\mathcal{S}_b} (\eta(\mathbf{s}_i) - \eta(\mathbf{s}_j))^2, \tag{2.20}$$

where $\mathcal{S}_b$ is a set of location pairs falling into a vector difference bin of choice and n_b is the size of this set (i.e., number of pairs in the bin). Often, $2\hat{\gamma}(\mathbf{s}_i, \mathbf{s}_j)$ is calculated for a set of bins, usually over a range of distances. Also note that, if the spatial process η is not observed directly, the residuals $\mathbf{e}$ resulting from a fit of the independent error model (2.16) are used instead. The empirical variogram is a moment-based estimator that is often credited to Matheron (1963) (though see Cressie 1990 for a discussion of the history of geostatistics).†

Estimated only as a function of Euclidean distance between observation locations, the empirical semivariogram for the raw temperature data is shown in Figure 2.7a, whereas the semivariogram for the residuals (after regressing temperature on longitude and latitude) is shown in Figure 2.7b. In Figure 2.7, both semivariograms generally increase to an asymptote, but the semivariogram for raw temperature has

* The "exp" is an element-wise exponential, exponentiating each element of the matrix on the inside of the parentheses.

† The term "semivariogram" is often used in the spatial statistics literature and refers to $\gamma(d)$, differing from the variogram by a factor of 2. Spatial statistical software often computes the semivariance directly.

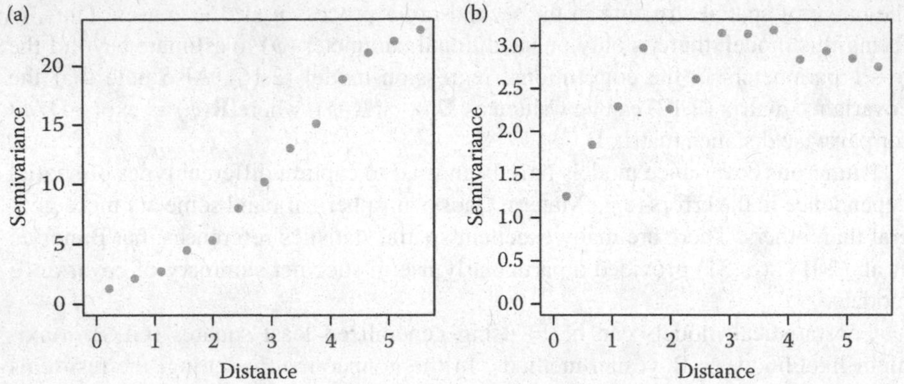

FIGURE 2.7 Empirical semivariogram for temperature data (a) and (b) residuals after regressing temperature on longitude and latitude. The range of the x-axis is half of the maximum distance in the spatial domain.

a much larger asymptote and reaches it at larger distances. The maximum semivariance is smaller for the residuals because most of the variation has been accounted for by the covariates (i.e., longitude and latitude). Also, the point at which the semivariogram levels off for the residuals occurs at a smaller distance because the range of spatial structure in the raw temperatures includes the major north–south trend.

In layman's terms, the two key assumptions in geostatistical modeling can be intuited as

1. *Stationarity*: The spatial structure of η does not vary with location.
   a. *Intrinsic stationarity*: The variance of the process for any pair of locations is only a function of the vector difference (i.e., $\mathbf{s}_i - \mathbf{s}_j$) between locations.*
   b. *Second-order stationarity*: The covariance of the process at any pair of locations is only a function of the vector difference between locations.
2. *Isotropy*: The spatial structure of η does not vary with direction; the process only depends on the distance between locations.

A second-order stationary process is also intrinsically stationary. Thus, the necessary condition for the variogram is stricter than that for the covariogram. However, if the process is intrinsically stationary, then the empirical variogram's advantage is that it does not involve estimation of the mean. In practice, it is common to estimate the spatial dependence in η with the empirical variogram regardless of the true stationarity of the process.

Even though we only observe them at a finite set of locations, CSPs are continuous. Thus, the estimated covariance is valid for any locations in the spatial domain, not just where the data were collected. This is the fundamental element that allows for

* A vector difference, $\mathbf{s}_i - \mathbf{s}_j$, for example, is a vector of the same dimension as the position vectors that contains information about distance and direction from $\mathbf{s}_i$ to $\mathbf{s}_j$.

prediction at locations that were unobserved. The continuity in space is one reason why spatial maps are often referred to as "processes."*

Before we turn to prediction, we describe the covariance modeling that is used in many applications of geostatistics. After the empirical variogram is estimated and plotted against d for a set of distance bins (e.g., Figure 2.7), it allows us to visualize the spatial structure in the process. It is often critical to find a parametric form for this covariance so that (1) the covariance matrix can be used for further inference and (2) we can learn about the covariance at distances other than those used in the empirical variogram. The ability to calculate covariance for all locations in the spatial domain facilitates prediction. Thus, we must find a parametric model that fits the empirical variogram well. Like the covariance models discussed earlier (2.18), there is a suite of parametric variogram models that are related to the covariance models through $c(d) = c(0) - \gamma(d)$. Weighted least squares is a common method for fitting the parametric variogram model to the empirical variogram and yields parameter estimates for σ^2 and ϕ (and others if the model contains more). The covariance parameter estimates can be substituted into the covariance matrix $\hat{\boldsymbol{\Sigma}}$, which can then be used for estimating $\boldsymbol{\beta}$ from the linear regression model (2.16) using GLS:

$$\hat{\boldsymbol{\beta}}_{GLS} = (\mathbf{X}'\hat{\boldsymbol{\Sigma}}^{-1}\mathbf{X})^{-1}\mathbf{X}'\hat{\boldsymbol{\Sigma}}^{-1}\mathbf{y}. \tag{2.21}$$

In principle, this process is iterated such that the covariance matrix is estimated based on the new residuals $\mathbf{e} = \mathbf{y} - \mathbf{X}\hat{\boldsymbol{\beta}}_{GLS}$ and then Equation 2.21 is used again to update the regression coefficient estimates. In practice, we have found that the iteratively reweighted least squares procedure requires few iterations to converge to stable estimates.

There are several alternatives to the iterative reweighted least squares estimation procedure, including maximum likelihood and Bayesian methods. In the case of maximum likelihood, we begin with the fully parametric model and seek to find the parameter values that maximize

$$L(\boldsymbol{\beta}, \sigma^2, \phi) \propto |\boldsymbol{\Sigma}(\sigma^2, \phi)|^{-1/2} \exp\left(-\frac{1}{2}(\mathbf{y} - \mathbf{X}\boldsymbol{\beta})'\boldsymbol{\Sigma}(\sigma^2, \phi)^{-1}(\mathbf{y} - \mathbf{X}\boldsymbol{\beta})\right), \tag{2.22}$$

where $\boldsymbol{\Sigma}(\sigma^2, \phi)$ makes it explicit that the covariance matrix depends on the parameters σ^2 and ϕ. In the Bayesian framework, we specify priors for the model parameters ($\boldsymbol{\beta}$, σ^2, and ϕ) and find the joint posterior distribution of these parameters given the data.

In cases where there may be small-scale variability or sources of measurement error, the geostatistical model is modified slightly to include an uncorrelated error term (often referred to as a "nugget" effect in the spatial statistics literature) such that

$$\mathbf{y} = \mathbf{X}\boldsymbol{\beta} + \boldsymbol{\eta} + \boldsymbol{\varepsilon}, \tag{2.23}$$

* A Gaussian process, for example, is a continuous random process arising from a normal distribution (perhaps in many dimensions) with covariance structure.

where $\boldsymbol{\varepsilon} \sim \text{N}(\mathbf{0}, \sigma_\varepsilon^2 \mathbf{I})$. This generalization adds an additional parameter to the model that needs to be estimated, but provides a way for the error in the spatial process to arise from correlated and uncorrelated sources.

Using the temperature data as an example, the semivariogram for the temperature residuals (Figure 2.7b) suggests a nugget may be useful for describing the covariance structure because the semivariance is larger than zero at very small distances.

2.2.2 Prediction

Optimal prediction of the response variable y, in the spatial context, is referred to as "Kriging," named after the mining engineer D.L. Krige (see Cressie 1990 for details). Given that response variables (i.e., data) are considered random variables until they are observed, for prediction, we seek the conditional distribution of unobserved response variables given those that were observed. That is, for a set of observed data $\mathbf{y}_o$ and a set of unobserved data $\mathbf{y}_u$, we wish to characterize the distribution $[\mathbf{y}_u|\mathbf{y}_o]$, or at least moments of this probability distribution, which is referred to as the predictive distribution.* In the case of interpolation (prediction within the space of the data), we are often interested in obtaining the predictions $\hat{\mathbf{y}}_u = E(\mathbf{y}_u|\mathbf{y}_o)$. A tremendously useful feature of the Gaussian distribution is that it has analytically tractable marginal and conditional distributions that are also Gaussian. To see this, consider the joint distribution of the observed and unobserved data, such that

$$\begin{pmatrix} \mathbf{y}_o \\ \mathbf{y}_u \end{pmatrix} \sim N\left(\begin{pmatrix} \mathbf{X}_o \\ \mathbf{X}_u \end{pmatrix} \boldsymbol{\beta}, \begin{pmatrix} \boldsymbol{\Sigma}_{o,o} & \boldsymbol{\Sigma}_{o,u} \\ \boldsymbol{\Sigma}_{u,o} & \boldsymbol{\Sigma}_{u,u} \end{pmatrix} \right), \tag{2.24}$$

where the o subscript is used to denote correspondence with the observed data set and u with the unobserved data set and the associated covariance and cross-covariance matrices are indicated by the ordering of their subscripts. Then, using properties of the multivariate normal distribution, the conditional distribution of the unobserved data ($\mathbf{y}_u$) given the observed data ($\mathbf{y}_o$) is

$$\mathbf{y}_u|\mathbf{y}_o \sim N(\mathbf{X}_u\boldsymbol{\beta} + \boldsymbol{\Sigma}_{u,o}\boldsymbol{\Sigma}_{o,o}^{-1}(\mathbf{y}_o - \mathbf{X}_o\boldsymbol{\beta}), \boldsymbol{\Sigma}_{u,u} - \boldsymbol{\Sigma}_{u,o}\boldsymbol{\Sigma}_{o,o}^{-1}\boldsymbol{\Sigma}_{o,u}). \tag{2.25}$$

When the parameters are all known, Equation 2.25 is the exact predictive distribution of the unobserved data; thus, the Kriging predictions are obtained using

$$\hat{\mathbf{y}}_u = \mathbf{X}_u\boldsymbol{\beta} + \boldsymbol{\Sigma}_{u,o}\boldsymbol{\Sigma}_{o,o}^{-1}(\mathbf{y}_o - \mathbf{X}_o\boldsymbol{\beta}), \tag{2.26}$$

which is also known as the best linear unbiased predictor (BLUP). The BLUP is a well-known statistical concept used in many forms of prediction.

---

* As previously mentioned, brackets used as $[\cdot]$ denote a probability distribution. Originally, the bracket notation used in this way (Gelfand and Smith 1990) represented a PDF or PMF, but more recently, it has been adopted as a space-saving notation for probability distributions in general (Hobbs and Hooten 2015).

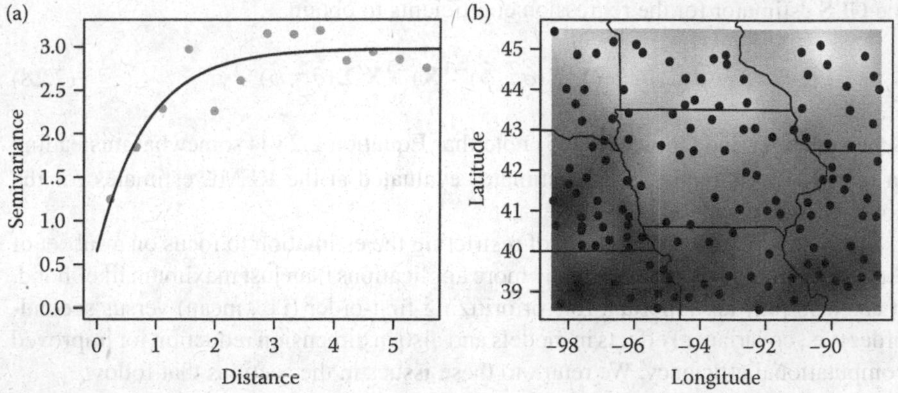

FIGURE 2.8 (a) Empirical (points) and fitted (line) semivariogram for the residuals after regressing temperature on longitude and latitude. Fitted semivariogram is based on an exponential covariance model with nugget effect. (b) Spatial predictions for temperature using Kriging (darker is warmer). U.S. state boundaries overlaid as lines and observation locations shown as points.

In the Midwestern U.S. temperature example (Figure 2.5), we fit a covariance model to the residuals (Figure 2.8a) and performed Kriging to obtain optimal predictions of temperature for the entire spatial domain (Figure 2.8b). The resulting temperature prediction field is much more flexible than a regression on the covariates (e.g., longitude and latitude) could provide alone.

2.2.3 Restricted Maximum Likelihood

We have described how model parameters in Equation 2.22 could be estimated using least squares techniques. However, alternative model-based approaches can also be used. One commonly used alternative method for fitting the geostatistical model in Equation 2.23 is referred to as restricted maximum likelihood (REML). Originally described by Patterson and Thompson (1971), REML is a method where the data are transformed such that the regression coefficients $\boldsymbol{\beta}$ are removed from the likelihood in a way that leaves only structure orthogonal to $\mathbf{X}$, and then the remaining "profile likelihood"* is maximized with respect to the covariance parameters. Using REML, consider the transformed data $\mathbf{Gy}$, where $\mathbf{G} = \mathbf{I} - \mathbf{X}(\mathbf{X}'\mathbf{X})^{-1}\mathbf{X}'$, then the profile likelihood for $\mathbf{Gy}$ reduces to

$$L(\sigma^2, \phi) \propto |\mathbf{G}\boldsymbol{\Sigma}(\sigma^2, \phi)\mathbf{G}'|^{-1/2} \exp\left(-\frac{1}{2}(\mathbf{Gy})'(\mathbf{G}\boldsymbol{\Sigma}(\sigma^2, \phi)\mathbf{G}')^{-1}\mathbf{Gy}\right) \quad (2.27)$$

because $E(\mathbf{Gy}) = (\mathbf{I} - \mathbf{X}(\mathbf{X}'\mathbf{X})^{-1}\mathbf{X}')\mathbf{X}\boldsymbol{\beta} = \mathbf{0}$. After Equation 2.27 is maximized to obtain $\hat{\sigma}^2$ and $\hat{\phi}$, these covariance parameter estimates can be substituted back into

* Sometimes called a residual likelihood in other literature.

the GLS estimator for the regression coefficients to obtain

$$\hat{\boldsymbol{\beta}}_{REML} = (\mathbf{X}'\boldsymbol{\Sigma}(\hat{\sigma}^2, \hat{\phi})^{-1}\mathbf{X})^{-1}\mathbf{X}'\boldsymbol{\Sigma}(\hat{\sigma}^2, \hat{\phi})^{-1}\mathbf{y}. \tag{2.28}$$

Schabenberger and Gotway (2005) note that Equation 2.28 is somewhat misleading in that $\hat{\boldsymbol{\beta}}_{REML}$ is really a GLS estimator evaluated at the REML estimates for the covariance parameters.

In the big picture, this concept of restricting the estimation to focus on a subset of the larger parameter space has many more applications than just maximum likelihood. It can also play an important role prioritizing first-order (i.e., mean) versus second-order (i.e., covariance) effects in models and also in dimension reduction for improved computational efficiency. We return to these issues in the sections that follow.

2.2.4 Bayesian Geostatistics

The Bayesian version of the previously described geostatistical model (2.23) is essentially the same except for (1) the formal treatment of model parameters as random quantities and (2) a formal mechanism to incorporate prior information. Even though the Bayesian geostatistical model is very similar in spirit, the two differences just mentioned are not subtle. In fact, many statisticians would view both Bayesian requirements as advantages in terms of how they allow one to account for uncertainty in a rigorous statistical modeling framework (Hobbs and Hooten 2015). Regardless of one's particular viewpoint about all parameters being random and the use of prior information, it is undeniable that Bayesian methods are useful and rapidly becoming popular in scientific studies. In fact, many contemporary statistical models for animal movement are specified in a Bayesian framework. We return to these in later chapters.

As this is the first description of specific Bayesian models in this book, we take this opportunity to introduce some helpful notation. Bayesian methods primarily involve the specification of probability distributions for quantities we observe (i.e., data) and for those quantities we wish to learn about (e.g., parameters and missing data) as well as the ability to find required conditional distributions; thus, we refer to numerous PDFs (as well as probability mass functions PMFs). Any excess symbols used to denote these types of functions can quickly become tedious to manage in large expressions; thus, we employ the Bayesian bracket notation. In doing so, let the PDF (or PMF) of a random variable θ be denoted as $[\theta]$ (as opposed to $f(\theta)$, $P(\theta)$, $p(\theta)$, or $\pi(\theta)$). Then, conditional distributions can be conveyed using the traditional "|" notation; for example, the conditional distribution of y given θ is written as $[y|\theta]$.*

Bayesian models are specified conditionally, in pieces, such that the data arise from a distribution that depends on other process or state variables (and perhaps parameters), and then those, in turn, have a distribution that depends on parameters that also have a distribution. Thus, in formulating a Bayesian model, we need only write the conditional distributions for each of the components. For a geostatistical

* It appears that Gelfand and Smith (1990) were the first to employ such notation and we thank them for it every time we write a posterior or full-conditional distribution using this notation because it is streamlined and uncluttered.

model without a nugget effect, the data portion of the model can be written as

$$\mathbf{y} \sim \mathrm{N}(\mathbf{X}\boldsymbol{\beta}, \boldsymbol{\Sigma}) \equiv [\mathbf{y}|\boldsymbol{\beta}, \sigma^2, \phi], \tag{2.29}$$

where the covariance matrix is parameterized as before, $\boldsymbol{\Sigma} \equiv \sigma^2\mathbf{R}(\phi)$. To complete the model specification, we provide prior distributions containing any understanding we might have of the model parameters before the data were collected. In this case, one potential prior specification could be

$$\boldsymbol{\beta} \sim \mathrm{N}(\boldsymbol{\mu}_\beta, \boldsymbol{\Sigma}_\beta) \equiv [\boldsymbol{\beta}],$$

$$\sigma^2 \sim \mathrm{IG}(\alpha_1, \alpha_2) \equiv [\sigma^2],$$

$$\phi \sim \mathrm{Gamma}(\gamma_1, \gamma_2) \equiv [\phi],$$

where "IG" refers to the inverse gamma distribution and is parameterized as

$$[\sigma^2] \equiv \frac{1}{\alpha_1^{\alpha_2}\Gamma(\alpha_2)}(\sigma^2)^{\alpha_2 - 1} e^{-(1/\alpha_1\sigma^2)}.$$

To fit the model, we find the conditional distribution of the unknowns (parameters) given the knowns (data). This distribution, $[\boldsymbol{\beta}, \sigma^2, \phi|\mathbf{y}]$, is known as the posterior distribution. Using Bayes' law, we can write out the posterior distribution as a function of the model distributions (i.e., data model and parameter models)

$$[\boldsymbol{\beta}, \sigma^2, \phi|\mathbf{y}] = c(\mathbf{y}) \cdot [\mathbf{y}|\boldsymbol{\beta}, \sigma^2, \phi][\boldsymbol{\beta}][\sigma^2][\phi], \tag{2.30}$$

where the product of priors is used because we are assuming the parameters are independent *a priori*. The constant $c(\mathbf{y})$ is actually a function of the data $\mathbf{y}$ and is a single number that allows the left-hand side of Equation 2.30 to integrate to 1, as required of all PDFs. We could attempt to integrate the right-hand side of Equation 2.30 directly to find $c(\mathbf{y})$; however, in this case, the integral is not analytically tractable.* Thus, we rely on one of many Bayesian computational methods to find the posterior distribution for the parameters of interest.

As we noted in Chapter 1, MCMC is an incredibly useful computational method for fitting Bayesian models and has the advantage of being relatively intuitive and easy to program (as compared with many other methods). The basic idea underpinning MCMC is to sample a single parameter (or subset of parameters) from the conditional distribution, given everything else (termed the "full-conditional distribution," and denoted as [parameter$|\cdot$]), assuming that everything else in the model is actually known (i.e., data and other parameters). For the parameter vector $\boldsymbol{\beta}$, the full-conditional distribution is $[\boldsymbol{\beta}|\cdot] \equiv [\boldsymbol{\beta}|\mathbf{y}, \sigma^2, \phi]$. After a sample, $\boldsymbol{\beta}^{(k)}$, is obtained, for the kth MCMC iteration, we sample the next parameter, $(\sigma^2)^{(k)}$, from its full-conditional distribution $[\sigma^2|\cdot]$, and then sample the remaining parameter $\phi^{(k)}$ from

* An analytically intractable expression cannot be written in closed form (i.e., pencil and paper).

its full-conditional distribution $[\phi|\cdot]$. After we have sequentially sampled all parameters from their full-conditionals using the latest sampled values of each parameter being conditioned on, we loop back to the first parameter and sample each parameter again such that we are always conditioning on the most recent values for parameters in the loop. MCMC theory shows that these sequences of samples, called Markov chains, will eventually produce a sample from the correct joint posterior distribution, given enough iterations of the MCMC algorithm. Hobbs and Hooten (2015) provide additional insight about MCMC that solidifies the quick introduction presented here.

After the samples have been obtained, various point and interval estimates (among other important quantities) for other parameters can be approximated by computing sample statistics on the Markov chains themselves. For example, we could find the posterior mean of the regression coefficients by averaging the set of MCMC samples

$$E(\boldsymbol{\beta}|\mathbf{y}) \approx \frac{\sum_{k=1}^{K} \boldsymbol{\beta}^{(k)}}{K},$$

where $k = 1, \ldots, K$ represent the iterations in the MCMC algorithm and the total number of MCMC iterations K is large enough that the posterior mean is well approximated. Posterior summarization is trivial (i.e., taking various sample averages of the MCMC output) because the sampling-based method for approximating integrals, called Monte Carlo (MC) integration, has excellent properties. For example, one could approximate any integral using MC samples (independent and identically distributed) $\theta^{(k)} \sim [\theta]$ for $k = 1, \ldots, K$ with

$$E_\theta(g(\theta)) = \int g(\theta)[\theta]d\theta \approx \frac{\sum_{k=1}^{K} g(\theta^{(k)})}{K}, \tag{2.31}$$

for some PDF $[\theta]$ and function of theta $g(\theta)$. Therefore, coupling MC integration with MCMC output from Bayesian model fitting yields an incredibly powerful tool for finding posterior quantities of nearly any function of model parameters. Trying to provide such inference under non-Bayesian paradigms, if possible at all, requires complicated procedures such as the delta method (e.g., Ver Hoef 2012) or further computational burden, such as bootstrapping.

In practice, MCMC algorithms may require some "burn-in" period where the samples are still converging to the correct posterior distribution, and thus, a set of initial samples (often the first fourth or half) are discarded before computing posterior summary statistics. Furthermore, it may not always be easy to assess whether an MCMC algorithm has converged, and although some statistics and guidelines exist, it is an ongoing challenge to assess convergence in high-dimensional settings.

MCMC algorithms are surprisingly easy to construct in a statistical programming language such as R (R Core Team 2013), but there are also several automated MCMC sampling softwares available (e.g., BUGS, JAGS, INLA, and STAN; Lunn et al. 2000; Plummer 2003; Lindgren and Rue 2015; Carpenter et al. 2016). Furthermore, we emphasize that, even though MCMC has led to numerous breakthroughs in statistics and science, and has served as a catalyst for Bayesian methods and studies in general, new Bayesian computational approaches are regularly being developed. Depending on the desired inference and model, some alternative computational approaches have advantages over MCMC. However, as previously mentioned, few, if any, alternatives are as robust, intuitive, and as easy to implement as MCMC.

One of the primary advantages of MCMC and the Bayesian approach to geostatistics in general is that uncertainty can properly be accounted for in both parameter estimation and prediction. In fact, where many of the non-Bayesian approaches to geostatistics involve a sequential set of estimation procedures (i.e., first obtain ordinary least square [OLS] coefficient estimates, calculate residuals, estimate variogram, then find GLS coefficient estimates), parameter estimation and prediction can all be done simultaneously under the Bayesian paradigm using MCMC.

In MCMC, one only needs to sample from the full-conditional distribution for each unknown quantity of interest given everything else in the model. For prediction, we only need to sample $\mathbf{y}_u^{(k)}$ from its full-conditional $[\mathbf{y}_u^{(k)}|\cdot]$ with the other parameters in an MCMC algorithm. Basic linear algebra leads to the necessary full-conditional distribution, which turns out to be the predictive distribution for $\mathbf{y}_u$ we described previously (2.25):

$$[\mathbf{y}_u|\cdot] = N(\mathbf{X}_u\boldsymbol{\beta} + \boldsymbol{\Sigma}_{u,o}\boldsymbol{\Sigma}_{o,o}^{-1}(\mathbf{y}_o - \mathbf{X}_o\boldsymbol{\beta}), \boldsymbol{\Sigma}_{u,u} - \boldsymbol{\Sigma}_{u,o}\boldsymbol{\Sigma}_{o,o}^{-1}\boldsymbol{\Sigma}_{o,u}). \qquad (2.32)$$

Therefore, it is trivial to obtain MCMC samples from Equation 2.32 inside of a larger MCMC algorithm and, using the output, we can easily find the Bayesian Kriging predictions $\mathrm{E}(\mathbf{y}_u|\mathbf{y}_o)$ by averaging the MCMC samples for $\mathbf{y}_u$ according to Equation 2.31. Furthermore, the sample variance (2.31) of the MCMC samples for $\mathbf{y}_u$ approximates the posterior Kriging variance $\mathrm{Var}(\mathbf{y}_u|\mathbf{y}_o)$ while incorporating the uncertainty involved in the estimation of model parameters. The Bayesian approach to geostatistics is probably the most coherent method for performing prediction while properly accommodating uncertainty.

2.3 DISCRETE SPATIAL PROCESSES

We considered processes where the phenomenon of interest theoretically varies continuously over some spatial domain in the previous section. Now consider a discrete (i.e., areal) spatial domain composed of spatial units $\mathcal{A}_i$ for $i = 1, \ldots, n$. These spatial units could be regions, lines, or points, but they are countable. Assuming that we could measure $y(\mathcal{A}_i)$ for all n units of interest, we seek to (1) describe the amount of spatial structure in the observed process and/or (2) model the process in terms of some linear predictor, as we did with the CSPs.

We need to define potential relationships among spatial units $\mathcal{A}_i$ for $i = 1, \ldots, n$. Given that we are often concerned with a finite set of discrete spatial units, it is common to create an $n \times n$ "proximity" matrix $\mathbf{W}$, where row i contains zeros and ones with the ones corresponding to the neighbors of unit i in the spatial domain. There are alternative specifications of $\mathbf{W}$; for example, instead of ones, the proportion of shared boundary of neighboring units could be used. For conventional reasons, we

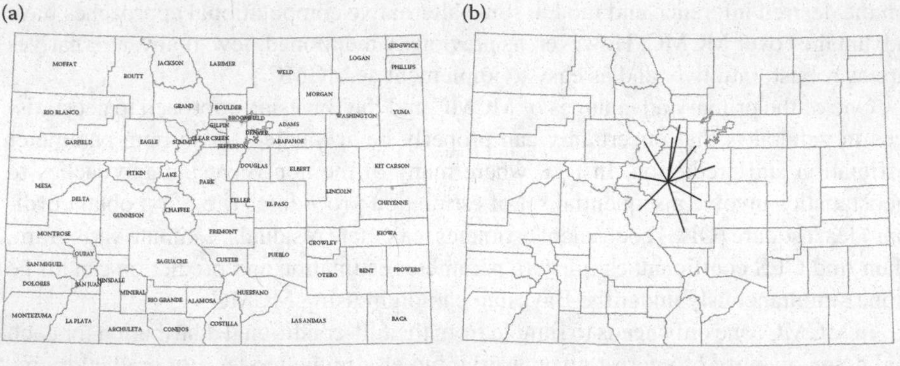

FIGURE 2.9 (a) Map of Colorado counties and (b) connections (straight black lines) between Park county and the neighboring counties in Colorado within 100 km of Park county.

retain the binary proximity matrix such that

$$w_{ij} = \begin{cases} 0 & \text{if } \mathcal{A}_j \notin \mathcal{N}_i \\ 1 & \text{if } \mathcal{A}_j \in \mathcal{N}_i \end{cases},$$

where w_{ij} are the elements of $\mathbf{W}$ and $\mathcal{N}_i$ indicates the neighborhood of unit $\mathcal{A}_i$. In a regular grid, the nearest neighbors (i.e., north, south, east, west) of grid cell $\mathcal{A}_i$ could comprise the neighborhood $\mathcal{N}_i$. The *a priori* specification of $\mathbf{W}$ is akin to the choice of parametric covariance function in geostatistics. Thus, for irregularly located regions, it is common to define the neighborhood $\mathcal{N}_i$ as all other units within some prespecified distance d.*

Consider the U.S. state of Colorado, for example (Figure 2.9). There are 64 counties in the state of Colorado, each irregularly sized and shaped (Figure 2.9a). The set of counties within (and other political or ecological regions) have discrete spatial support.

2.3.1 Descriptive Statistics

Exploratory data analysis with areal data often involves assessing spatial clustering and regularity. Clustered spatial processes are characterized by nearby spatial regions having responses with similar values giving the appearance of smooth maps (Figure 2.10c). In contrast, regularity in spatial processes is exhibited by large differences in the spatial process for nearby regions. Regular areal data often resemble a checkerboard (Figure 2.10a). Areal processes without any apparent clustering or regularity will appear as randomly arranged regions (Figure 2.10b).

The two most commonly used descriptive statistics for discrete spatial processes are called the Moran's I and Geary's C statistic. Both indicate the degree of positive

---

* The distances d_{ij} are often calculated based on (1) Euclidean distance between unit centroids $\mathbf{c}_i$ and $\mathbf{c}_j$ or (2) minimum distance between $\mathcal{A}_i$ and $\mathcal{A}_j$.

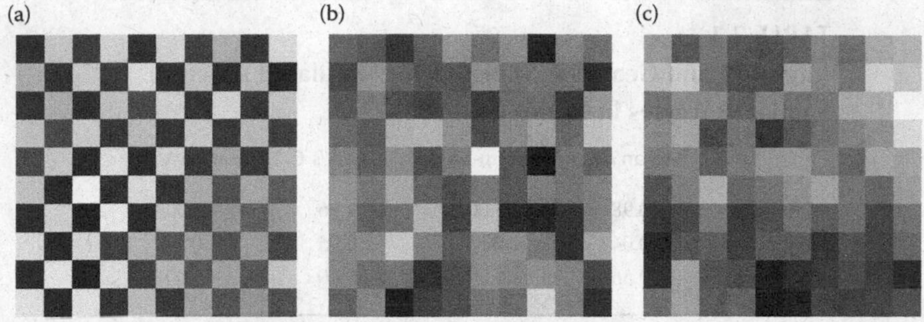

FIGURE 2.10 Simulated areal data on a regular grid arising from (a) a regular process, (b) a random process, and (c) a clustered process.

or negative spatial structure in a process but have subtle, yet important, differences. We describe each in what follows.

For the sample variance $\hat{\sigma}^2$ of the data $\mathbf{y} \equiv (y_1, \ldots, y_n)'$, the Moran's I statistic

$$\frac{n}{(n-1)\hat{\sigma}^2 \sum_i \sum_j w_{ij}} \sum_i \sum_j w_{ij}(y_i - \bar{y})(y_j - \bar{y}) \tag{2.33}$$

is the discrete-space analog to the correlation function used in continuous space. Note that we have used a subscript index to simplify the notation; that is, $y_i \equiv y(\mathcal{A}_i)$. The Moran's I statistic ranges from –1 to 1 and, under certain assumptions, the mean of the Moran's I statistic is $-1/(n-1)$. Values of Moran's I close to 1 indicate spatial clustering (or similarity for neighboring units) while values closer to –1 indicate spatial regularity (or dissimilarity for neighboring units).

The Geary's C statistic

$$\frac{1}{2\hat{\sigma}^2 \sum_i \sum_j w_{ij}} \sum_i \sum_j w_{ij}(y_i - y_j)^2 \tag{2.34}$$

is more similar to the variogram in geostatistics.* Ranging from 0 to 2 with a mean of 1, large Geary's C values (i.e., >1) correspond to regularity (or negative autocorrelation) while small values (i.e., <1) correspond to clustering (or positive autocorrelation). Despite the fact that Geary's C enjoys similar properties as the variogram in not requiring the precalculation of the sample mean $\bar{y}$, the Moran's I statistic is more popular and heavily used in summarizing spatial data.

To help decide whether spatial structure exists, hypothesis testing can be conducted using Moran's I or Geary's C under Gaussian assumptions or through the use of Monte Carlo tests. In these cases, the null hypothesis is often "no spatial structure." Furthermore, one common use for the Moran's I and Geary's C statistics is to check modeling assumptions by examining the remaining structure in the data after accounting for desired first-order effects. Different versions of the Moran's I

---

* Geary's C is known as the Durbin–Watson statistic in time series.

TABLE 2.1
Moran's I and Geary's C Statistics for Simulated Discrete Spatial Processes in Figure 2.10

	Moran's I	Moran p-Value	Geary's C	Geary p-Value
(a) Regular	−0.98	<0.001	1.96	<0.001
(b) Random	0.04	0.52	0.95	0.53
(c) Clustered	0.56	<0.001	0.40	<0.001

statistic have been developed for investigating spatial structure in the residuals of a linear model: $\mathbf{e} = \mathbf{y} - \hat{\mathbf{y}}$, where $\hat{\mathbf{y}} = \mathbf{X}\hat{\boldsymbol{\beta}}$. In matrix notation, the Moran's I statistic for residuals is

$$\frac{n}{\sum_{ij} w_{ij}} \frac{(\mathbf{Ge})'\mathbf{W}(\mathbf{Ge})}{(\mathbf{Ge})'\mathbf{Ge}}, \tag{2.35}$$

where $\mathbf{G} = \mathbf{I} - \mathbf{X}(\mathbf{X}'\mathbf{X})^{-1}\mathbf{X}'$ as in Equation 2.27.

Using our simulated discrete spatial data (Figure 2.10) and a proximity matrix based on first-order spatial neighbors (i.e., North, South, East, and West), the Moran's I and Geary's C statistics and associated p-values for a two-sided hypothesis test (i.e., assuming the null hypothesis of no spatial structure) are shown in Table 2.1.

The Moran's I and Geary's C statistics both suggest significant regularity in Figure 2.10a, clustering in Figure 2.10c, and a lack of evidence for discrete spatial structure in Figure 2.10b.

As an example involving real data, consider the counties in the U.S. state of Colorado (Figure 2.9). Based on state records, the total avian species richness (ever occurring in the Colorado) by county is shown in Figure 2.11a. The potential correlates (i.e., log(minimum elevation), log(human population), and total area in square kilometers) with avian richness are shown in Figure 2.11b–d.

The associated Moran's I and Geary's C statistics (and p-values) for the Colorado county discrete spatial data presented in Figure 2.11 are shown in Table 2.2. The statistics in Table 2.2 are based on a binary proximity matrix where neighboring counties are defined to be within a 100 km radius (based on county centroids).

TABLE 2.2
Moran's I and Geary's C Statistics for Colorado County Discrete Spatial Processes in Figure 2.11

	Moran's I	Moran p-Value	Geary's C	Geary p-Value
(a)	0.33	<0.001	0.87	0.158
(b)	0.62	<0.001	0.32	<0.001
(c)	0.58	<0.001	0.59	<0.001
(d)	0.16	0.005	0.66	0.002

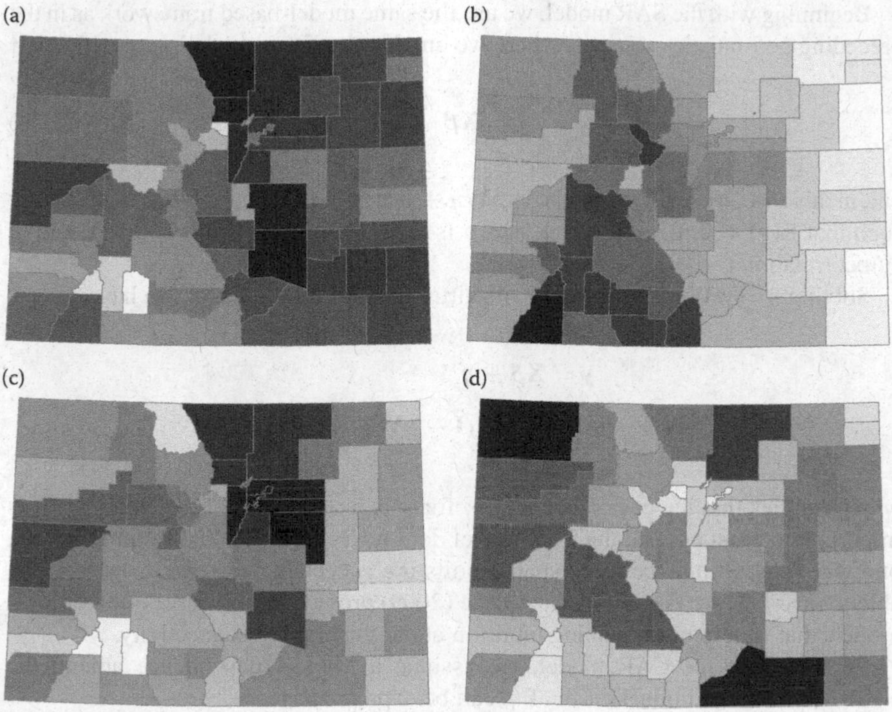

FIGURE 2.11 Maps (darker corresponds to larger values) of (a) avian species richness, (b) log(minimum elevation), (c) log(human population), and (d) total area in square kilometers.

The Moran's I and Geary's C statistics for the avian species richness data and associated covariates in Colorado suggest that all of these discrete spatial data are clustered. The only exception occurs for avian species richness itself (Figure 2.11a), which only has a significant Moran's I statistic (no confirmation from Geary's C).

2.3.2 Models for Discrete Spatial Processes

Two main types of models are commonly used to formally account for first- and second-order spatial processes in discrete spatial data. These are simultaneous autoregressive (SAR) models and conditional autoregressive (CAR) models.* As with the Moran's I and Geary's C statistics, there are also subtle, but important, differences between SAR and CAR models and we describe these in what follows. However, to foreshadow, we note that the differences between SAR and CAR models may be somewhat of a moot point because one can always find a CAR specification that is equivalent to a given SAR specification.

* Note that some refer to CAR models as Besag models, after their early development by Besag (1974).

Beginning with the SAR model, we use the same model-based framework as in the preceding geostatistics sections, where we employ the linear modeling specification

$$\mathbf{y} = \mathbf{X}\boldsymbol{\beta} + \boldsymbol{\eta}, \tag{2.36}$$

but, in this case, we let the errors, $\boldsymbol{\eta} = \rho\mathbf{W}\boldsymbol{\eta} + \boldsymbol{\nu}$, depend on themselves stochastically such that $\text{E}(\boldsymbol{\nu}) = \mathbf{0}$, $\text{E}(\boldsymbol{\nu}\boldsymbol{\nu}') = \sigma^2\mathbf{I}$, and ρ is a parameter that controls the degree of autocorrelation ($-1 < \rho < 1$).

Solving $\boldsymbol{\eta} = \rho\mathbf{W}\boldsymbol{\eta} + \boldsymbol{\nu}$ for $\boldsymbol{\eta}$ and substituting into Equation 2.36, we have

$$\begin{aligned}\mathbf{y} &= \mathbf{X}\boldsymbol{\beta} + \boldsymbol{\eta}\\ &= \mathbf{X}\boldsymbol{\beta} + (\mathbf{I} - \rho\mathbf{W})^{-1}\boldsymbol{\nu},\end{aligned}$$

which implies that the covariance matrix for $\boldsymbol{\eta}$ is $\sigma^2(\mathbf{I} - \rho\mathbf{W})^{-1}(\mathbf{I} - \rho\mathbf{W}')^{-1}$. It is important to point out that the SAR model does not require $\mathbf{W}$ to be symmetric (i.e., one-way relationships between spatial units are acceptable) but we do need to be able to invert $(\mathbf{I} - \rho\mathbf{W})$. LeSage and Pace (2009) provide a solid description of SAR models that is helpful for gaining intuition about the implied connectivity.

To formulate the CAR model, we assume a Markov dependence among the errors η_i.* The conditional mean of η_i can be expressed as

$$\text{E}(\eta_i | \{\eta_j, j \in \mathcal{N}_i\}) = \sum_{j \in \mathcal{N}_i} c_{ij}\eta_j \tag{2.37}$$

and

$$\text{Var}(\eta_i | \{\eta_j, j \in \mathcal{N}_i\}) = \sigma_i^2, \tag{2.38}$$

where c_{ij} are weights based on the proximity with neighbors and σ_i^2 varies with i, imparting nonstationary in the spatial process.† An interesting and critical result for CAR models is that they can be written jointly, using matrix notation such as SAR models. Thus, let the CAR model be defined as

$$\mathbf{y} = \mathbf{X}\boldsymbol{\beta} + \boldsymbol{\eta}, \tag{2.39}$$

where $\boldsymbol{\eta} \sim \text{N}(\mathbf{0}, \sigma^2(\mathbf{I} - \rho\mathbf{W})^{-1})$ and the proximity matrix $\mathbf{W}$ must be symmetric. It is common to reparameterize the covariance matrix of the CAR model such that $\boldsymbol{\eta} \sim \text{N}(\mathbf{0}, \sigma^2(\text{diag}(\mathbf{W1}) - \rho\mathbf{W})^{-1})$, where $\text{diag}(\mathbf{W1})$ is a diagonal matrix with the

---

* This Markov assumption implies that, given the neighbors, the process at a location is independent of all other nonneighboring locations.

† Statistical models for discrete spatial processes do not have the same assumptions as those for continuous spatial processes.

row sums of $\mathbf{W}$ on the diagonal and zero elsewhere. In this latter specification, the correlation parameter ρ is bounded between –1 and 1.

It has become common to fix $\rho = 1$ in CAR models and refer to them as "intrinsic" CAR models (ICARs). The recent popularity of ICAR specifications is due to several reasons:

1. Most real data scenarios yield processes with positive autocorrelation ($\rho > 0$).
2. Only very large values of ρ (i.e., $\rho \rightarrow 1$) impose strong visible positive autocorrelation in $\boldsymbol{\eta}$.
3. $\rho = 1$ simplifies the precision matrix in the CAR model.
4. $\rho = 1$ facilities computationally efficient fitting algorithms (details in what follows).

The implementation of SAR and CAR models is similar to the former geostatistical models where they can be implemented in either a maximum likelihood or Bayesian paradigm. The CAR specification naturally pairs with an MCMC algorithm because the full-conditional distributions for η_i are Gaussian and can be readily simulated from sequentially.

To demonstrate the differences in inference resulting from the regular linear model and the CAR model, we fit both models to the Colorado avian species richness data (i.e., Figure 2.11). As a typical variance stabilizing transformation, we used the natural log of species richness for a response variable. We used the standardized natural log of county population size from the 2010 census and standardized county area as covariates. We specifically left out elevation as a potentially important "missing covariate." Heuristically, we expect greater species richness in counties with more people and in larger counties. We might also expect there to be latent spatial structure in the residuals from a regular multiple linear regression model fit.

A maximum likelihood analysis of these data confirms our hypotheses that log county population size and county area are positive predictors of recorded avian diversity (Table 2.3). While the parameter estimates resulting from the regular linear

TABLE 2.3
Parameter Estimates and p-Values for Avian Log Species Richness Based on the Standardized Natural Log of County Population Size from the 2010 Census and Standardized County Area as Covariates (Figure 2.11) in the Regular Linear Model (LM) and CAR Model

Covariate	LM Estimate	LM p-Value	CAR Estimate	CAR p-Value
Intercept	1.733	<0.001	1.743	<0.001
log(population)	0.019	<0.001	0.024	<0.001
Area	0.011	0.008	0.005	0.185

model fit are both positive and apparently important (i.e., small p-values), a Moran's I test of the residuals suggested that there may be remaining unaccounted for spatial dependence in the errors (p-value < 0.001). Thus, fitting a CAR model to the same data (using a neighborhood structure based on centroids within a 100-km radius), we find that the county log population covariate still seems significant, whereas the county area covariate is no longer a significant predictor of log richness (Table 2.3). A Moran's I test of the CAR residuals indicated no remaining evidence of spatial structure after accounting for correlated errors (p-value $= 0.671$).

The results of the Colorado avian species richness analysis illustrate an important reason to account for latent dependence in data. Assuming independent errors when they are actually positively correlated can cause parameter estimates to be overly narrow, inflating the chance of inferring a significant first-order effect. When we added the spatial dependence to the regression model using a CAR structure, the county area p-value increased, leading us to downplay its importance in explaining avian species richness. Furthermore, and most importantly, because the assumptions of the linear model were not met in this example, it cannot be used to provide statistical inference, whereas the CAR model results can be used.

Finally, recall that our model did not include the elevation covariate. However, the CAR model we fit did include a positively correlated spatial random effect. Thus, the spatial random effect helped account for the missing covariate of elevation, at least to some extent. Figure 2.12 provides a visual perspective of how spatial structure helps to account for the missing elevation covariate. Notice that the opposite pattern appears in the (a) and (b) panels of Figure 2.12. Heuristically, we expect higher elevations to negatively affect avian species richness. Thus, the spatial random effect needs to appear as the opposite pattern of log(minimum elevation) to influence the model in the same way as the actual covariate. In this case, the estimated spatial random effect does indeed have a pattern similar to that expected based on our prior understanding of the system. Thus, the spatial random effect is capable of accounting for the same type of spatial structure that appears in the topography of Colorado.

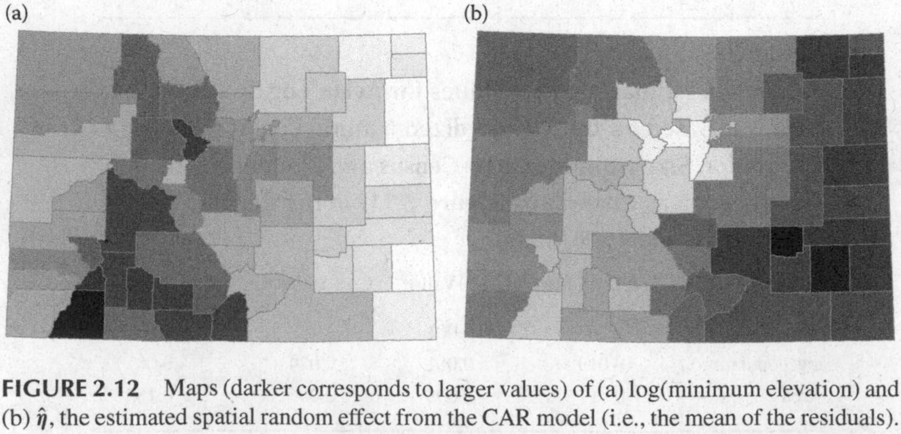

FIGURE 2.12 Maps (darker corresponds to larger values) of (a) log(minimum elevation) and (b) $\hat{\eta}$, the estimated spatial random effect from the CAR model (i.e., the mean of the residuals).

2.4 SPATIAL CONFOUNDING

The concept of spatial confounding arises from the fact that first- and second-order predictors can be inadvertently correlated and has received quite a bit of attention in the recent literature (e.g., Hodges and Reich 2010; Paciorek 2010; Hanks et al. 2015b). To illustrate the concept of confounding, consider the linear mixed model specification that could arise in any of the explicit spatial models above (e.g., geostatistical, SAR, CAR) and note that the spatial random effect can be parameterized as $\boldsymbol{\eta} = \mathbf{H}\boldsymbol{\alpha}$, then

$$\begin{aligned} \mathbf{y} &= \beta_0 + \mathbf{X}\boldsymbol{\beta} + \boldsymbol{\eta} + \boldsymbol{\varepsilon} \\ &= \beta_0 + \mathbf{X}\boldsymbol{\beta} + \mathbf{H}\boldsymbol{\alpha} + \boldsymbol{\varepsilon}. \end{aligned} \tag{2.40}$$

The presence of $\mathbf{H}$ makes it clear that there are, in fact, two "design" matrices in this model, one for the fixed effects ($\mathbf{X}$) and one for the random effects ($\mathbf{H}$). Under this specification (2.40), assume $\boldsymbol{\eta} \sim \text{N}(\mathbf{0}, \sigma_\eta^2 \mathbf{Q}^{-1})$ (for either a continuous or discrete spatial process), where $\mathbf{Q}^{-1}$ is the spatial correlation matrix whose inverse can be decomposed as $\mathbf{Q} = \mathbf{H}\boldsymbol{\Lambda}\mathbf{H}'$, then $\boldsymbol{\alpha} \sim \text{N}(\mathbf{0}, \sigma_\eta^2 \boldsymbol{\Lambda}^{-1})$ such that $\boldsymbol{\Lambda}$ is a diagonal matrix. Just as multicollinear covariates can bias the estimates of $\boldsymbol{\beta}$ in the standard linear model, it can also influence the regression coefficients in the mixed model framework, which includes the spatial models we have described when a nugget effect ($\boldsymbol{\eta}$) is used.

If the columns of $\mathbf{H}$ are linearly independent of the columns of $\mathbf{X}$, these models perform as expected for parameter estimation. However, when the columns of $\mathbf{H}$ are not linearly independent of $\mathbf{X}$, one may wish to consider remedial measures. Hodges and Reich (2010) and Hughes and Haran (2013) present a restriction approach for forcing the first-order process to take precedence over the second-order process in CAR models and Hanks et al. (2015b) developed a similar method for geostatistical models. In fact, the restriction is essentially the same idea that is used in REML estimation where the second-order process is restricted to the residual space of the first-order process. Following Hodges and Reich (2010), we describe the basic restricted spatial regression approach and follow up in the next section with the modification presented by Hughes and Haran (2013).

To arrive at one set of orthogonal basis vectors for $\mathbf{H}$ (2.40), consider the spectral decomposition of the matrix $\mathbf{G} = \mathbf{H}\boldsymbol{\Lambda}\mathbf{H}'$.* In this decomposition, which is also known as the eigen decomposition, the columns of $\mathbf{H}$ are the eigenvectors, while the diagonal elements of the diagonal matrix $\boldsymbol{\Lambda}$ are the eigenvalues of the residual operator $\mathbf{G} = \mathbf{I} - \mathbf{X}(\mathbf{X}'\mathbf{X})^{-1}\mathbf{X}'$.† The corresponding model for the spectral coefficients $\boldsymbol{\alpha}$ is $\boldsymbol{\alpha} \sim \text{N}(\mathbf{0}, \sigma_\eta^2 (\mathbf{H}'\mathbf{Q}\mathbf{H})^{-1})$. This restricted spatial regression will guarantee that the point estimates for $\boldsymbol{\beta}$ are the same as those resulting from the nonspatial model (i.e., the model without $\boldsymbol{\eta}$ in Equation 2.40).

---

* The continuous version of this decomposition is referred to as a Karhunen–Loeve expansion and is the basis for principal components analysis.

† Technically, the matrix $\mathbf{H}$ needs to be truncated so that it only contains the first $n - \text{rank}(\mathbf{X})$ eigenvectors of $\mathbf{G}$.

Nothing prevents one from using this same restriction approach on the covariates for the fixed effects in the model ($\mathbf{X}$) if certain collinear covariates are believed to have a priority over others. This procedure is probably not wise to apply as a blanket approach in all analyses (Paciorek 2010; Hanks et al. 2015b). Serious consideration should be given to the covariates used in a model. However, in most ecological studies, we try to collect information on the factors we feel are most relevant (i.e., suspected to be causal) for the response variables we observe. Thus, few ecologists would have much hesitation about giving their carefully selected fixed effect covariates priority in a model over second-order spatial structure.

2.5 DIMENSION REDUCTION METHODS

When the observed data are high-dimensional (i.e., many spatial locations; on the order $\mathcal{O}(10^5)$ or more), then the otherwise trivial calculation (2.26) becomes very computationally demanding due to the inverse of the covariance matrix $\mathbf{\Sigma}_{o,o}$ for the observed data.* In fact, given that the covariance matrix for observed data in the likelihood (2.22) contains unknown parameters, it needs to be inverted at every step in an optimization routine. Thus, it is often of interest, even necessary, to find computationally efficient ways to deal with such large calculations. Out of necessity, dimension reduction methods are becoming popular in spatial and spatio-temporal statistics. We highlight some of the approaches to dimension reduction in what follows.

2.5.1 Reducing Necessary Calculations

Disregarding the aforementioned issues that could arise due to spatial confounding for now (we return to that topic later), we present a way to reduce the number of calculations made in spatial regression models without actually reducing the dimensionality of the spatial process itself. Consider the spatially explicit model with both first- and second-order effects

$$\mathbf{y} = \beta_0 + \mathbf{X}\boldsymbol{\beta} + \boldsymbol{\eta} + \boldsymbol{\varepsilon} \tag{2.41}$$

$$= \beta_0 + \mathbf{X}\boldsymbol{\beta} + \mathbf{H}\boldsymbol{\alpha} + \boldsymbol{\varepsilon}. \tag{2.42}$$

As before, suppose that $\boldsymbol{\eta} \sim \text{N}(\mathbf{0}, \sigma_\eta^2\mathbf{Q}^{-1})$ and $\boldsymbol{\varepsilon} \sim \text{N}(\mathbf{0}, \sigma_\varepsilon^2\mathbf{I})$. If $\mathbf{H}$ is an $n \times n$ matrix of orthonormal basis functions (e.g., Fourier basis functions, wavelets), then $\boldsymbol{\alpha}$ are spectral coefficients whose implied distribution will be $\boldsymbol{\alpha} \sim N(\boldsymbol{\mu}_\alpha, \sigma_\alpha^2\mathbf{\Lambda})$, where $\mathbf{\Lambda}$ is a diagonal matrix and, typically, $\boldsymbol{\mu}_\alpha = \mathbf{0}$ and $\sigma_\alpha^2 = \sigma_\eta^2$. Note that this is the same basic idea as that discussed in the preceding sections, however, we can use fast computational algorithms to calculate the necessary transformation $\boldsymbol{\eta} = \mathbf{H}\boldsymbol{\alpha}$ (and inverse transformation, $\mathbf{H}'\boldsymbol{\eta} = \boldsymbol{\alpha}$; recall that, if the columns of $\mathbf{H}$ are orthogonal, we have $\mathbf{H}'\mathbf{H} = \mathbf{I}$). As an example, Wikle (2002) employs the discrete cosine transform, whereas Hooten et al. (2003) used the fast Fourier transform, to get back

* Note that, if the data are of dimension $\mathcal{O}(10^5)$, then the covariance matrix is on the order of $\mathcal{O}(10^{10})$ elements, a frighteningly large number of values to store in the computer, let alone do calculations with.

and forth between $\boldsymbol{\alpha}$ and $\boldsymbol{\eta}$. In the Bayesian generalized linear mixed model setting (which includes the linear mixed model), an advantage of the orthogonality is that the full-conditional distribution for $\boldsymbol{\alpha}$ is

$$[\boldsymbol{\alpha}|\cdot] = \text{N}\left(\left(\mathbf{H}'\mathbf{H} + \frac{\boldsymbol{\Lambda}^{-1}}{\sigma_\alpha^2}\right)^{-1}\left(\mathbf{H}'(\mathbf{y} - \beta_0 - \mathbf{X}\boldsymbol{\beta}) + \frac{\boldsymbol{\Lambda}^{-1}}{\sigma_\alpha^2}\boldsymbol{\mu}_\alpha\right),\right.$$

$$\left.\left(\mathbf{H}'\mathbf{H} + \frac{\boldsymbol{\Lambda}^{-1}}{\sigma_\alpha^2}\right)^{-1}\right), \tag{2.43}$$

where the inner product, $\mathbf{H}'\mathbf{H} = \mathbf{I}$, and the covariance matrix in Equation 2.43 is the inverse of a diagonal matrix (because both $\mathbf{H}'\mathbf{H}$ and $\boldsymbol{\Lambda}$ are diagonal). This, by itself, can dramatically reduce the number of calculations required in an MCMC algorithm and speed up model fitting.

A disadvantage to using this approach is that the matrix $\mathbf{H}$ should only have to be calculated once or the savings gained in computing the full-conditional in Equation 2.43 are tempered by having to recalculate $\mathbf{H}$ repeatedly. The matrix $\mathbf{Q}$ must be known in advance because $\mathbf{H}$ is often computed as a direct function of $\mathbf{Q}$. In the geostatistical setting, we often assume the correlation matrix $Q_{ij}^{-1} \equiv \exp(-d_{ij}/\phi)$. In this case, the distances d_{ij}, between locations i and j, are easily calculated (and thus, known) but the parameter ϕ is almost always unknown. A practical, yet perhaps unfulfilling, remedial approach for empirically fixing $\mathbf{Q}$ is to either use a separate set of data to estimate ϕ (and then fix its value in $\mathbf{Q}$) or, similarly, use the same data set to estimate ϕ. In the latter case, the approach is referred to as "empirical Bayes." If $\mathbf{Q}$ is known and conditioned on, the expansion matrix $\mathbf{H}$ can be easily calculated, allowing the reparameterization in Equation 2.42 to be advantageous computationally. In the same spirit, the ICAR model specification (2.39) implies that $\mathbf{Q} = (\text{diag}(\mathbf{W1}) - \mathbf{W})$, where $\mathbf{W}$ is typically a binary proximity matrix indicating which spatial regions are neighbors of each other. This proximity matrix is often fixed by the researcher, and thus, $\mathbf{Q}$ is fixed and can be used to compute $\mathbf{H}$. Thus, the number of calculations can be reduced in both continuous and discrete spatial process modeling using the first-order reparameterization in Equation 2.42.

2.5.2 Reduced-Rank Models

We described a reparameterized version of the explicit spatial model (2.42) from the preceding section. This model specification is surprisingly simple, yet can be useful from many different perspectives. In essence, the main idea is to find a set of basis functions* $\mathbf{H}$ that result in a computational advantage while still providing the intended spatial structure in the model (Hefley et al. 2016a).

While the reparameterized spatial model is useful in its own right, further computational efficiency can be achieved by choosing $\mathbf{H}$ carefully (e.g., Wikle 2010a). In

* More appropriately, these should probably be referred to as basis vectors because they are represented as a discrete set of values in practice.

particular, the expansion matrix $\mathbf{H}$, in the previous section, is not technically reducing dimensionality, but rather, reducing the number of required computations. More formally, the previously specified matrix $\mathbf{H}$ is a full-rank $n \times n$ matrix. If, instead, we consider a lower-rank matrix $\tilde{\mathbf{H}}$ that has dimension $n \times p$, where $p \ll n$, we arrive at the following modification of the spatial model (2.42):

$$\begin{aligned} \mathbf{y} &= \beta_0 + \mathbf{X}\boldsymbol{\beta} + \boldsymbol{\eta} + \boldsymbol{\varepsilon} \\ &\approx \beta_0 + \mathbf{X}\boldsymbol{\beta} + \tilde{\mathbf{H}}\tilde{\boldsymbol{\alpha}} + \boldsymbol{\varepsilon}, \end{aligned} \tag{2.44}$$

where the coefficient vector is distributed as $\tilde{\boldsymbol{\alpha}} \sim \text{N}(\boldsymbol{\mu}_\alpha, \boldsymbol{\Sigma}_\alpha)$ now has dimension $p \times 1$ and Equation 2.44 is often thought of as an approximation to the true intended model (2.42). Thus, the term $\tilde{\mathbf{H}}\tilde{\boldsymbol{\alpha}}$ is an approximation of $\boldsymbol{\eta}$ and is distributed multivariate Gaussian with mean $\tilde{\mathbf{H}}\boldsymbol{\mu}_\alpha$ and covariance matrix $\tilde{\mathbf{H}}\boldsymbol{\Sigma}_\alpha\tilde{\mathbf{H}}'$. The matrix of basis vectors $\tilde{\mathbf{H}}$ is typically parameterized using distances between locations and a small set of parameters (i.e., $\boldsymbol{\phi}$) or arises from a decomposition (e.g., spectral) of a parameterized covariance matrix. Furthermore, the $\boldsymbol{\mu}_\alpha$ is often a zero vector and $\boldsymbol{\Sigma}_\alpha$ is diagonal, simplifying the distribution of $\tilde{\mathbf{H}}\tilde{\boldsymbol{\alpha}}$.

In fitting the reduced-rank model using a Bayesian framework and MCMC, we arrive at a familiar form for the full-conditional distribution of the coefficients

$$[\tilde{\boldsymbol{\alpha}}|\cdot] = N\left(\left(\tilde{\mathbf{H}}'\tilde{\mathbf{H}} + \boldsymbol{\Sigma}_\alpha^{-1}\right)^{-1}\left(\tilde{\mathbf{H}}'(\mathbf{y} - \beta_0 - \mathbf{X}\boldsymbol{\beta}) + \boldsymbol{\Sigma}_\alpha^{-1}\boldsymbol{\mu}_\alpha\right), \left(\tilde{\mathbf{H}}'\tilde{\mathbf{H}} + \boldsymbol{\Sigma}_\alpha^{-1}\right)^{-1}\right). \tag{2.45}$$

The full-conditional distribution in Equation 2.45 has a computational advantage over Equation 2.43 because the precision matrix $(\tilde{\mathbf{H}}'\tilde{\mathbf{H}} + \boldsymbol{\Sigma}_\alpha^{-1})$ is only of dimension $p \times p$, which is much smaller than the $n \times n$ precision matrix in Equation 2.43. The dimension reduction implies that the precision matrix will be much easier to invert within an MCMC algorithm (or a maximum likelihood optimization).

Furthermore, for Bayesian Kriging, the full-conditional predictive distribution for unobserved data $\mathbf{y}_u$

$$[\mathbf{y}_u|\cdot] = \text{N}(\mathbf{X}_u\boldsymbol{\beta} + \boldsymbol{\eta}_u, \sigma_\varepsilon^2\mathbf{I}) \tag{2.46}$$

is sampled from, in an MCMC setting, to learn about the posterior predictive distribution, $[\mathbf{y}_u|\mathbf{y}]$. The predictive distribution in Equation 2.46 relies on $\boldsymbol{\eta}_u \equiv \tilde{\mathbf{H}}_u\tilde{\boldsymbol{\alpha}}$, the correlated random field at the unobserved locations of interest. The matrix $\tilde{\mathbf{H}}_u$ contains the basis functions at the locations where predictions are desired.

An alternative approach for fitting the reduced-rank model (2.44) is to use an integrated likelihood approach. Using a process called "Rao-Blackwellization," we integrate the random effects $\tilde{\boldsymbol{\alpha}}$ out of the product of the data and process models to yield the integrated likelihood

$$[\mathbf{y}|\boldsymbol{\phi}, \boldsymbol{\mu}_\alpha, \boldsymbol{\Sigma}_\alpha, \boldsymbol{\beta}, \sigma_\varepsilon^2] = \int [\mathbf{y}|\boldsymbol{\phi}, \tilde{\boldsymbol{\alpha}}, \boldsymbol{\beta}, \sigma_\varepsilon^2][\tilde{\boldsymbol{\alpha}}|\boldsymbol{\mu}_\alpha, \boldsymbol{\Sigma}_\alpha]d\tilde{\boldsymbol{\alpha}}. \tag{2.47}$$

When $\boldsymbol{\mu}_\alpha \equiv \mathbf{0}$, the integrated likelihood in Equation 2.47 will be multivariate Gaussian with mean $\mathbf{X}\boldsymbol{\beta}$ and covariance $\boldsymbol{\Sigma}_y \equiv \tilde{\mathbf{H}}\boldsymbol{\Sigma}_\alpha\tilde{\mathbf{H}}' + \boldsymbol{\Sigma}_\varepsilon$. The integrated likelihood (2.47) does not contain the vector $\tilde{\boldsymbol{\alpha}}$; thus, we do not have to sample $\tilde{\boldsymbol{\alpha}}$ when fitting the model using MCMC. MCMC algorithms that require samples for $\tilde{\boldsymbol{\alpha}}$ can be slow to converge, while MCMC algorithms based on the integrated likelihood often show improved convergence and mixing. However, to fit the model based on the integrated likelihood (2.47) using MCMC, we do have to invert the covariance matrix $\boldsymbol{\Sigma}_y$, which is now $n \times n$.

If the sample size, n, is large, the inversion of $\boldsymbol{\Sigma}_y$ can be computationally prohibitive. Fortunately, the Sherman–Morrison–Woodbury identity allows us to invert special matrices of the form $\mathbf{A} + \mathbf{BCD}$ using

$$(\mathbf{A}+\mathbf{BCD})^{-1} = \mathbf{A}^{-1} - \mathbf{A}^{-1}\mathbf{B}(\mathbf{C}^{-1}+\mathbf{DA}^{-1}\mathbf{B})^{-1}\mathbf{DA}^{-1}. \tag{2.48}$$

For the reduced-rank model based on the integrated likelihood, if $\boldsymbol{\Sigma}_\alpha = \sigma_\alpha^2\mathbf{I}$ and $\boldsymbol{\Sigma}_\varepsilon = \sigma_\varepsilon^2\mathbf{I}$, then

$$\boldsymbol{\Sigma}_y^{-1} = \frac{\mathbf{I}}{\sigma_\varepsilon^2} - \frac{\tilde{\mathbf{H}}}{\sigma_\varepsilon^2}\left(\frac{\mathbf{I}}{\sigma_\alpha^2} + \frac{\tilde{\mathbf{H}}'\tilde{\mathbf{H}}}{\sigma_\varepsilon^2}\right)^{-1}\frac{\tilde{\mathbf{H}}'}{\sigma_\varepsilon^2}. \tag{2.49}$$

Thus, $(\mathbf{I}/\sigma_\alpha^2) + (\tilde{\mathbf{H}}'\tilde{\mathbf{H}}/\sigma_\varepsilon^2)$ is only a $p \times p$ matrix and can be inverted quickly. Furthermore, if the basis vectors in $\tilde{\mathbf{H}}$ are orthogonal (e.g., eigenvectors), then $\tilde{\mathbf{H}}'\tilde{\mathbf{H}}$ is often diagonal, further reducing the required computation to compute the precision matrix $\boldsymbol{\Sigma}_y^{-1}$ and sample the model parameters in an MCMC algorithm. For large data sets (i.e., n greater than a few hundred), the integrated likelihood method is useful for constructing fast and stable MCMC algorithms to fit the reduced-rank geostatistical model (2.44).

Bayesian Kriging based on the integrated likelihood model is achieved by sampling from the predictive full-conditional distribution

$$[\mathbf{y}_u|\cdot] = \text{N}(\mathbf{X}_u\boldsymbol{\beta} + \boldsymbol{\Sigma}_{y,u,o}\boldsymbol{\Sigma}_y^{-1}(\mathbf{y}-\mathbf{X}\boldsymbol{\beta}), \boldsymbol{\Sigma}_{y,u}^{-1} - \boldsymbol{\Sigma}_{y,u,o}\boldsymbol{\Sigma}_y^{-1}\boldsymbol{\Sigma}_{y,o,u}), \tag{2.50}$$

where $\boldsymbol{\Sigma}_{y,u,o}$ are the cross-covariance matrices between the unobserved and observed spatial locations and $\boldsymbol{\Sigma}_{y,o,u} \equiv \boldsymbol{\Sigma}'_{y,u,o}$. Sampling from the predictive full-conditional (2.50) does not affect the model fit; thus, it can be performed during or after the remainder of the MCMC samples are obtained for model parameters. The predictive full-conditional (2.50) also depends on $\boldsymbol{\Sigma}_y^{-1}$; thus, predictive samples can be obtained quickly using the Sherman–Morrison–Woodbury identity (2.49).

2.5.3 Predictive Processes

Another approach to dimension reduction in spatially explicit models that is rapidly growing in popularity is referred to as the "predictive process" approach (Banerjee et al. 2008). The basic idea underpinning the predictive process involves using a prediction of the correlated spatial field, $\hat{\boldsymbol{\eta}}$, rather than the true field itself, $\boldsymbol{\eta}$, in a geostatistical model, such that

$$\begin{aligned}\mathbf{y} &= \mathbf{X}\boldsymbol{\beta} + \boldsymbol{\eta} + \boldsymbol{\varepsilon} \\ &\approx \mathbf{X}\boldsymbol{\beta} + \hat{\boldsymbol{\eta}} + \boldsymbol{\varepsilon}.\end{aligned} \tag{2.51}$$

To obtain the predictions $\hat{\boldsymbol{\eta}}$, consider a set of m knot locations $\tilde{\mathbf{S}}$ that exist in the space of the n data locations $\mathbf{S}$, where $m \ll n$. If $\boldsymbol{\eta} \sim \text{N}(\mathbf{0}, \boldsymbol{\Sigma}_\eta)$ and $\boldsymbol{\varepsilon} \sim \text{N}(\mathbf{0}, \sigma_\varepsilon^2\mathbf{I})$, then a reasonable approach to obtain $\hat{\boldsymbol{\eta}}$ is with the linear predictor

$$\hat{\boldsymbol{\eta}} \equiv \tilde{\boldsymbol{\Sigma}}_\eta \tilde{\boldsymbol{\Sigma}}^{-1} \tilde{\boldsymbol{\eta}}, \tag{2.52}$$

where $\tilde{\boldsymbol{\Sigma}}_\eta$ is the $n \times m$ cross-covariance matrix between data locations $\mathbf{S}$ and knot locations $\tilde{\mathbf{S}}$, $\tilde{\boldsymbol{\Sigma}}$ is the $m \times m$ covariance matrix for the knot locations $\tilde{\mathbf{S}}$, and $\tilde{\boldsymbol{\eta}} \sim \text{N}(\mathbf{0}, \tilde{\boldsymbol{\Sigma}})$ is an $m \times 1$ correlated random vector. Substituting Equation 2.52 into Equation 2.51 and disregarding the fact that the predictive process is an approximation, we have

$$\begin{aligned}\mathbf{y} &= \mathbf{X}\boldsymbol{\beta} + \hat{\boldsymbol{\eta}} + \boldsymbol{\varepsilon} \\ &= \mathbf{X}\boldsymbol{\beta} + \tilde{\boldsymbol{\Sigma}}_\eta \tilde{\boldsymbol{\Sigma}}^{-1} \tilde{\boldsymbol{\eta}} + \boldsymbol{\varepsilon}.\end{aligned} \tag{2.53}$$

Thus, rather than needing to invert the large covariance matrix, $\boldsymbol{\Sigma}_\eta$, at every step in a statistical computer algorithm (e.g., MCMC), we only need to invert the $m \times m$ matrix $\tilde{\boldsymbol{\Sigma}}$ and sample an m-dimensional correlated random vector $\tilde{\boldsymbol{\eta}}$. If the number of knots (m) is small relative to the sample size (n), then the predictive process procedure can be very computationally advantageous.

An interesting and relevant note is that the predictive process specification takes the same form as the other reduced-rank specifications we described previously (2.44). That is, $\tilde{\boldsymbol{\Sigma}}_\eta \tilde{\boldsymbol{\Sigma}}^{-1} \tilde{\boldsymbol{\eta}} = \tilde{\mathbf{H}}\tilde{\boldsymbol{\alpha}}$, where $\tilde{\boldsymbol{\Sigma}}_\eta \tilde{\boldsymbol{\Sigma}}^{-1}$ is the matrix of basis functions and $\tilde{\boldsymbol{\eta}}$ represents the process on a lower-dimensional manifold. The key difference between the predictive process and more conventional methods for dimension reduction is in the choice and properties of basis functions. One could argue that the predictive process approach is heuristically more tangible than other spectral approaches for defining basis functions because, in the predictive process procedure, the knot locations are in the same space as the data locations. The associated basis functions in $\tilde{\boldsymbol{\Sigma}}_\eta \tilde{\boldsymbol{\Sigma}}^{-1}$ can be visualized in Euclidean space and the coefficients $\tilde{\boldsymbol{\eta}}$ are the values of the spatial process at the set of knots.

Furthermore, the predictive process was originally intended for use when the covariance matrices (i.e., $\boldsymbol{\Sigma}_\eta$, $\tilde{\boldsymbol{\Sigma}}_\eta$, and $\tilde{\boldsymbol{\Sigma}}$) are functions of unknown parameters, and thus, must be computed and inverted at each step of a statistical algorithm. For example, the elements of each of the covariance matrices might be modeled geostatistically as an exponential function $\sigma_\eta^2 \exp(-d_{ij}/\phi)$ where d_{ij} represents the distance between any two points i and j such that these points could be either in the data locations, knot locations, or both.

Additionally, the parameter ϕ could be estimated using an auxiliary source of data or using empirical Bayes with the same source of data. This allows the expansion matrix $\tilde{\boldsymbol{\Sigma}}_\eta \tilde{\boldsymbol{\Sigma}}^{-1}$ to be fixed *a priori*, resulting in a further speedup of computation. Regardless of the approach, as with any method, the devil is in the details in terms of how to actually implement this method. In the former dimension reduction approaches, one estimates the correlation matrix *a priori* and then spectrally decomposes and truncates it to obtain $\tilde{\mathbf{H}}$, whereas in the predictive process approach, one must select the number and locations of knots. This is relatively easy, but a poor choice of knot locations can dramatically misrepresent the spatial structure in $\boldsymbol{\eta}$.

Finally, Banerjee et al. (2008) actually use an integrated likelihood based on the same procedure described in the previous section. The integrated likelihood approach using a set of basis vectors based on the predictive process yields a fast and stable MCMC algorithm for fitting a reduced-rank geostatistical model.

Returning to the Midwestern U.S. temperature data described in the earlier sections, suppose we wish to model the temperature field using a Bayesian framework (as opposed to the non-Bayesian framework we employed to predict the temperature in Section 2.2). We considered the same measurements from the previous analysis for the response variable (i.e., average maximum temperature in the Midwestern United States) and covariates (i.e., latitude and longitude).

We fit two Bayesian geostatistical models to the temperature data, the first using the standard linear mixed model specification

$$\mathbf{y} = \mathbf{X}\boldsymbol{\beta} + \boldsymbol{\eta} + \boldsymbol{\varepsilon}, \tag{2.54}$$

where $\mathbf{y}$ represents the temperature measurements and $\mathbf{X}$ contains the covariates at the measurement locations. We modeled the correlated random effects $\boldsymbol{\eta}$ as Gaussian random fields with exponential covariance structure.* We used the predictive process formulation

$$\mathbf{y} = \mathbf{X}\boldsymbol{\beta} + \hat{\boldsymbol{\eta}} + \boldsymbol{\varepsilon} \tag{2.55}$$

for the second model. In this case, we still used an exponential covariance model, but with the predictive process basis function expansion $\hat{\boldsymbol{\eta}} \equiv \tilde{\boldsymbol{\Sigma}}_\eta \tilde{\boldsymbol{\Sigma}}^{-1} \tilde{\boldsymbol{\eta}}$.

Figure 2.13 compares the predictions from the first (i.e., full-rank) and second (i.e., predictive process) Bayesian geostatistical models described previously. The full-rank Bayesian geostatistical model required approximately 30 s to fit on a 2 × 2.93 Ghz processor machine with 32 GB of memory, whereas the reduced-rank predictive process model required only 8 s based on 15 evenly spaced knots throughout the prediction domain. The surfaces are similar, with the predictive process resulting in a slightly smoother predicted temperature field.

Spatial statistics provide a rich source of tools that can be used in many other fields and are critical for many types of animal movement inference. For example, in Chapter 4, we delve deeper into SPP models and, in Chapter 6, we rely on geostatistical models for movement trajectories. However, given that telemetry data are

* Gaussian processes and Gaussian random fields are the same thing; they both are realizations of a continuous Gaussian distribution with correlation structure (i.e., a nondiagonal covariance matrix).

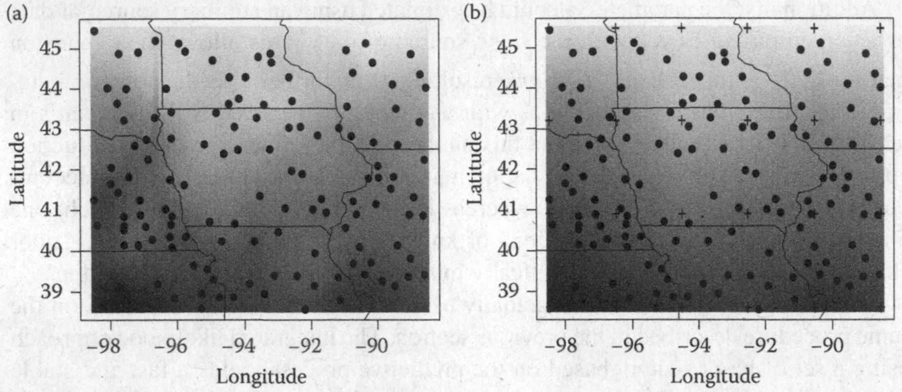

FIGURE 2.13 Predictive surfaces (darker corresponds to warmer temperatures) resulting from Bayesian model fits and Kriging based on the Midwestern temperature data. Panel (a) displays the temperature predictions based on the full-rank model and panel (b) displays the temperature predictions based on the predictive process reduced-rank model. Points correspond to measurement locations and crosses correspond to the knot locations that anchor the basis functions.

also explicitly temporal, we summarize fundamental statistics for time series in the following chapter.

2.6 ADDITIONAL READING

The use of spatial statistics in ecological modeling is increasing and new and useful methodological developments are appearing regularly. Highly readable overall references for spatial statistics include Schabenberger and Gotway (2005), Waller and Gotway (2004), Cressie (1990), Chapter 4 of Cressie and Wikle (2011), Banerjee et al. (2014), and the edited volume by Gelfand et al. (2010) for recent developments. More specifically, in terms of point processes, Møller and Waagepetersen (2004) was traditionally referred to, but newer references have become more popular (e.g., Illian et al. 2008; Baddeley and Turner 2000). Diggle and Ribeiro (2007) provide a nice summary of model-based geostatistics and Rue and Held (2005) provide a technical overview of Markov random fields, including specific subjects such as CAR models. Ver Hoef et al. (In Review) provide an overview of CAR and SAR models.

With increasing need to analyze "big data," dimension reduction has become popular in the spatial statistics literature recently. Aside from the excellent overview in Wikle (2010a), see some of the specific new developments by Rue et al. (2009), Datta et al. (2016), and Katzfuss (2016) for cutting-edge dimension reduction ideas for massive data sets.

For a primer on the use of Bayesian statistical methods for analyzing ecological data, see Hobbs and Hooten (2015), and for more detailed references in specific application areas, see Royle and Dorazio (2008) and Clark (2007). For a more technical overview of Bayesian methods, see Gelman et al. (2014). Bolker (2008) provides a good refresher on all other statistical methods in ecology.

3 Statistics for Temporal Data

Animal telemetry data usually consist of time-indexed spatial locations, and can be thought of as multivariate time series. Thus, a foundation in the statistical treatment of time series data is important for modeling animal movement. This chapter provides a useful set of tools and concepts that one may wish to apply to telemetry data.

3.1 UNIVARIATE TIME SERIES

We begin our review of time series by introducing notation and terminology in the univariate context and then move to the multivariate context in the section that follows. The essential premise in statistics for time series involves the assessment and modeling of dependence in temporally indexed data and processes. Much like in spatial statistics, we consider a variable y_t for time index $t = 1, \ldots, T$ that exhibits temporal variation. This variation can be a function of first- and/or second-order processes. A first-order process, in the time series context, corresponds to a model for the mean of y_t such that $\text{E}(y_t) = \mu_t$ and the mean process might also vary over time as a function (either deterministic or stochastic) of some other variables. For example, a regression formulation for the mean process can be specified as $\mu_t = \mathbf{x}_t'\boldsymbol{\beta}$, where $\mathbf{x}_t$ is a vector of covariates for time index t and $\boldsymbol{\beta}$ are the usual coefficients that link the covariate to the mean.

The second-order perspective in time series is concerned with the covariance of the data or process (i.e., $\text{cov}(y_t, y_\tau)$ for all t and τ). As in spatial statistics, for time series, we need ways to assess the first- and second-order structure, as well as ways to model it. The assessment and estimation of first-order processes are fairly straightforward and could entail an examination of the correlations among y_t and $\mathbf{x}_t$ for all t. However, it is the second-order structure for which we need new machinery. One might argue that time series data are merely a simplification of spatial data, and thus, we could apply all of the same descriptive methods we described in Chapter 2. Translating spatial statistical methods into the time series context, we would associate variogram approaches with continuous temporal processes and Moran's I statistics with discrete temporal processes, for example. In fact, versions of these methods exist for time series and were created somewhat independently of their spatial counterparts.*

Why are we concerned with temporal dependence? Aside from the fact that we are interested in studying a naturally dynamic process in animal movement, as in spatial statistics, there can be consequences associated with using a first-order-only

* One could argue that time is strictly a forward process. Regardless of this fact (at least the way we experience time as humans, with the potential exception of Kurt Vonnegut), statistical approaches make use of information on both sides of the time point of interest, as we demonstrate in what follows.

model for a first- and second-order process. For example, most ecologists are primarily interested in first-order effects (i.e., things that influence the mean of the process under study). Thus, we should avoid making invalid model assumptions that lead to erroneous inference. To illustrate how invalid assumptions about second-order dependence can lead to erroneous first-order inference, consider the following contrived example.

Suppose the following two sets of data are collected from known distributions:

1. *Independent error:* $y_1, y_2 \sim [y|\mu, \sigma^2]$, where $\text{cov}(y_1, y_2) = 0$,
2. *Dependent error:* $z_1, z_2 \sim [z|\mu, \sigma^2]$, where $\text{cov}(z_1, z_2) \neq 0$,

where μ and σ^2 correspond to the mean and variance, and note that the square bracket notation $[\cdot]$ refers to a probability distribution, as before. These distributions could be Gaussian, but need not be in this example. Suppose we are interested in estimating the first-order mean μ in this setting. The usual estimator for a population mean is the sample mean, which is $\hat{\mu}_y = (y_1 + y_2)/2$ and $\hat{\mu}_z = (z_1 + z_2)/2$. As an estimator, the sample mean enjoys the excellent properties of unbiasedness and known variance in the case where σ^2 is known. To see how these properties arise, we derive the first two moments for the distribution of $\hat{\mu}$ in detail. First, the expectation of the sample mean for y is

$$\begin{aligned}
\text{E}(\hat{\mu}_y) &= \text{E}\left(\frac{y_1 + y_2}{2}\right) \\
&= \frac{1}{2}\text{E}\left(y_1 + y_2\right) \\
&= \frac{1}{2}\left(\text{E}(y_1) + \text{E}(y_2)\right) \\
&= \frac{1}{2}\left(\mu + \mu\right) \\
&= \frac{1}{2}(2\mu) \\
&= \mu.
\end{aligned}$$

Thus, the sample mean is an unbiased estimator of μ. The same procedure can be applied to show that $\hat{\mu}_z$ is also unbiased. Thus, we have an unbiased estimator for a homogeneous mean in both cases (i.e., the independent error (1) or dependent error (2) case).

Proceeding with the variance, we take a similar approach, but using variance and covariance operators. To find the variance of $\hat{\mu}_y$, we start by considering the variance as a covariance of that quantity and itself and then expand the covariance term as

$$\begin{aligned}
\text{Var}(\hat{\mu}_y) &= \text{cov}(\hat{\mu}_y, \hat{\mu}_y) \\
&= \frac{1}{2^2}\text{cov}\left(y_1 + y_2, y_1 + y_2\right)
\end{aligned}$$

$$
\begin{aligned}
&= \frac{1}{2^2}\left(\text{cov}(y_1, y_1) + \text{cov}(y_1, y_2) + \text{cov}(y_2, y_1) + \text{cov}(y_2, y_2)\right) \\
&= \frac{1}{2^2}\left(\text{Var}(y_1) + 2\text{cov}(y_1, y_2) + \text{Var}(y_2)\right) \\
&= \frac{1}{2^2}\left(\sigma^2 + 2\text{cov}(y_1, y_2) + \sigma^2\right) \\
&= \frac{1}{2^2}(2\sigma^2) \\
&= \frac{\sigma^2}{2}.
\end{aligned}
$$

Thus, the variance of $\hat{\mu}_y$ is the population variance divided by the sample size (as we recall from our first statistics course). But, what is the variance of $\hat{\mu}_z$? The variance of $\hat{\mu}_z$ can be found using the same procedure, except notice that the $2\text{cov}(z_1, z_2)$ (on line 5 in the above derivation) is not zero, implying that

$$\text{Var}(\hat{\mu}_z) = \frac{\sigma^2}{2} + \frac{\text{cov}(z_1, z_2)}{2}. \tag{3.1}$$

The variance of the estimator for μ is either larger or smaller for z than it is for y. In this case, $\text{Var}(\hat{\mu}_z)$ will be larger when the covariance between z_1 and z_2 is positive and smaller when negative. Thus, positive dependence in time series data,* if unaccounted for, will lead to confidence intervals that are too narrow, inflating the chance of a type 1 error in decision making based on first-order effects. Thus, in what follows, we provide the background to assess, and then account for, temporal dependence in data and processes.

3.1.1 Descriptive Statistics

Some of the most commonly used methods for assessing dependence in time series data are autocorrelation functions (ACFs) and partial autocorrelation functions (PACFs). The term "autocorrelation" refers to a set of random variables that are correlated with themselves. For a time series $\boldsymbol{\eta} \equiv (\eta_1, \ldots, \eta_t, \ldots, \eta_T)'$, we can express covariance in terms of an autocovariance function $\gamma(t, \tau) = \text{E}((\eta_t - \text{E}(\eta_t))(\eta_\tau - \text{E}(\eta_\tau)))$.† If the temporal process is stationary with homogeneous mean equal to zero, we write the autocovariance as a function of the distance between time points Δt

$$\gamma(\Delta t) = \text{E}(\eta_t \eta_{t-\Delta t}), \tag{3.2}$$

and the resulting ACF is $\rho(\Delta t) = \gamma(\Delta t)/\gamma(0)$.

Returning to the concept of stationarity for temporal processes: If the covariance only depends on Δt for all η_t and η_τ, then it is stationary. Typically, an assumption

---

* Or any other data, especially spatial data.

† We use η, instead of ε, because we will add an error term to the model structure in later sections.

of homogeneous mean is also required for temporal stationarity, but we mention that separately because we intend to model variation in the mean (i.e., first-order) elsewhere. The covariance function for time series is the analog to the covariogram in spatial statistics and the stationarity assumption has a similar interpretation as well; specifically, that the temporal process behaves according to the same dependence throughout the entire time series.

We need a way to estimate the covariance function for time series. Thus, we estimate $\gamma(\Delta t)$ and $\rho(\Delta t)$ with

$$\hat{\gamma}(\Delta t) = \frac{\sum_{t=1+\Delta t}^{T}(\eta_t - \bar{\eta})(\eta_{t-\Delta t} - \bar{\eta})}{T - \Delta t}, \tag{3.3}$$

and

$$\hat{\rho}(\Delta t) = \frac{\hat{\gamma}(\Delta t)}{\hat{\gamma}(0)}. \tag{3.4}$$

Using these estimators in a large sample situation, if η is not correlated,* approximately 95% of $\hat{\rho}(\Delta t)$ should fall in the interval $(-1.96/\sqrt{T-\Delta t}, 1.96/\sqrt{T-\Delta t})$. This provides a way to test if an observed time series meeting the aforementioned assumptions is uncorrelated.

The other useful statistic for assessing temporal dependence is called the PACF. The PACF provides inference about the correlation between η_t and $\eta_{t-\Delta t}$ with the dependence from the time points between removed. The PACF is estimated as

$$\hat{\rho}(\Delta t, \Delta t) = \begin{cases} \hat{\rho}(1) & \text{if } \Delta t = 1 \\ \dfrac{\hat{\rho}(\Delta t) - \sum_{j=1}^{\Delta t-1} \hat{\rho}(\Delta t - 1, j)\hat{\rho}(\Delta t - j)}{1 - \sum_{j=1}^{\Delta t-1} \hat{\rho}(\Delta t - 1, j)\hat{\rho}(j)} & \text{if } \Delta t = 2, 3, \ldots \end{cases}, \tag{3.5}$$

where $\hat{\rho}(\Delta t, j) = \hat{\rho}(\Delta t - 1, j) - \hat{\rho}(\Delta t, \Delta t)\hat{\rho}(\Delta t - 1, \Delta t - j)$ for $j = 1, \ldots, \Delta t - 1$. In the next section, we use the ACF and PACF together to help identify potential model structures for temporal processes.

Another approach for assessing autocorrelation in time series data is to specify a model that explicitly contains it. For now, we assume that the model for autocorrelation is nonparametric (i.e., not naming the distribution that provides the stochasticity explicitly). When we suspect a linear trend in the data, we can specify a model such that

$$y_t = \beta_0 + \beta_1 t + \varepsilon_t, \tag{3.6}$$

where ε_t are independent and normally distributed errors, and we might be interested in removing that first-order structure and then assess the second-order structure by computing a statistic based on the residuals $e_t = y_t - \beta_0 - \beta_1 t$. The Durbin–Watson

---

* An uncorrelated stationary temporal process is typically referred to as "white noise" in the time series literature.

statistic then is computed as

$$\hat{d} = \frac{\sum_{t=2}^{T}(e_t - e_{t-1})^2}{\sum_{t=1}^{T} e_t^2}. \tag{3.7}$$

One can then find the associated confidence interval for this statistic and gauge whether lag 1 autocorrelation is evident in the data beyond the first-order trend. It is important to note that this statistic only examines lag 1 autocorrelation. That is, it is concerned only with e_t and $e_{t-\Delta t}$, where $\Delta t = 1$ for all t; although it can be adapted to assess autocorrelation at larger lags.

Consider the four simulated time series in Figure 3.1. Panels (a–c) in Figure 3.1 show time series with increasing amounts of positive temporal dependence, whereas

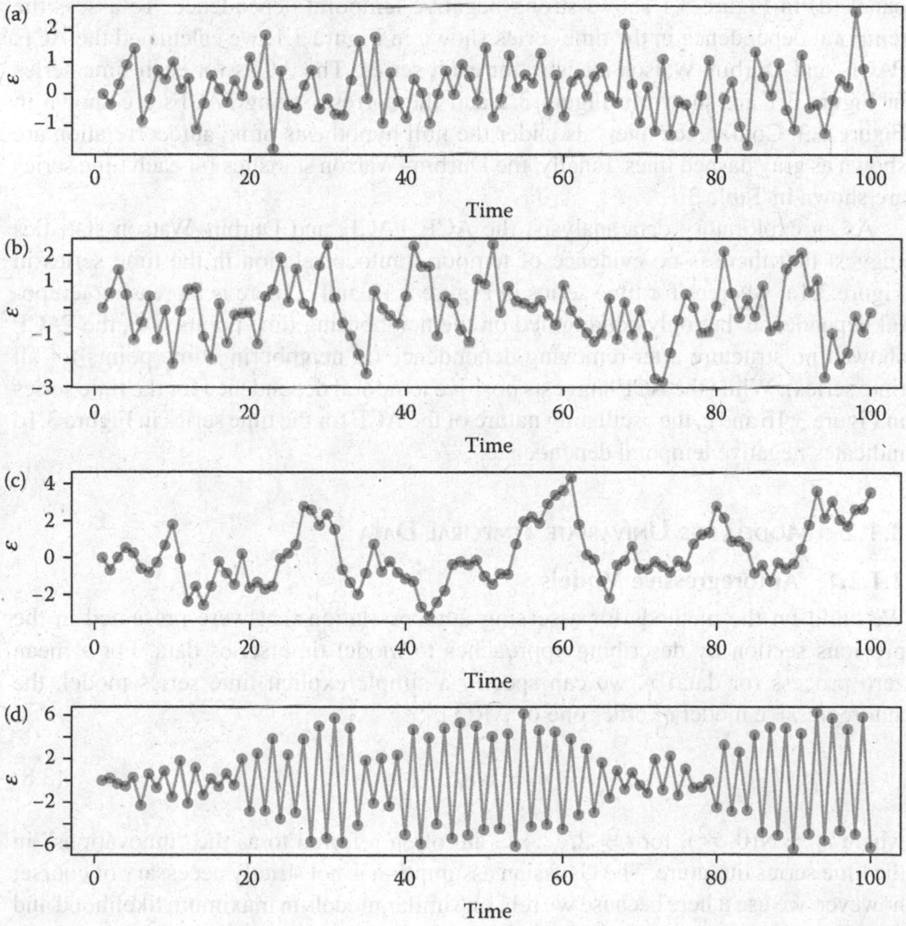

FIGURE 3.1 Simulated time series with mean zero, variance equal to 1, and (a) no temporal dependence, (b) moderate positive temporal dependence, (c) strong positive temporal dependence, and (d) strong negative temporal dependence.

TABLE 3.1
Durbin–Watson Statistics and p-Values at Lag 1 Unit of Time for a Two-Sided Hypothesis Test for the Time Series in Figure 3.1

	a	b	c	d
Durbin–Watson	2.31	1.18	0.37	3.88
p-Value	0.12	<0.001	<0.001	<0.001

Note: The null hypothesis is no autocorrelation.

panel (d) in Figure 3.1 shows strong negative temporal dependence. To assess the temporal dependence in the time series shown in Figure 3.1, we calculated the ACF, PACF, and Durbin–Watson statistic for each series. The ACFs for each time series in Figure 3.1 are shown in Figure 3.2 and the corresponding PACFs are shown in Figure 3.3. Confidence intervals under the null hypothesis of no autocorrelation are shown as gray dashed lines. Finally, the Durbin–Watson statistics for each time series are shown in Table 3.1.

As an exploratory data analysis, the ACF, PACF, and Durbin–Watson statistics suggest that there is no evidence of temporal autocorrelation in the time series in Figure 3.1a, whereas for time series in Figure 3.1b and c, there is increasing temporal dependence, but only conditioned on the neighboring time points (i.e., the PACF showed no structure after removing dependence on neighboring time points for all time series). While the ACF suggests positive temporal dependence for the time series in Figure 3.1b and c, the oscillating nature of the ACF for the time series in Figure 3.1d indicates negative temporal dependence.

3.1.2 Models for Univariate Temporal Data

3.1.2.1 Autoregressive Models

We build on the methods for assessing autocorrelation that were presented in the previous section by describing approaches to model time series data. For a mean zero process (or data) η, we can specify a simple explicit time series model, the autoregressive model of order one or AR(1):

$$\eta_t = \alpha\eta_{t-1} + \varepsilon_t, \tag{3.8}$$

where $\varepsilon_t \sim \text{N}(0, \sigma^2)$, for $t = 2, \ldots, T$, are often referred to as the "innovations" in the time series literature. The Gaussian assumption is not strictly necessary of course; however, we use it here because we rely on similar models in maximum likelihood and Bayesian contexts later. In this case, the phrase "order one" refers to the fact that the time lag $\Delta t = 1$ and the parameter α controls the dynamics in the model, essentially influencing the amount of either positive ($\alpha > 0$) or negative ($\alpha < 0$) dependence. The AR(1) process (3.8) will be well behaved (i.e., not explosive) if $-1 < \alpha < 1$.

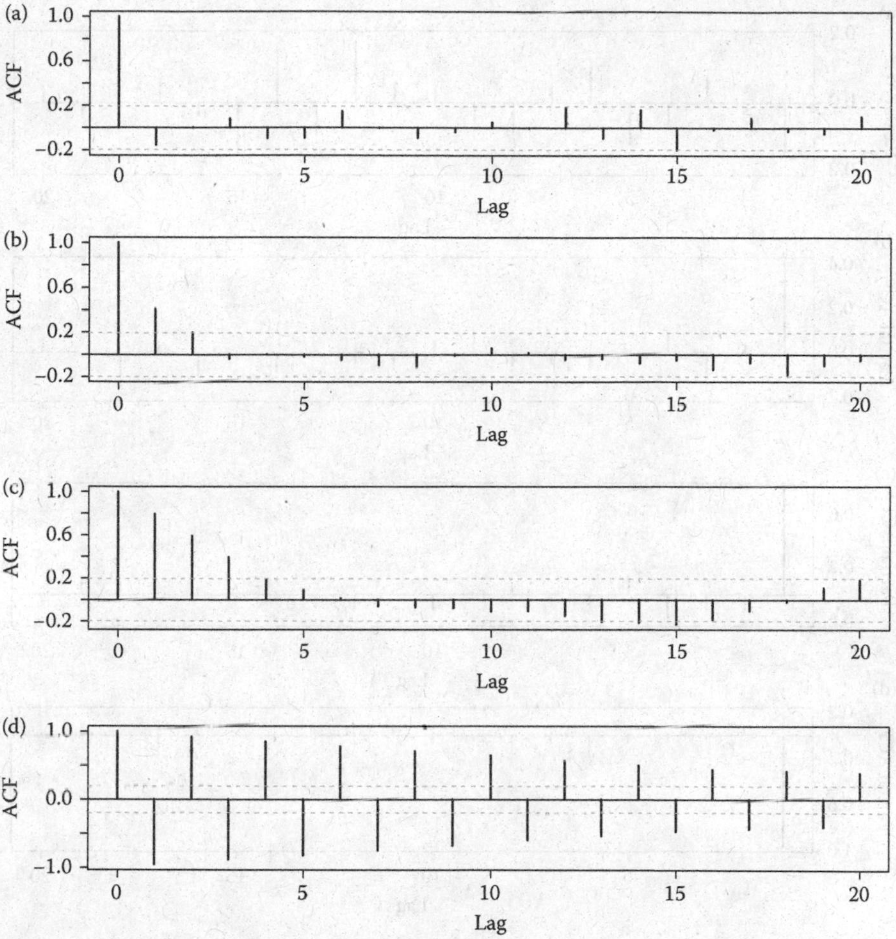

FIGURE 3.2 ACF for simulated time series with mean zero, variance equal to 1, and (a) no temporal dependence, (b) moderate positive temporal dependence, (c) strong positive temporal dependence, and (d) strong negative temporal dependence. Gray dashed lines show a 95% confidence interval under the null hypothesis.

At $\alpha = 0$, the time series is a white noise process (i.e., independent) with mean zero and variance σ^2. However, for $\alpha \neq 0$, the process is often referred to as a "random walk." That is, each step in the time series is a step of random length in a random direction away from the previous location (in η space). When $\alpha = 1$, the random walk is not stationary and can wander anywhere it wants in the real numbers. The random walk can be used as a model for temporal dependence where strong autocorrelation is present or desired.

The AR(1) model is naturally conditional, that is, η_t depends on η_{t-1}, but it can also be coerced into a joint model such that $\boldsymbol{\eta} \sim \text{N}(\mathbf{0}, \boldsymbol{\Sigma})$. In the joint model, the precision (i.e., inverse covariance) matrix $\boldsymbol{\Sigma}^{-1}$ is tri-diagonal with $(1-\alpha)/\sigma^2$ for the first and

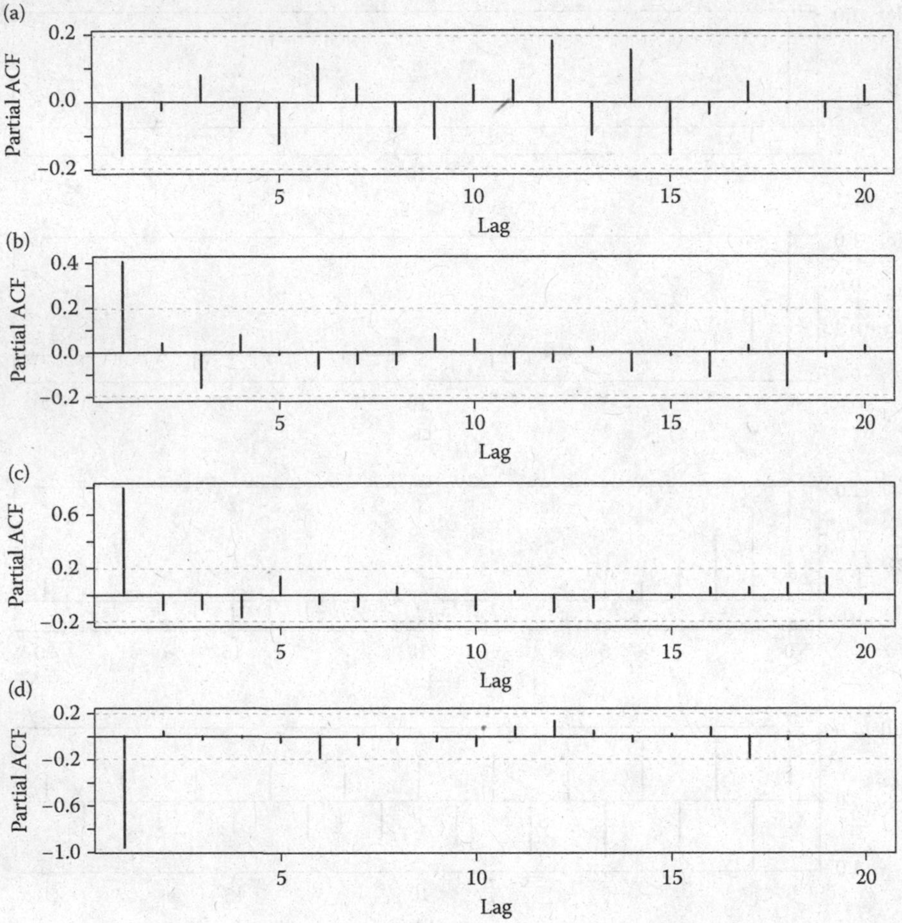

FIGURE 3.3 PACF for simulated time series with mean zero, variance equal to 1, and (a) no temporal dependence, (b) moderate positive temporal dependence, (c) strong positive temporal dependence, and (d) strong negative temporal dependence. Gray dashed lines show a 95% confidence interval under the null hypothesis.

last diagonal elements, $(2 - \alpha)/\sigma^2$ on the diagonal elements for $t = 2, \ldots, T - 1$, α/σ^2 on the first off-diagonals, and zero elsewhere. This is the same result discussed in the CAR models for areal spatial processes, but where the discrete areal support is one-dimensional (1-D) in time (i.e., instead of 2-D in space).

Higher-order autoregressive models can be specified similar to the AR(1). For example, the AR(2) can be specified as

$$\eta_t = \alpha_1 \eta_{t-1} + \alpha_2 \eta_{t-2} + \varepsilon_t, \tag{3.9}$$

where the second autoregressive coefficient controls the dependence at time lag $\Delta t = 2$. Models with dependence at higher-order lags are often referred to as AR(p) models, where p denotes the highest-order lag in the model. It should be noted that, outside of the field of economics, AR models of higher order than 2 are not common. One of the reasons higher-order models are not common in ecology is that they can be difficult to interpret.

There are several extensions we might want to make for this type of autoregressive model. The first, and perhaps most obvious, is to allow for a trend in the process. Thus, we denote y_t as the response variable to clarify that we are now specifying models for something other than a mean zero stationary process. Consider a scenario where there exists a heterogeneous temporal trend in the data and we wish to account for it in the model. We have many options in that case; however, to express the general idea of how to set up such a model, we limit ourselves to only AR(1) dynamics. A univariate autoregressive temporal model with linear heterogeneous trend is specified as

$$y_t = \mathbf{x}_t'\boldsymbol{\beta} + \alpha y_{t-1} + \varepsilon_t, \tag{3.10}$$

where $\varepsilon_t \sim \text{N}(0, \sigma^2)$ and $\mathbf{x}_t$ represents a vector of temporally referenced covariates with corresponding regression coefficients $\boldsymbol{\beta}$. This heterogeneous time series model (3.10) includes the dynamics directly for the y_t process. An alternative specification includes the dynamics on the "error" process from Equation 3.8 such that

$$y_t = \mathbf{x}_t'\boldsymbol{\beta} + \eta_t, \tag{3.11}$$

where $\eta_t \sim \text{N}(\alpha\eta_{t-1}, \sigma^2)$. Substituting $\alpha\eta_{t-1} + \varepsilon_t$ into Equation 3.11, for η_t, implies that

$$\begin{aligned}
y_t &= \mathbf{x}_t'\boldsymbol{\beta} + \eta_t \\
&= \mathbf{x}_t'\boldsymbol{\beta} + \alpha\eta_{t-1} + \varepsilon_t \\
&= \mathbf{x}_t'\boldsymbol{\beta} + \alpha(y_{t-1} - \mathbf{x}_{t-1}'\boldsymbol{\beta}) + \varepsilon_t \\
&= \mathbf{x}_t'\boldsymbol{\beta} + \alpha y_{t-1} - \alpha\mathbf{x}_{t-1}'\boldsymbol{\beta} + \varepsilon_t \\
&= (\mathbf{x}_t - \alpha\mathbf{x}_{t-1})'\boldsymbol{\beta} + \alpha y_{t-1} + \varepsilon_t,
\end{aligned} \tag{3.12}$$

where $\varepsilon_t \sim \text{N}(0, \sigma^2)$ as in Equation 3.8. The difference in specifications (3.11) and (3.12) is that the new covariates $(\mathbf{x}_t - \alpha\mathbf{x}_{t-1})$ are now a weighted lag 1 difference of the original covariates $(\mathbf{x}_t)$. These different specifications affect the inference obtained from fitting such models. In the situation where $\alpha = 0$, the nondynamic temporal regression model results (i.e., no AR(1) component), as we might expect it to. Conversely, when $\alpha = 1$ (i.e., strong autocorrelation), the model becomes

$$y_t = (\mathbf{x}_t - \mathbf{x}_{t-1})'\boldsymbol{\beta} + y_{t-1} + \varepsilon_t, \tag{3.13}$$

where the dynamic component is a random walk and the new covariates $(\mathbf{x}_t - \mathbf{x}_{t-1})$ are a velocity vector in covariate space describing the change in covariates during that time period. Thus, the stronger the autocorrelation in $\boldsymbol{\eta}$, the more the inference about $\boldsymbol{\beta}$ shifts away from a direct effect of $\mathbf{x}_t$ on y_t, and shifts toward the effect of a change in covariates (over time) on the associated change in the response variable.* Thus, the model with autocorrelated errors (3.11) can be thought of as a form of discretized differential equation model and is a very important topic that will arise in Chapter 6, when modeling animal movement.

Each of the time series in Figure 3.1 were simulated from an AR(1) process with mean zero and variance 1 using model (3.8). In Figure 3.1, panel (a) used an autocorrelation parameter $\alpha = 0$, panel (b) used $\alpha = 0.5$, panel (c) used $\alpha = 0.9$, and panel (d) used $\alpha = -0.9$. The fact that we used AR(1) models to simulate each of the data sets was suggested by the ACF, PACF, and Durbin–Watson statistics.

To simulate a higher-order AR time series model with a trend (Figure 3.4), we use

$$y_t = \beta_0 + \beta_1 x_t + \eta_t, \tag{3.14}$$

where $\beta_0 = 1$, $\beta_1 = 1$, and $\eta_t \sim \text{N}(\alpha_1 \eta_{t-1} + \alpha_2 \eta_{t-2}, \sigma^2)$ with $\alpha_1 = 0.1$, $\alpha_2 = 0.5$, and $\sigma^2 = 1$. The temporal covariate x_t is simulated from a sine function (Figure 3.4).

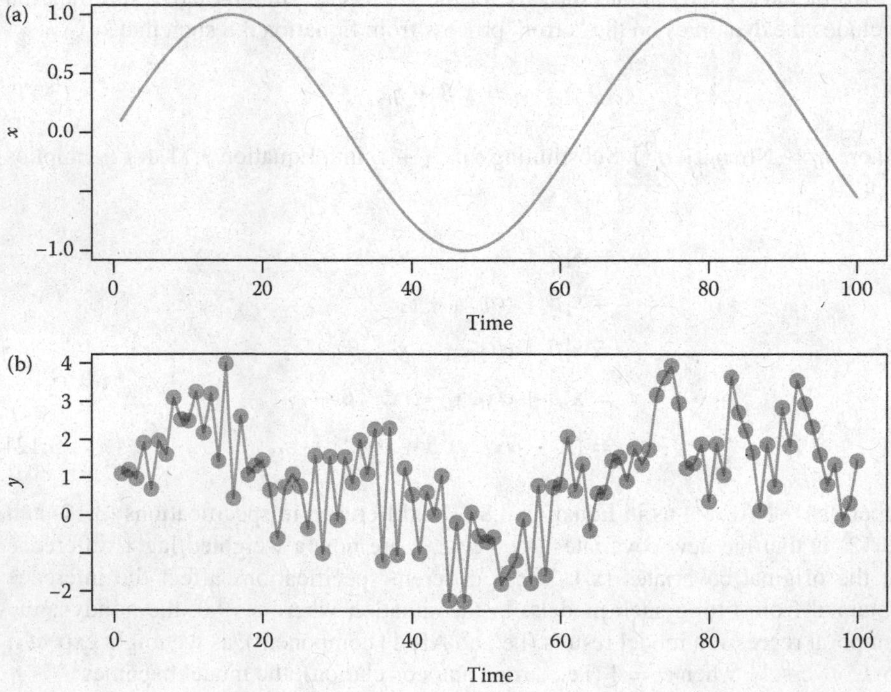

FIGURE 3.4 Simulated time series with heterogeneous trend specified as Equation 3.14. (a) The temporal covariate x_t and (b) the time series y_t.

* Note that this can also be written as $(y_t - y_{t-1}) = (\mathbf{x}_t - \mathbf{x}_{t-1})'\boldsymbol{\beta} + \varepsilon_t$. So, the change in "position" (y) is a function of a change in the driving factors ($\mathbf{x}$). This will be an important formulation in some of the animal movement models that follow.

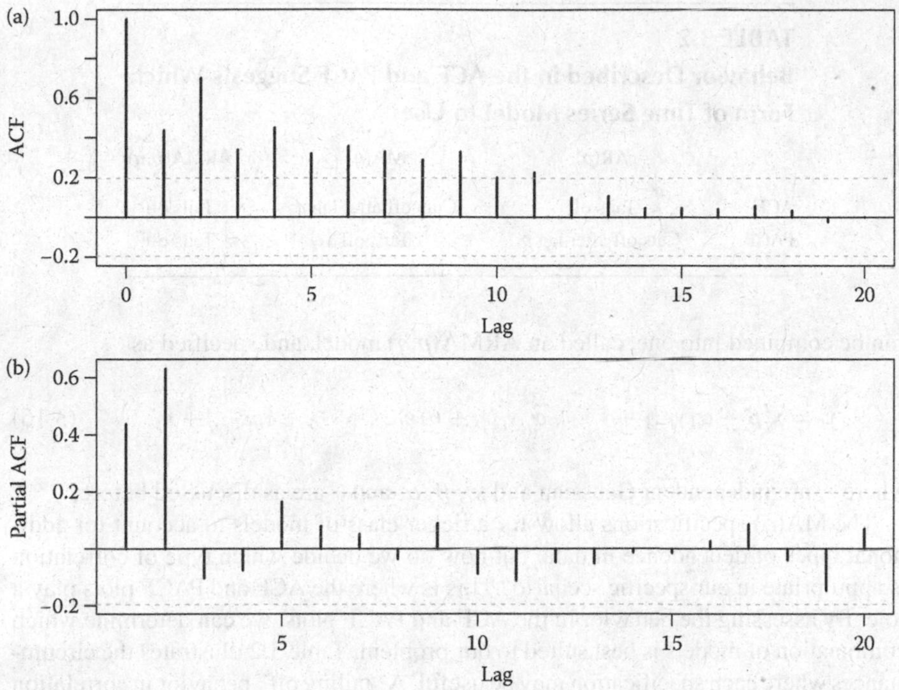

FIGURE 3.5 ACF (a) and PACF (b) for the simulated time series with heterogeneous trend specified as Equation 3.14. Gray dashed lines show a 95% confidence interval under the null hypothesis.

The ACF and PACF plots shown in Figure 3.5 indicate longer range dependence in the time series and dependence at time lag 2 after accounting for dependence at lag 1.

3.1.2.2 Moving Average Models

Autoregressive structures are not the only form of temporal dependence commonly used; there are also a suite of moving average models that can be useful for characterizing temporal structure in data. Moving average models (i.e., MA(q), for order q) differ from AR(p) models in that they are a regression of the response variable on previous error terms. These previous error terms in the model are often referred to as "shocks" and are assumed to arise from an error distribution independently. The unique characteristic of these shocks is that the response is only affected by shocks at the specified number of lags (q) in the model.

A basic MA(1) model, with a linear first-order trend, can be written as

$$y_t = \mathbf{x}_t'\boldsymbol{\beta} + \theta\varepsilon_{t-1} + \varepsilon_t, \tag{3.15}$$

where $\varepsilon_t \sim \text{N}(0, \sigma^2)$ for all t and θ is the MA(1) regression coefficient. Note that the difference between Equation 3.15 and the former model (3.12) is that the errors (ε_t) are uncorrelated in Equation 3.15. Furthermore, the two types of time series models

TABLE 3.2
Behavior Described in the ACF and PACF Suggests Which Form of Time Series Model to Use

	AR(p)	MA(q)	ARMA(p,q)
ACF	Tails off	Cuts off after lag q	Tails off
PACF	Cuts off after lag p	Tails off	Tails off

can be combined into one, called an ARMA(p,q) model, and specified as

$$y_t = \mathbf{x}_t'\boldsymbol{\beta} + \alpha_1 y_{t-1} + \cdots + \alpha_p y_{t-p} + \theta_1 \varepsilon_{t-1} + \cdots + \theta_q \varepsilon_{t-q} + \varepsilon_t, \tag{3.16}$$

where ε_t are independent Gaussian and $\mathbf{x}_t$, $\boldsymbol{\beta}$, α, and θ are as discussed before.

The MA(q) specifications allow for a richer class of models to account for additional types of dependence in data, but how do we decide which type of correlation is appropriate in our specific scenario? This is where the ACF and PACF plots play a role. By assessing the behavior in the ACF and PACF plots, we can determine which combination of models is best suited to our problem. Table 3.2 illustrates the circumstances where each specification may be useful. A "tailing off" behavior in correlation function plot refers to the smooth decrease in magnitude (either positive or negative) of the function at increasing lags, whereas "cutting off" behavior can be seen when the function abruptly reduces to some negligible magnitude (either positive or negative). Thus, the combination of ACF and PACF plots provides insight about which types of shocks are influencing the system under study (Shumway and Stoffer 2006).

To illustrate the behavior of an ARMA time series process, we simulate using an ARMA(1,1) model with no covariate-based trend

$$y_t = \alpha_1 y_{t-1} + \theta_1 \varepsilon_{t-1} + \varepsilon_t, \tag{3.17}$$

where $\varepsilon_t \sim \text{N}(0, 1)$ are independent, $\alpha_1 = 0.9$, and $\theta_1 = 0.9$ (Figure 3.6). Following the guidance in Table 3.2, the ACF and PACF in Figure 3.6 indicate that the time series does show characteristics of an ARMA time series, as it should, because both the ACF and PACF tail off rather than cut off after a certain lag.

3.1.2.3 Backshift Notation

A common and useful notation used in the time series literature is the "backshift" operator B.* The backshift operator is used as a function on the temporal variable at time point t as $By_t = y_{t-1}$. The backshift operator can be used sequentially to simplify notation as well; for example, $B^2 y_t = BBy_t = By_{t-1} = y_{t-2}$. Using the backshift operator, we can reformulate each of the previously discussed time series models.

* Sometimes also referred to as a "lag" operator.

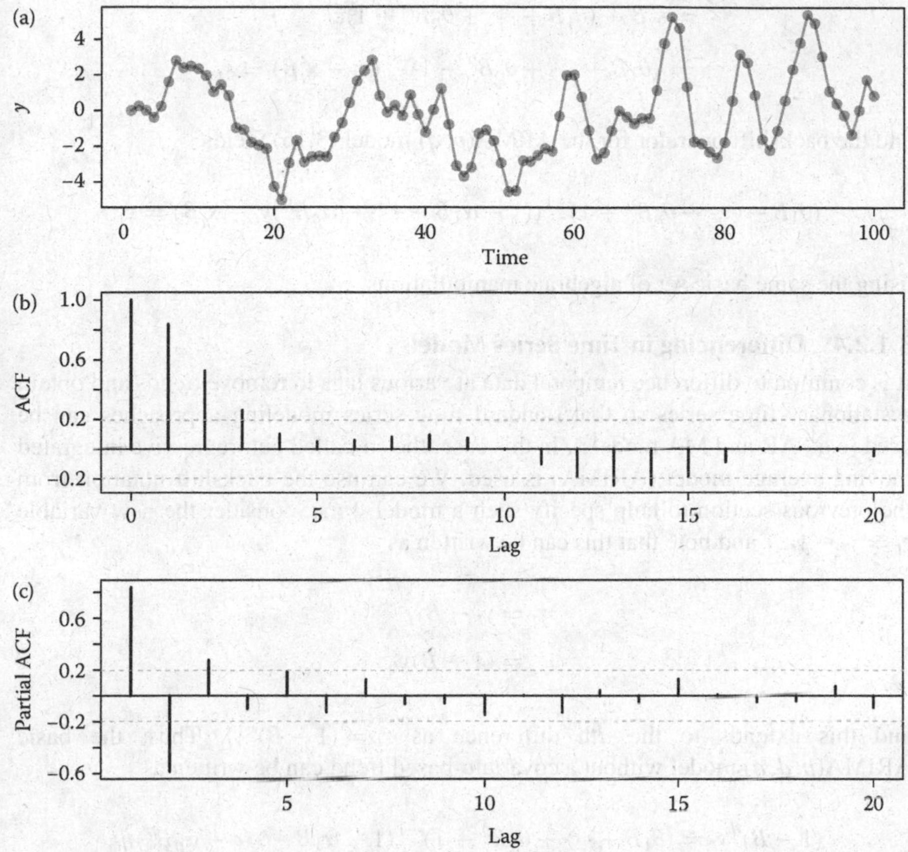

FIGURE 3.6 Simulated ARMA(1,1) time series (a) based on Equation 3.17, ACF (b), and PACF (c) for the simulated ARMA(1,1) time series. Gray dashed lines show a 95% confidence interval under the null hypothesis.

Thus, the AR(p) model can be written as

$$\begin{aligned} y_t &= \mathbf{x}_t'\boldsymbol{\beta} + \alpha_1 y_{t-1} + \cdots + \alpha_p y_{t-p} + \eta_t \\ &= \mathbf{x}_t'\boldsymbol{\beta} + \alpha_1 B y_t + \cdots + \alpha_p B^p y_t + \eta_t \\ &= \mathbf{x}_t'\boldsymbol{\beta} + (\alpha_1 B + \cdots + \alpha_p B^p) y_t + \eta_t \\ &\Rightarrow (1 - \alpha_1 B - \cdots - \alpha_p B^p) y_t - \mathbf{x}_t'\boldsymbol{\beta} = \eta_t. \end{aligned}$$

Similarly, for the MA(q), the use of a backshift operator implies

$$\begin{aligned} y_t &= \mathbf{x}_t'\boldsymbol{\beta} + \theta_1 \varepsilon_{t-1} + \cdots + \theta_p \varepsilon_{t-p} + \varepsilon_t \\ &= \mathbf{x}_t'\boldsymbol{\beta} + \theta_1 B \varepsilon_t + \cdots + \theta_p B^p \varepsilon_t + \varepsilon_t \end{aligned}$$

$$
\begin{aligned}
&= \mathbf{x}_t'\boldsymbol{\beta} + (\theta_1 B + \cdots + \theta_p B^p + 1)\varepsilon_t \\
&\Rightarrow (\theta_1 B + \cdots + \theta_p B^p + 1)^{-1}(y_t - \mathbf{x}_t'\boldsymbol{\beta}) = \varepsilon_t,
\end{aligned}
$$

and the backshift operator for the ARMA(p, q) model (3.16) yields

$$(\theta_1 B + \cdots + \theta_p B^p + 1)^{-1}((1 - \alpha_1 B - \cdots - \alpha_p B^p)y_t - \mathbf{x}_t'\boldsymbol{\beta}) = \varepsilon_t,$$

using the same basic set of algebraic manipulations.

3.1.2.4 Differencing in Time Series Models

It is common to difference temporal data at various lags to remove trends and obtain a stationary time series so that standard time series modeling approaches can be used (e.g., AR and MA models). In this case, the so-called autoregressive integrated moving average model (ARIMA) is used. We can use the backshift notation from the previous section to help specify such a model. First, consider the new variable $z_t = y_t - y_{t-1}$ and note that this can be written as

$$
\begin{aligned}
z_t &= y_t - By_t \\
&= (1 - B)y_t,
\end{aligned}
$$

and this extends to the dth difference as $z_t = (1 - B)^d y_t$. Thus, the basic ARIMA(p, d, q) model without a covariate-based trend can be written as

$$(1 - B)^d y_t = (\theta_1 B + \cdots + \theta_p B^p + 1)^{-1}(1 - \alpha_1 B - \cdots - \alpha_p B^p)\eta_t,$$

where d corresponds to the chosen order of difference. If there is a need for further trend explanation with covariates, then the form of ARIMA(p, d, q) is

$$(1 - B)^d y_t = \mathbf{x}_t'\boldsymbol{\beta} + (\theta_1 B + \cdots + \theta_p B^p + 1)^{-1}(1 - \alpha_1 B - \cdots - \alpha_p B^p)\eta_t.$$

The above model can be rewritten as $y_t = g_1(y_1, \ldots, y_{t-1}) + g_2(\mathbf{x}_t) + g_3(\eta_1, \ldots, \eta_t)$, where the component functions are described as

- g_1: linear combination of previous observations (differencing)
- g_2: linear combination of covariates at time t (regression model)
- g_3: function of random shocks η in terms of $\boldsymbol{\alpha}$ and $\boldsymbol{\theta}$ (ARMA dependence)

3.1.2.5 Fitting Time Series Models

Regardless of the specification, there are several approaches that are useful for fitting AR, MA, ARMA, and ARIMA models. We focus mainly on the AR models here, but these approaches can be modified for the other situations. There are four general approaches one can use to fit time series models:

1. Ordinary least squares
2. Yule–Walker estimation
3. Maximum likelihood
4. Bayesian estimation

Beginning with OLS, we recognize the form of the basic AR(p) model as a regression model

$$y_t = \alpha_0 + \alpha_1 y_{t-1} + \cdots + \alpha_p y_{t-p} + \varepsilon_t, \tag{3.18}$$

for $t = 1, \ldots, T$ and where we only have a homogeneous trend (i.e., an intercept, α_0) for now. Using matrix notation, the model (3.18) can be written as $\mathbf{y} = \mathbf{Y}\boldsymbol{\alpha} + \boldsymbol{\varepsilon}$, where $\mathbf{y} \equiv (y_T, \ldots, y_{p+1})'$ and $\boldsymbol{\varepsilon} \equiv (\varepsilon_T, \ldots, \varepsilon_{p+1})'$. The "design" matrix $\mathbf{Y}$ then is a $(T-p) \times (p+1)$ matrix containing a first column of ones and then remaining columns are lagged data vectors. That is, column two of $\mathbf{Y}$ is $(y_{T-1}, \ldots, y_p)'$ and column three is $(y_{T-2}, \ldots, y_{p-1})'$ and so on.

Then, we minimize the objective function $(\mathbf{y} - \mathbf{Y}\boldsymbol{\alpha})'(\mathbf{y} - \mathbf{Y}\boldsymbol{\alpha})$ with respect to $\boldsymbol{\alpha}$ to obtain $\hat{\boldsymbol{\alpha}}$. This results in the usual OLS estimator for the autoregressive coefficient vector $\hat{\boldsymbol{\alpha}} = (\mathbf{Y}'\mathbf{Y})^{-1}\mathbf{Y}'\mathbf{y}$.

The second method, called Yule–Walker estimation, hinges on the method of moments approach.* We know, from the section on ACFs, that $\gamma(\Delta t) = \text{cov}(y_t, y_{t-\Delta t})$; thus, we can obtain the following Yule–Walker equations:

$$\gamma(\Delta t) = \alpha_1 \gamma(\Delta t - 1) + \cdots + \alpha_p \gamma(\Delta t - p),$$

$$\sigma^2 = \gamma(0) - \alpha_1 \gamma(1) - \cdots - \alpha_p \gamma(p),$$

for $\Delta t = 1, \ldots, p$. Then, in matrix notation, we can write $\boldsymbol{\Gamma\alpha} = \boldsymbol{\gamma}$ and $\sigma^2 = \gamma(0) - \boldsymbol{\alpha}'\boldsymbol{\gamma}$, where the $p \times p$ matrix $\boldsymbol{\Gamma}$ is symmetric containing $\gamma(0)$ on the diagonals and $\gamma(j)$ on the jth off-diagonal (up to $p-1$). Thus, $\hat{\boldsymbol{\alpha}} = \hat{\boldsymbol{\Gamma}}^{-1}\hat{\boldsymbol{\gamma}}$, where the $\gamma(\Delta t)$ are estimated with Equation 3.3. If T is large, we can obtain 95% confidence intervals for $\boldsymbol{\alpha}$ using

$$\hat{\alpha}_j \pm 1.96\sqrt{\frac{\hat{\sigma}^2 \hat{\Gamma}_{jj}^{-1}}{n}}.$$

The maximum likelihood approach to model fitting is different from OLS and Yule–Walker in that we typically make Gaussian assumptions about the errors and then specify the joint distribution for all of the data as a product of the conditional distributions

$$[y_1, \ldots, y_T] = f(y_T | y_1, \ldots, y_{T-1}) f(y_{T-1} | y_1, \ldots, y_{T-2}) \cdots f(y_2 | y_1) f(y_1). \tag{3.19}$$

* Method of moments is the process of equating population moments (i.e., often means and variances) in the data generating probability distribution with sample moments and then solving algebraically for the parameters in the distribution.

Thus, the likelihood in terms of the parameters $\boldsymbol{\alpha}$ and σ^2 can be written as

$$L(\boldsymbol{\alpha}, \sigma^2) = \prod_{t=p+1}^{T} \text{N}(y_t | \alpha_0 + \alpha_1 y_{t-1} + \cdots + \alpha_p y_{t-p}, \sigma^2), \tag{3.20}$$

if we condition on $y_1, \ldots, y_p$. At this point, we maximize the function (3.20) with respect to $\boldsymbol{\alpha}$ and σ^2 to obtain the MLEs. This likelihood-based approach has the added benefit of being able to perform model comparison using Akaike Information Criterion (AIC), but it is not the only method available for comparing models (see Hooten and Hobbs 2015).

Finally, a Bayesian AR(p) model might use the same likelihood as Equation 3.20, as well as prior distributions for the unknown parameters $\boldsymbol{\alpha}$ and σ^2. It can be shown that conjugate priors for this model, though not necessarily the best choice, are

$$\boldsymbol{\alpha} \sim \text{N}(\boldsymbol{\mu}_\alpha, \sigma_\alpha^2 \mathbf{I}),$$

$$\sigma^2 \sim \text{IG}(\omega_1, \omega_2),$$

where $\boldsymbol{\mu}_\alpha$, σ_α^2, ω_1, and ω_2 are hyperparameters that are fixed and known *a priori*. Then, the posterior distribution for this AR(p) model is

$$[\boldsymbol{\alpha}, \sigma^2 | \mathbf{y}] \propto \prod_{t=p+1}^{T} [y_t | \alpha_0 + \alpha_1 y_{t-1} + \cdots + \alpha_p y_{t-p}, \sigma^2][\boldsymbol{\alpha}][\sigma^2], \tag{3.21}$$

where the likelihood arises from Equation 3.20 and $[\boldsymbol{\alpha}]$ and $[\sigma^2]$ represent the prior distributions. The full-conditional distributions necessary to construct an MCMC algorithm for this model are multivariate Gaussian and inverse gamma and are trivial to sample from sequentially.

Consider the four simulated time series in Figure 3.1. Each of these time series were simulated from an AR(1) process and exploratory data analysis suggested that autocorrelation is present in the time series in panels (b–d) (Figure 3.1). The point estimates of α obtained from fitting the AR(1) time series model to each of the data sets in Figure 3.1 using four different methods are shown in Table 3.3. The Bayesian AR(1) assumed a Gaussian prior for α with mean zero and variance 1 and an inverse gamma prior for σ^2 with $q = 2$ and $r = 1$. Table 3.3 indicates that while all of the estimation methods provide similar inference, there are differences among them.

Overall, similar fitting approaches can be constructed for more complicated MA, ARMA, and ARIMA models as well. Regardless of the approach, we should proceed with a series of routine model-checking techniques, computing model residuals and constructing ACF and PACF plots for them, to assess whether various sources of dependence exist, beyond those accounted for by the model.

TABLE 3.3
Truth and Parameter Point Estimates for the Autocorrelation Parameter α in the AR(1) Model for Each of the Time Series in Figure 3.1a–d Using Four Different Estimation Methods: Ordinary Least Squares, Yule–Walker, Maximum Likelihood, and the Bayesian Posterior Mean

	a	b	c	d
Truth	0	0.5	0.9	−0.9
OLS	−0.157	0.406	0.830	−0.985
Y–W	−0.157	0.406	0.794	−0.957
MLE	−0.156	0.402	0.821	−0.974
Bayes	−0.156	0.403	0.828	−0.984

3.1.3 Forecasting

Much like in spatial statistics, we may be interested in obtaining predictions for the process of interest. The most commonly sought form of prediction in time series is a forecast, that is, a temporal extrapolation. In spatial statistics, we were mostly concerned with optimal interpolation and cautioned against extrapolation (e.g., by Kriging only inside a convex hull of the data). From a dynamic modeling perspective, we may also be interested in interpolation (to estimate missing values in the data), but a primary concern is forecasting. Strong assumptions about stationarity must be made to perform this type of extrapolation. With this in mind, we need a framework that we can use for prediction in the temporal setting. To begin, we consider the prediction of future responses.

Suppose we have data $\mathbf{y} \equiv (y_1, \ldots, y_T)'$ and desire a prediction for y_{T+1}. Recall that, for the linear regression model with independent errors (i.e., $\mathbf{y} = \mathbf{X}\boldsymbol{\beta} + \boldsymbol{\varepsilon}$), we seek $\hat{y}_{\text{pred}} = \text{E}(y_{\text{pred}}|\mathbf{y}) = \mathbf{x}'_{\text{pred}}\hat{\boldsymbol{\beta}}$, which has prediction error variance $\sigma^2(1 + \mathbf{x}'_{\text{pred}}(\mathbf{X}'\mathbf{X})^{-1}\mathbf{x}_{\text{pred}})$, where $\mathbf{x}_{\text{pred}}$ represent the set of covariates for the prediction of interest. The time series analog for the AR(1) model, $y_t = \alpha y_{t-1} + \varepsilon_t$, where data exist for $t = 1, \ldots, T$, would be a one-step-ahead prediction $\hat{y}_{T+1} = \alpha y_T$ when the coefficient α is known. In this setting, the prediction error variance $\hat{\sigma}^2_{T+1}$, for the one-step-ahead prediction, is

$$\hat{\sigma}^2_{T+1} = \sigma^2(1 + \alpha^2). \tag{3.22}$$

Note that predictions for higher-order (in time) time series models can be obtained in a similar fashion, as well as more complicated models like ARMA and ARIMA models (see Shumway and Stoffer 2006 for further details).

Figure 3.7 shows 10 step-ahead predictions and 95% prediction intervals for each of the time series presented in Figure 3.1. The predictions in Figure 3.7 were obtained by fitting the AR(1) model using maximum likelihood, yielding the results

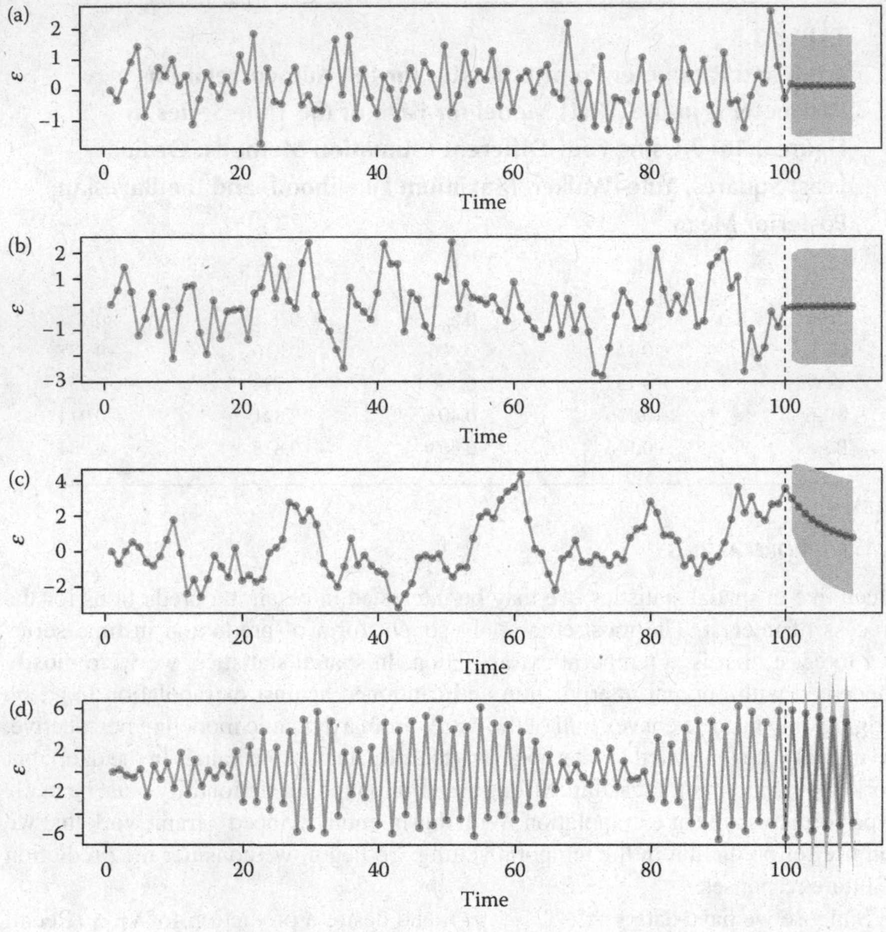

FIGURE 3.7 Predictions based on a maximum likelihood fit of the AR(1) model to each of the time series in Figure 3.1. Dashed vertical line represents the time point before which data exist. Gray region shows a 95% prediction interval.

in Table 3.3. Notice how the predictions converge to the overall mean of the time series and the prediction intervals widen. For time series with stronger temporal dependence, the predictions converge to the mean more slowly (Figure 3.7c and d).

The Bayesian perspective on forecasting with temporal data and a dynamic model is similar, but predictions and predictive distributions are embedded in the same computational procedure we use to fit the model (i.e., find the posterior distribution). To see this, consider the simple Bayesian AR(1) model

$$y_t \sim \text{N}(\alpha y_{t-1}, \sigma^2), \quad t = 1, \ldots, T,$$

$$\alpha \sim \text{N}(0, \sigma_\alpha^2),$$

$$\sigma^2 \sim \text{IG}(r, q).$$

When we fit the model, we seek the posterior distribution $[\alpha, \sigma^2|\mathbf{y}]$, which can be sampled from using MCMC.* To obtain a forecast for y_{T+1}, we find the posterior predictive distribution

$$[y_{T+1}|\mathbf{y}] = \int\int [y_{T+1}|\alpha, \sigma^2, \mathbf{y}][\alpha, \sigma^2|\mathbf{y}] d\alpha d\sigma^2, \tag{3.23}$$

using composition sampling in an MCMC algorithm.† For this model, the full-conditional predictive distribution is $[y_{T+1}|\alpha, \sigma^2, \mathbf{y}] = \mathrm{N}(\alpha y_T, \sigma^2)$; thus, we only need to be able to sample from a Gaussian distribution given our current values for α and σ^2 in the MCMC algorithm. After the samples $y_{T+1}^{(1)}, \ldots, y_{T+1}^{(k)}, \ldots, y_{T+1}^{(K)}$ (where $k = 1, \ldots, K$ refers to the MCMC iteration) have been obtained, they can be summarized using Monte Carlo integration just like any other model parameter. That is, we compute the sample average of those resulting predictive realizations to obtain the posterior predictive mean for the forecast.

3.1.4 Additional Univariate Time Series Notes

We now have a suite of time series methods we can apply to data that accommodate various sources of temporal dependence. However, in our brief overview of these methods, many important points are left out. As a catch-all for a few of these remaining issues, consider the following comments:

- The ARIMA(p, d, q) model can contain a large number of parameters ($p + q + 1 + \#$ of covariates), and thus, can be difficult to fit without a large amount of data.‡ Also, the scientific interpretation of the parameters is questionable; thus, many think of these phenomenological time series models as ways to temporally smooth the data or to provide a statistical framework for forecasting.
- The differencing aspect of ARIMA models can be helpful in making stationarity assumptions, but it can also remove important cyclical behavior. This latter effect can decrease ARIMA forecasting strength. Obvious periodicity in time series models ("seasonality" in the time series literature) should probably be modeled explicitly.
- Many forms of seasonality (i.e., periodicity) can be accommodated in ARMA models by parameterizing the trend (i.e., $\mathbf{X}\boldsymbol{\beta}$) in terms of cyclical functions. For example, a purely seasonal model with no dynamics

---

* In fact, the full-conditional distributions for this model are conjugate, meaning that a fully Gibbs MCMC algorithm can easily be constructed.

† Composition sampling involves taking the current MCMC values for parameters and substituting them into the full-conditional predictive distribution, then sampling from that iteratively, as you would sample any other parameters in the MCMC algorithm.

‡ It is not uncommon to see economic or financial ARIMA models with tens or even hundreds of parameters. Model selection is often employed to find the parameter combination that provides best predictive ability.

can be written as $y_t = \beta_0 + \beta_1 \cos(2\pi\omega t) + \beta_2 \sin(2\pi\omega t) + \varepsilon_t$. This combined sine/cosine formulation arises from the fact that $a\cos(2\pi\omega t + b) = a\cos(b)\cos(2\pi\omega t) - a\sin(b)\sin(2\pi\omega t)$, where a represents the amplitude of the wave, ω represents the frequency, and b represents the phase.

- ACF and PACF plots provide a good place to start for deciding which models to fit, but with real data, it is not always clear what the correct order (i.e., p and q) should be.
- As with any other type of modeling that relies on linear regressions and Gaussian errors, some transformation of y_t may be necessary to ensure homoskedasticity and normality are valid assumptions.*
- There are many other important characteristics of temporal processes and models that were not covered here, for example, invertibility, stationarity, redundancy, and causality. These are worth a review, but beyond the scope of this book.
- The methods we present herein are associated with the time domain. However, an entire subdiscipline of time series statistics is concerned with the frequency domain of temporal processes and data. The frequency domain can be very useful in studying periodic time series, but also tends to be less intuitive for non-statisticians.

3.1.5 Temporally Varying Coefficient Models

Not all temporal models need to be dynamic. For example, in Section 3.1.1, we described a simple situation involving a trend in the temporal data

$$y_t = \beta_0 + \beta_1 t + \varepsilon_t. \tag{3.24}$$

It is also possible that this trend can be explained by temporally varying covariates, in which case the model becomes

$$y_t = \mathbf{x}_t'\boldsymbol{\beta} + \varepsilon_t. \tag{3.25}$$

A natural question might be: How can we make these models more flexible? One approach that can be potentially useful is a form of semiparametric regression.† For example, to generalize the basic homogeneous trend model $y_t = \mu + \varepsilon_t$ would be to let the mean of the temporal process vary over time such that $y_t = \mu_t + \varepsilon_t$, where μ_t is unknown for all t. However, without repeated measurements at each time, the model is overparameterized (i.e., too many parameters and not enough data to learn about all of them). Thus, we need to reduce the dimensionality of the model so that the known-to-unknown ratio is greater.

The basic idea underpinning semiparametric regression is to project the time-varying quantity (in our case, μ_t) into a different (hopefully reduced dimensional)

* Models that directly account for heteroskedasticity are referred to as "stochastic volatility" models.

† Semiparametric regression is often referred to as additive modeling by the machine learning community. The term "additive" is used because we are "adding" up effects much like in regular linear regression.

space.* It really only means that we plan to model the temporal variation in a transformation of the temporal space. The transformation is provided by a set of basis functions that describe different portions of the temporal space ahead of time so that we do not have to figure out the mean of the process everywhere independently, but rather as a subset of the space. More formally, we can reparameterize the temporally varying mean model as

$$\begin{aligned} y_t &= \mu_t + \varepsilon_t \\ &= \mathbf{h}_t'\boldsymbol{\phi} + \varepsilon_t, \end{aligned} \tag{3.26}$$

where the vectors $\mathbf{h}_t$ contain information about a region in temporal space and $\boldsymbol{\phi}$ are the coefficients to be estimated. As with other regression models, this can be written in full matrix notation as $\mathbf{y} = \mathbf{H}\boldsymbol{\phi} + \boldsymbol{\varepsilon}$. Thus, when the set of basis vectors are known, it is trivial to estimate $\boldsymbol{\phi}$. In practice, there are a few issues with this model. First, the new "design" matrix of basis functions $\mathbf{H}$ is $T \times T$ and the coefficient vector $\boldsymbol{\phi}$ is $T \times 1$. Thus, under this full-rank scenario, we gain nothing in terms of dimension reduction. Second, we need to choose the specific form of basis functions in $\mathbf{H}$.

To reduce the dimension of the unknowns in the model, consider the approximation

$$\begin{aligned} \mathbf{y} &= \boldsymbol{\mu} + \boldsymbol{\varepsilon} \\ &= \mathbf{H}\boldsymbol{\phi} + \boldsymbol{\varepsilon} \\ &= \tilde{\mathbf{H}}\tilde{\boldsymbol{\phi}} + \boldsymbol{\varepsilon}, \end{aligned}$$

where the new matrix of basis vectors $\tilde{\mathbf{H}}$ is $T \times p$, and, similarly, the new coefficients $\tilde{\boldsymbol{\phi}}$ are $p \times 1$. If $p \ll T$, we gain a substantial amount of power for estimating μ_t.

The actual choice of $\mathbf{H}$ or $\tilde{\mathbf{H}}$ is somewhat arbitrary, like many choices we make in statistical modeling. Some have better support than others based on their characteristics and the specific application being considered. As a subset of the many forms we could use for $\mathbf{H}$, consider the following popular choices:

- *Piecewise constant:* For p contiguous subsets of the temporal domain $\mathcal{T}_j$ for $j = 1, \ldots, p$, let

$$h_{j,t} = \begin{cases} 0 & \text{if } t \notin \mathcal{T}_j \\ 1 & \text{if } t \in \mathcal{T}_j \end{cases},$$

- *Piecewise linear:* For p contiguous subsets of the temporal domain $\mathcal{T}_j$ for $j = 1, \ldots, p$, let

$$h_{j,t} = \begin{cases} 0 & \text{if } t \notin \mathcal{T}_j \\ t - \min(\mathcal{T}_j) & \text{if } t \in \mathcal{T}_j \end{cases},$$

* The phrase "project it into a reduced dimensional space" is commonly used in time series and spatial statistics (e.g., recall our discussion of reduced-rank models in the previous chapter).

where $\min(\mathcal{T}_j)$ is the minimum value (i.e., infimum) in the time set $\mathcal{T}_j$.

- *B-splines:* For p "knot" locations τ_j ($j = 1, \ldots, p$) in the temporal domain, let

$$h_{j,t}(l) = \frac{t - \tau_j}{\tau_{j+l-1} - \tau_j} h_{j,t}(l-1) + \frac{\tau_{j+l} - t}{\tau_{j+l} - \tau_{j+1}} h_{j+1,t}(l-1)$$

for $j = 1, \ldots, p + 2L - l$, where $l = 1, \ldots, L$ refers to the B-spline order and the first order is defined as

$$h_{j,t}(1) = \begin{cases} 1 & \text{if } \tau_j \leq t < \tau_{j+1} \\ 0 & \text{otherwise} \end{cases}.$$

The B-spline basis functions* are related to cubic splines and commonly used in semiparametric statistics. Despite their apparent complexity, as compared with piecewise constant or piecewise linear splines, B-splines are trivial to calculate using modern statistical software.

In semiparametric regression, the coefficients $\boldsymbol{\phi}$ sometimes lack an obvious interpretation unless the basis functions are somehow mechanistically informed. Furthermore, the types of basis functions described above are commonly referred to as "landmark" basis functions, meaning that they depend on knot locations or fixed and known regions of the temporal domain. Another type of basis function is the radial basis function, typically a real function that is centered at a knot and decays radially from it.

Using radial basis functions (specifically, thin plate spline basis functions), we fit the temporally varying coefficient model in Equation 3.26 to each of the time series from Figure 3.1 as an additive regression. Figure 3.8 shows the prediction and 95% prediction interval of the model fits to each of the time series. To fit each model, we used a technique known as regularization (Hooten and Hobbs 2015) to select the optimal predicting model.† Using 10 regularized radial basis functions, Figure 3.8a suggests that there is no discernible pattern in the time series other than a slight downward trend. Recall that the time series in Figure 3.8a arises from a Monte Carlo (i.e., independent) sample from a Gaussian distribution. For the time series in Figure 3.8b and c, the temporally varying coefficient model captures more of the pattern suggestive of autocorrelation in the data. The negatively autocorrelated time series in Figure 3.8d, however, is not amenable to the choice of basis functions and temporally varying coefficient model we used. The predictions do not capture the oscillating pattern in the negatively autocorrelated time series because there are important dynamics in the process that were ignored. Thus, naive semiparametric regression is best for relatively smooth time series. Also, temporally varying coefficient models like the type

---

* It is common to hear the terms "basis functions" and "basis vectors" used interchangeably, especially in statistics. However, "basis vectors" refer to the case where the functions themselves have been discretized for use in computation.

† Regularization involves penalizing the complexity of a model so that it is parsimonious enough to provide good predictions. Additive models are often penalized using generalized cross-validation (GCV).

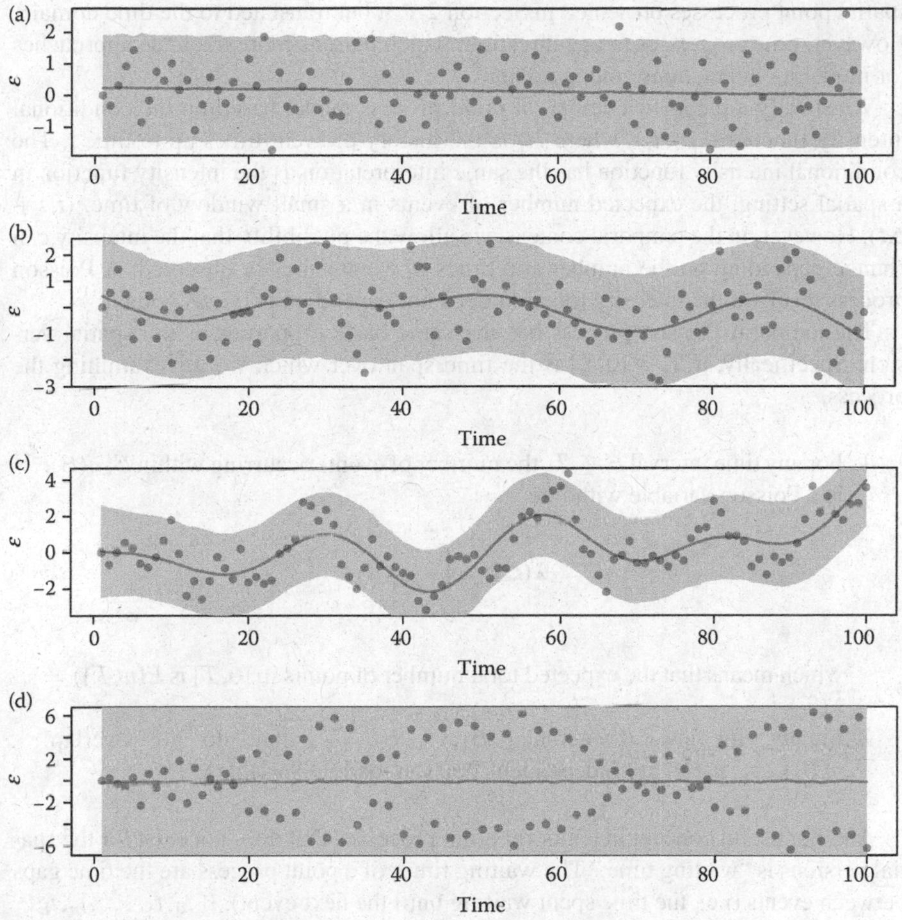

FIGURE 3.8 Predictions based on temporally varying coefficient model to each of the time series in Figure 3.1. The 95% prediction intervals are shaded in gray.

presented here are best for interpolated prediction rather than extrapolated prediction (i.e., forecasting).

3.1.6 Temporal Point Processes

A temporal point process is different than other processes described in this chapter. Similar to the spatial point processes presented in Chapter 2, the data for a temporal point process are the times at which an event occurs, rather than the value of a variable at a given time.* Thus, the models presented in this section are analogous to the

* Although, when both the times and characteristics of the event at each time are observed, it is referred to as a marked temporal point process (e.g., volume of water displaced during a sequence of geyser eruptions).

spatial point processes presented in Section 2.1.3, but translated to the time domain. However, time progresses in one direction, which permits more tractable approaches for modeling interactions among points.

We specify a parametric temporal point process model based on the conditional intensity function $\lambda(t|\mathcal{H}_t)$, where $\mathcal{H}_t$ is the history of event times up to time t. The conditional intensity function has the same interpretation as the intensity function in a spatial setting; the expected number of events in a small window of time, $(t, t+\Delta t)$. However, in the temporal context, we allow the possibility that the intensity can change depending on the number and times of events already observed. A Poisson process results if the intensity function does not depend on previous points.

The temporal Poisson process has the same basic properties as the spatial version. Specifically, if $\mathcal{T} \equiv [0, T]$ is the time span over which we are examining the process,

1. For any time interval $\mathcal{B} \subseteq \mathcal{T}$, the number of events occurring within $\mathcal{B}$, $n(\mathcal{B})$, is a Poisson variable with rate

$$\tilde{\lambda}(\mathcal{B}) = \int_{\mathcal{B}} \lambda(\tau)\, d\tau,$$

which means that the expected total number of points in $[0, T]$ is $E(n(\mathcal{T})) = \tilde{\lambda}(\mathcal{T})$.

2. Finally, for any J intervals, $\mathcal{B}_1, \ldots, \mathcal{B}_J \subseteq \mathcal{T}$, that do not overlap, $n(\mathcal{B}_1), \ldots, n(\mathcal{B}_J)$ are independent Poisson random variables.

Another useful concept in temporal point processes that does not exist for the spatial versions is "waiting time." The waiting times of a point process are the time gaps between events (i.e., the time spent waiting until the next event). If $t_0, t_1, \ldots, t_n, t_{n+1}$ (where $t_0 = 0$ and $t_{n+1} = T$) are the observed event times, then the associated waiting times are $t_i - t_{i-1}$ for $i = 1, \ldots, n$ (we revisit the truncated time, $T - t_n$, in what follows). Rather than specify a model for the event times themselves, we can specify a model for the waiting times and examine the resulting model for the event times. We show how to move between the two different model specifications.

We begin by slightly redefining the intensity function, $\lambda(t|\mathcal{H}_t)$. A point process is considered "orderly" if the chances of having more than one event in an interval approaches zero as the interval becomes very short. That is,

$$\lim_{\Delta t \to 0} \frac{P(n((t, t+\Delta t)) > 1)}{\Delta t} = 0. \tag{3.27}$$

If a point process is orderly, then the intensity function can be equivalently defined as the probability that a single event occurs in a very short interval

$$\lambda(t|\mathcal{H}_t) = \lim_{\Delta t \to 0} \frac{P(n((t, t+\Delta t)) = 1|\mathcal{H}_t)}{\Delta t}. \tag{3.28}$$

Thus, the probability of an event occurring in the interval $(t, t + \Delta t)$ is $P(n((t, t + \Delta t)) = 1|\mathcal{H}_t) \approx \lambda(t|\mathcal{H}_t)\Delta t$. Although this may sound like a strong assumption, most point processes fall into this category. This restriction is aimed at eliminating the chance that two events will occur simultaneously so that we can construct a proper density function.

To find the cumulative distribution function (CDF) and PDF of the waiting time given the intensity function and history $\mathcal{H}_t$, we need to find the CDF and PDF of the event time t_i given the previous events and the time since the last event. To accomplish this, we take a brief probability detour.

For any continuous random variable, X, we can write

$$P(X \in (x, x + \Delta x]|X > x) = \frac{F(x + \Delta x) - F(x)}{1 - F(x)}, \tag{3.29}$$

where F is the CDF of the random variable X. This results from the definitions of conditional probabilities and CDFs. If we divide each side of Equation 3.29 by Δx and let $\Delta x \to 0$, then

$$\lim_{\Delta x \to 0} \frac{P(X \in (x, x + \Delta x]|X > x)}{\Delta x} = \frac{f(x)}{1 - F(x)}, \tag{3.30}$$

where $f(x)$ is the PDF of the distribution of X, which results from the fact that $dF(x)/dx = f(x)$. In the context of event times, we obtain

$$\lim_{\Delta t \to 0} \frac{P(t_i \in (t, t + \Delta t]|\mathcal{H}_t)}{\Delta t} = \frac{f(t|\mathcal{H}_t)}{1 - F(t|\mathcal{H}_t)}. \tag{3.31}$$

Using Equations 3.27 and 3.28, we replace the left-hand side with $\lambda(t|\mathcal{H}_t)$, providing a way to calculate the intensity

$$\lambda(t|\mathcal{H}_t) = \frac{f(t|\mathcal{H}_t)}{1 - F(t|\mathcal{H}_t)}. \tag{3.32}$$

While Equation 3.32 provides a sense of the relationship between the intensity function and the CDF (or PDF) of the waiting time, it is not directly useful for model building. To further simplify the relationship,* we use

$$-f(t|\mathcal{H}_t) = \frac{d}{dt}(1 - F(t|\mathcal{H}_t)). \tag{3.33}$$

* Although it may not seem simple at first.

A positive function $H(x)$, such that $d \log H(x)/dx = h(x)/H(x)$, where $h(x) = dH(x)/dx$, together with Equation 3.33, provides a result in terms of just the waiting time CDF:

$$\lambda(t|\mathcal{H}_t) = -\frac{d}{dt}\log(1 - F(t|\mathcal{H}_t)). \tag{3.34}$$

Integrating each side and solving for the CDF results in the relationship

$$F(t|\mathcal{H}_t) = 1 - \exp\left(-\int_{t_{i-1}}^{t} \lambda(\tau|\mathcal{H}_\tau)d\tau\right). \tag{3.35}$$

Finally, by taking the derivative of Equation 3.35 with respect to t, we can find the waiting time PDF

$$f(t|\mathcal{H}_t) = \lambda(t|\mathcal{H}_t)\exp\left(-\int_{t_{i-1}}^{t} \lambda(\tau|\mathcal{H}_\tau)d\tau\right). \tag{3.36}$$

Now that we have derived the conditional PDFs for the event times given the previous event times, we form the full joint PDF for the entire set of events and obtain the likelihood for parameter estimation. Thus, we explicitly parameterize the intensity function as in Chapter 2 for spatial point processes (i.e., $\lambda(t|\boldsymbol{\beta}, \mathcal{H}_t)$, where $\boldsymbol{\beta}$ is a vector of parameters we wish to estimate). The joint likelihood is formed from the product of conditional PDFs; however, we must deal with the truncation between the last observed event, t_n, and the end of the study interval, T. We never see when event t_{n+1} occurs; we only know that $t_{n+1} > T$. To find the PDF of t_{n+1}, we need to find the probability that there are no events in the interval $(t_n, T]$, or that, given $t_n < T$, the unobserved t_{n+1} event happens at a time $>T$, which is equal to $1 - F(T|\mathcal{H}_T)$. Therefore, using Equation 3.35

$$\begin{aligned} f(t_{n+1}|\mathcal{H}_{t_{n+1}}) &= 1 - F(t_{n+1}|\boldsymbol{\beta}, \mathcal{H}_{t_{n+1}}) \\ &= \exp\left(-\int_{t_n}^{T} \lambda(\tau|\boldsymbol{\beta}, \mathcal{H}_\tau)d\tau\right). \end{aligned} \tag{3.37}$$

Finally, we have all the pieces to form the parametric model likelihood for a temporal point process

$$\begin{aligned} L(\boldsymbol{\beta}) &= \prod_{i=1}^{n+1} f(t_i|\boldsymbol{\beta}, \mathcal{H}_{t_i}) \\ &= \prod_{i=1}^{n+1} \lambda(t_i|\boldsymbol{\beta}, \mathcal{H}_{t_i})\exp\left(-\int_{t_{i-1}}^{t_i} \lambda(\tau|\boldsymbol{\beta}, \mathcal{H}_\tau)d\tau\right) \end{aligned}$$

$$= \left(\prod_{i=1}^{n} \lambda(t_i|\boldsymbol{\beta}, \mathcal{H}_{t_i})\right) \exp\left(-\sum_{i=1}^{n+1} \int_{t_{i-1}}^{t_i} \lambda(\tau|\boldsymbol{\beta}, \mathcal{H}_\tau) d\tau\right)$$

$$= \left(\prod_{i=1}^{n} \lambda(t_i|\boldsymbol{\beta}, \mathcal{H}_{t_i})\right) \exp\left(-\int_{0}^{T} \lambda(\tau|\boldsymbol{\beta}, \mathcal{H}_\tau) d\tau\right). \quad (3.38)$$

In Chapter 2, we showed the identical likelihood form for the spatial version of the point process. However, a notion of temporal dependence has been incorporated by conditioning on the history, $\mathcal{H}_t$. Therefore, the intensity function changes over the interval $[0, T]$ depending on when events occur.

The waiting time concept in temporal point processes is very similar to that of survival modeling based on "time to events" or failures. In survival modeling, the "hazard" function is mathematically equivalent to our conditional intensity function, $\lambda(t|\mathcal{H}_t)$. The waiting times are equivalent to the "failure" times that are modeled. Therefore, many of the parametric survival models are available for us to use in this context. One of the most popular survival models that incorporates covariates into the hazard function is the Cox proportional hazards (CPH) model (Cox and Oakes 1984). The CPH intensity function is given by

$$\lambda(t|\mathcal{H}_t) = \lambda_0(t - \Delta_t) \exp(\mathbf{x}'(t)\boldsymbol{\beta}), \quad (3.39)$$

where $\lambda_0(t - \Delta_t)$ is a baseline intensity function that depends only on the time since the last event, Δ_t. The time-indexed covariates in Equation 3.39 are denoted as the vector $\mathbf{x}(t)$ and $\boldsymbol{\beta}$ are the coefficients to be estimated. The term $\exp(\mathbf{x}'(t)\boldsymbol{\beta})$ scales the base intensity depending on the time series of covariate values. If we substitute the CPH intensity function back into the likelihood (3.38), the resulting log-likelihood is

$$\ell(\boldsymbol{\beta}) = \sum_{i=1}^{n} \left(\log \lambda_0(t_i - t_{i-1}) + \mathbf{x}'(t_i)\boldsymbol{\beta} - \int_{t_{i-1}}^{t_i} \exp(\log \lambda_0(\tau - t_{i-1}) + \mathbf{x}'(\tau)\boldsymbol{\beta}) d\tau\right). \quad (3.40)$$

To evaluate the log-likelihood, one can employ a trick similar to the Berman–Turner device (Berman and Turner 1992) we introduced in Chapter 2. In the temporal context, we select $J_i + 1$ quadrature points,* $u_{i,0}, \ldots, u_{i,J_i}$ within the interval $[t_{i-1}, t_i]$ (where $u_{i,0} = t_{i-1}$ and $u_{i,J_i} = t_i$). Then the log-likelihood can be approximated by

$$\ell(\boldsymbol{\beta}) \approx \sum_{i=1}^{n} \sum_{j=1}^{J_i} z_{i,j}(\log(u_{i,j} - u_{i,j-1}) + \log \lambda_0(u_{i,j} - t_{i-1}) + \mathbf{x}'(u_{i,j})\boldsymbol{\beta}) \quad (3.41)$$
$$- \exp(\log(u_{i,j} - u_{i,j-1}) + \log \lambda_0(u_{i,j} - t_{i-1}) + \mathbf{x}'(u_{i,j})\boldsymbol{\beta}),$$

* Recall the description of quadrature from Section 2.1.3.

where $z_{i,j} = 1$ if $j = J_i$ and zero for all other times. The log-likelihood function (3.41) is the same if the $z_{i,j}$ were treated as independent Poisson random variables with rates $\exp(\log(u_{ij} - u_{i,j-1}) + \log\lambda_0(u_{i,j} - t_{i-1}) + \mathbf{x}'(u_{i,j})\boldsymbol{\beta})$. This approximation was initially proposed by Holford (1980). Thus, if the log baseline intensity function $\log\lambda_0(\cdot)$ is linear in its parameters, one can use standard GLM software to fit a Poisson regression model with offsets equal to $\log(u_{i,j} - u_{i,j-1})$.

There are several different forms of baseline intensity that will produce different effects from events clustered together in time to events that are more regularly spaced than what would be expected from pure randomness. A very flexible class of waiting time distributions is the Weibull distribution. The PDF for the Weibull distribution is

$$f(t|\phi,\alpha) = \frac{\alpha}{\phi}\left(\frac{t}{\phi}\right)^{\alpha-1} e^{-(t/\phi)^{\alpha}}, \tag{3.42}$$

and the CDF is

$$F(t|\phi,\alpha) = 1 - e^{-(t/\phi)^{\alpha}}, \tag{3.43}$$

thus, if we model the waiting times with a Weibull distribution, the conditional intensity function is

$$\lambda(t|\mathcal{H}_t) = \frac{\alpha}{\phi}\left(\frac{t-\Delta_t}{\phi}\right)^{\alpha-1}, \tag{3.44}$$

where Δ_t is the time of the last observed event prior to t. For the Weibull intensity,

$$\log\lambda_0(t|\mathcal{H}_t) = \log\left(\frac{\alpha}{\phi}\right) - \log(\phi) + (\alpha-1)\log(t-\Delta_t), \tag{3.45}$$

which can be reparameterized as $\log\lambda_0(t|\mathcal{H}_t) = \beta_0 + \beta_1\log(t-\Delta_t)$ and is linear with respect to the parameters. Therefore, the Weibull baseline intensity can be used within a GLM in the Poisson approximation of the temporal point process likelihood. Depending on the value of β_1, one can obtain different behaviors in the clustering of events. For example, if $\beta_1 < 0$, the intensity decreases with $t - \Delta_t$, implying that events often occur close together in time. However, if $\beta_1 > 0$, the intensity increases with increasing waiting times; therefore, events tend to be more spread out and regular. A special case occurs at $\beta_1 = 0$, where the intensity is constant over all waiting times, $\lambda(t|\mathcal{H}_t) = \lambda(t) = \exp(\beta_0)$. At $\beta_1 = 0$, the intensity does not depend on the past history, and is therefore a Poisson process. Thus, at $\beta_1 = 0$, we obtain a process that is completely random (i.e., there is neither clustering of events or inhibition of events in time). Figure 3.9 shows realizations of a temporal point process for (a) $\beta_1 = -0.5$, (b) $\beta_1 = 0$, and (c) $\beta_1 = 1$. In Figure 3.9a, it is clearly visible that the events tend to cluster together, while in panel (b), the events seem to have no pattern in the times they occur, and finally, in panel (c), the events occur at a more regular time schedule.

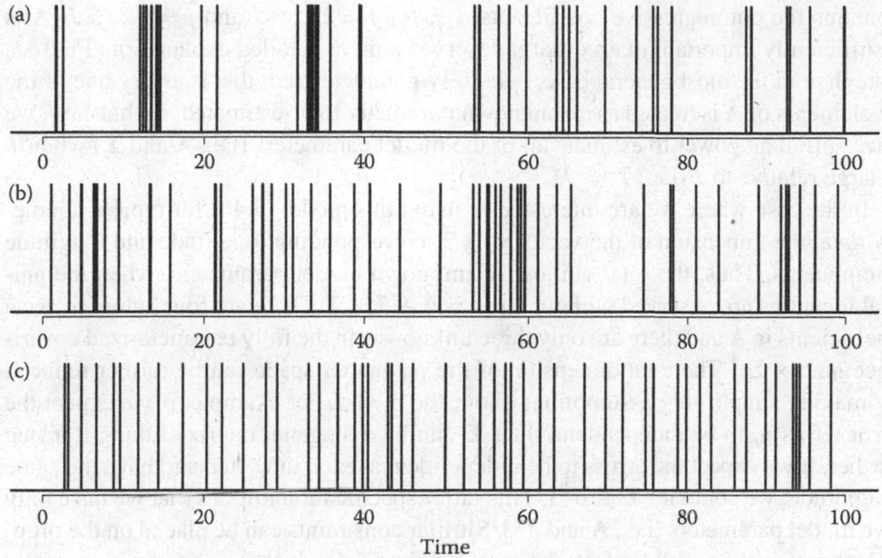

FIGURE 3.9 Example of a Weibull point process. Plot (a) illustrates clustering behavior when $\beta_1 < 0$ (or $\alpha < 1$ in the original parameterization), plot (b) illustrates constant intensity when $\beta_1 = 0$ ($\alpha = 1$), and finally, a more regularly spaced pattern is illustrated in plot (c) when $\beta_1 > 0$ ($\alpha > 1$).

3.2 MULTIVARIATE TIME SERIES

3.2.1 VECTOR AUTOREGRESSIVE MODELS

The previous section provided the basic terminology and background concerning temporal modeling from dynamic and nondynamic perspectives. However, it is difficult to imagine animal movement data as arising from a 1-D spatio-temporal process (i.e., a univariate time series). Thus, we extend the models already introduced to the 2-D setting. In doing so, this section provides the foundation for many of the specific animal movement process models in later chapters.

We begin by generalizing the notation. We index our former response variable y_t by j to denote which dimension of the multivariate process we are referring to. Representing the set of these variables in a vector, we have $\mathbf{y}_t = (y_{1,t}, \ldots, y_{j,t}, \ldots, y_{J,t})'$, for $j = 1, \ldots, J$ and $t = 1, \ldots, T$. Then the most straightforward way to introduce the concept of a multivariate dynamic statistical model is with an autoregressive specification. In these models, matrix notation becomes critical to reduce the complexity of necessary mathematical expressions. Thus, we write the centered (i.e., mean zero) vector autoregressive model of order one (VAR(1)) as

$$\mathbf{y}_t = \mathbf{A}\mathbf{y}_{t-1} + \boldsymbol{\eta}_t, \tag{3.46}$$

where the error vector is distributed as $\boldsymbol{\eta}_t \sim \text{N}(\mathbf{0}, \boldsymbol{\Sigma})$ with $J \times J$ error covariance matrix $\boldsymbol{\Sigma}$. The $J \times J$ matrix $\mathbf{A}$ is called the propagator, or transition, matrix and contains the autoregressive coefficients $\alpha_{j,\tilde{j}}$, for $j = 1, \ldots, J$ and $\tilde{j} = 1, \ldots, J$. $\mathbf{A}$ is a sufficiently important matrix that it deserves a more detailed explanation. First we note that, in the most general case, $\mathbf{A}$ is fully parameterized; that is, every one of the J^2 elements of $\mathbf{A}$ is treated as an unknown parameter to be estimated. In that case, we have sufficient power to estimate all of the model parameters (i.e., $\mathbf{A}$ and $\boldsymbol{\Sigma}$) when T is large relative to J (i.e., $T \gg J$).

In the case where we are interested in using this model (3.46) for typical telemetry data, the dimension of the vector $\mathbf{y}_t$ is 2, corresponding to latitude and longitude components. Thus, the total number of unknown model parameters, when the animal locations are observed without error, is $4 + 3 = 7$. There are four autoregressive coefficients in $\mathbf{A}$ and there are only three unknowns in the fully parameterized covariance matrix $\boldsymbol{\Sigma}$.* The total dimension of the parameter space can be further reduced by making simplifying assumptions about the model. For example, if we expect the error terms $\boldsymbol{\eta}_t$ to be independent, then $\boldsymbol{\Sigma}$ can be a diagonal matrix. Taking it a step further, if we expect the errors to be independent at each time step and have the same magnitude, we could let $\boldsymbol{\Sigma} \equiv \sigma^2\mathbf{I}$. This latter specification implies that we have only five model parameters (i.e., $\mathbf{A}$ and σ^2). Similar constraints can be placed on the propagator matrix $\mathbf{A}$ as well. In fact, the most parsimonious dynamic specification would involve $\mathbf{A} \equiv \mathbf{I}$, the identity matrix, with ones on the diagonal and zeros elsewhere. In this case, the VAR(1) model becomes

$$\begin{aligned} \mathbf{y}_t &= \mathbf{A}\mathbf{y}_{t-1} + \boldsymbol{\eta}_t \\ &= \mathbf{I}\mathbf{y}_{t-1} + \boldsymbol{\eta}_t \\ &= \mathbf{y}_{t-1} + \boldsymbol{\eta}_t, \end{aligned} \tag{3.47}$$

a random walk of order 1. Under this random walk specification for the animal telemetry data scenario, if $\boldsymbol{\Sigma} \equiv \sigma^2\mathbf{I}$, then the current position of the animal ($\mathbf{y}_t$) will be very close to the last position ($\mathbf{y}_{t-1}$) if the error variance σ^2 is small. For example, two simulated multivariate time series are shown in Figure 3.10. An initial position of $\mathbf{y}_1 = (0, 0)'$ is assumed for both time series and $\sigma^2 = 1$ in Figure 3.10a, whereas Figure 3.10b assumes $\sigma^2 = 2$.

It is important that even though we use the traditional term "error variance," the "error" terms $\boldsymbol{\eta}_t$ are really just a component of a stochastic temporal process (i.e., the individual's position). The simple random walk (3.47) is our first mechanistic animal movement model. Despite its simplicity, the random walk is useful (especially as a null model) and we return to it in the chapters that follow.

The random walk is a very simple dynamic model; it can be generalized by letting $\mathbf{A} \equiv \text{diag}(\boldsymbol{\alpha})$.† When the covariance $\boldsymbol{\Sigma}$ is also diagonal, fitting the VAR(1) model

---

* The free parameters in $\boldsymbol{\Sigma}$ are the two diagonal elements, representing the variances for each dimension, and a single off-diagonal element that controls the correlation between the dimensions. Covariance matrices need to be symmetric; thus, the upper right element in a 2×2 covariance matrix is the same as the lower left element.

† The "diag" function places the vector $\boldsymbol{\alpha}$ along the diagonal of a square matrix with zeros for all off-diagonal elements.

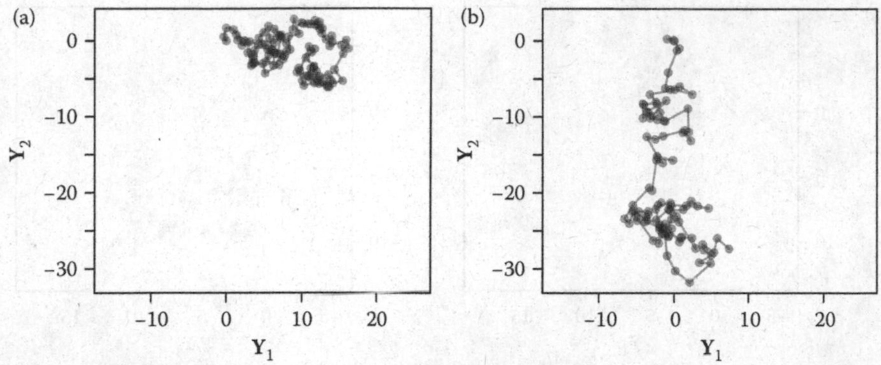

FIGURE 3.10 Simulated 2-D processes arising from Equation 3.47 plotted in 2-D space. Panel (a) assumes $\sigma^2 = 1$ and panel (b) assumes $\sigma^2 = 2$.

(3.46) essentially implies we are really fitting J independent univariate time series models (one for each dimension of $\mathbf{y}_t$). This makes it slightly more robust than the independent random walk model, but it does not harness the real utility of the general VAR(1).

In fact, the VAR(1) can provide surprisingly general dynamic behavior. The mechanism allowing for the flexibility in dynamics lies in the off-diagonal elements of $\mathbf{A}$ (Wikle and Hooten 2010). The off-diagonal elements control the interactions within the process from one time point to the next. As an example, consider the VAR(1) model for a 2-D dynamic process with homogeneous trend $\boldsymbol{\mu}$:

$$\mathbf{y}_t = \boldsymbol{\mu} + \mathbf{A}\mathbf{y}_{t-1} + \boldsymbol{\eta}_t, \tag{3.48}$$

which is a biased random walk. If $\mathbf{A}$ is parameterized such that it has $\alpha_{1,1} = \alpha_{2,2} = \alpha$ as diagonal elements with $\alpha_{1,2}$ and $\alpha_{2,1}$ on the off-diagonals, then the mean of the first element $y_{1,t}$ is $\text{E}(y_{1,t}|\mathbf{y}_{t-1}) = \mu_1 + \alpha_{1,1}y_{1,t-1} + \alpha_{1,2}y_{2,t-1}$. The conditional mean of $y_{1,t}$ indicates that $y_{1,t}$ will tend to be close to some weighted average of the global mean for that dimension (μ_1), the previous location in that dimension ($y_{1,t-1}$), and the previous location in the other dimension ($y_{2,t-1}$). A similar expression can be found for the conditional mean of the other dimension of the process. If the off-diagonal autoregressive coefficients $\alpha_{1,2}$ and $\alpha_{2,1}$ in $\mathbf{A}$ approach zero, we return to the independent random walk model, but as they increase, we see an increasing influence of one dimension of the process on the other in the dynamics. This range of possible interactions among dimensions allows for realistic behavior in a dynamic process such as animal movement.

Figure 3.11 shows two simulated multivariate time series arising from the VAR(1) model in Equation 3.48 in 2-D space. The mean was $\boldsymbol{\mu} = (0, 0)'$ and variance parameter was $\sigma^2 = 1$ for both simulations, but the propagator matrix $\mathbf{A}$ was specified to have 0.9 on the diagonal elements, with 0.1 on the off-diagonals in Figure 3.11a and -0.1 on the off-diagonal in Figure 3.11b. The elliptical shape of the time series in

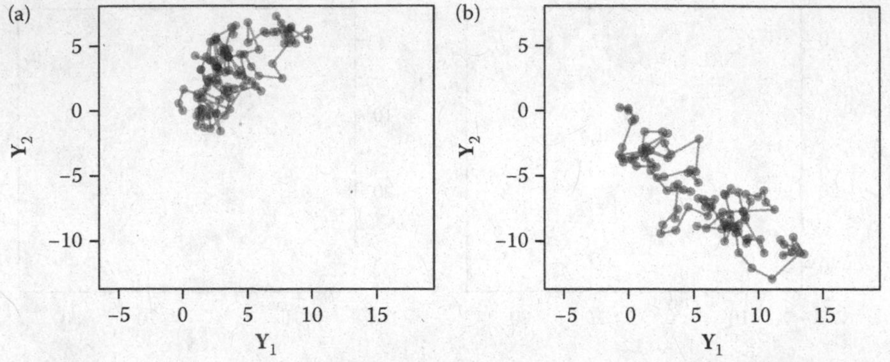

FIGURE 3.11 Simulated 2-D processes arising from Equation 3.48 plotted in 2-D space. Panels (a) and (b) assume $\boldsymbol{\mu} = (0, 0)'$ and $\sigma^2 = 1$. The propagator matrix **A** contains positive off-diagonal elements in panel (a), whereas panel (b) relies on a propagator matrix with negative off-diagonal elements.

2-D space in Figure 3.11 arises from the interaction between the directions in the dynamic process.

More mechanistic parameterizations of **A** are also possible and can be useful.* We describe several specific parameterizations of VAR models for animal telemetry data in the later chapters but, before we leave this topic, we note that higher-order autoregressive models are possible and potentially useful for multivariate processes.

Recall our description of univariate ARIMA models in the previous section. The same sort of temporal differencing can be used in the multivariate setting, but its interpretation may vary. For example, it might have additional utility beyond detrending a time series. It is possible that the differencing could be motivated by a discretized derivative used to relate velocities in the multivariate process (rather than locations). To see this, consider the integrated VAR(1) model on the quantity $\boldsymbol{\delta}_t = \mathbf{y}_t - \mathbf{y}_{t-1}$, where

$$\boldsymbol{\delta}_t = \mathbf{A}\boldsymbol{\delta}_{t-1} + \boldsymbol{\eta}_t. \tag{3.49}$$

Using substitution and algebra with this model shows that it is actually a VAR(2) model on the original location vectors $\mathbf{y}_t$. To see this, substitute $\mathbf{y}_t - \mathbf{y}_{t-1}$ into Equation 3.49 for $\boldsymbol{\delta}_t$ and rearrange terms with y_t on the left-hand side and all other terms on the right-hand side of the equality. The result is

$$\begin{aligned}\mathbf{y}_t &= \mathbf{A}(\mathbf{y}_{t-1} - \mathbf{y}_{t-2}) + \mathbf{y}_{t-1} + \boldsymbol{\eta}_t \\ &= (\mathbf{A} + \mathbf{I})\mathbf{y}_{t-1} - \mathbf{A}\mathbf{y}_{t-2} + \boldsymbol{\eta}_t,\end{aligned} \tag{3.50}$$

* Wikle and Hooten (2010) and Cressie and Wikle (2011) provide much more detailed descriptions of multivariate dynamic models, their utility, and implementation, especially as they pertain to spatio-temporal processes.

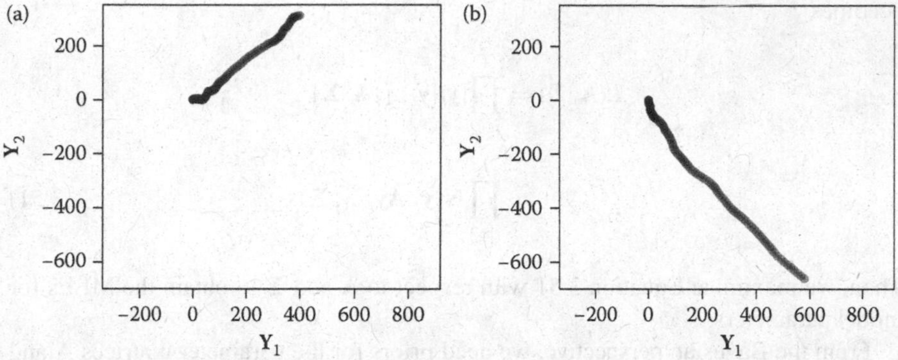

FIGURE 3.12 Simulated 2-D processes arising from Equation 3.50 plotted in 2-D space. Panels (a) and (b) assume $\sigma^2 = 1$. The propagator matrix **A** contains positive off-diagonal elements in panel (a), whereas panel (b) relies on a propagator matrix with negative off-diagonal elements.

where the two propagator matrices in the VAR(2) model are $(\mathbf{A} + \mathbf{I})$ and $-\mathbf{A}$. Thus, a particular parameterization of a VAR(2) implies integrated VAR(1) dynamics.*

Figure 3.12 shows two simulated multivariate time series arising from the VAR(1) model in Equation 3.50 in 2-D space. As in the preceding nonintegrated time series from Figure 3.11, the variance parameter was $\sigma^2 = 1$ for both simulations, but the propagator matrix **A** was specified to have 0.9 on the diagonal elements, with 0.1 on the off-diagonals in Figure 3.12a and −0.1 on the off-diagonal in Figure 3.12b. Thus, the simulated time series based on Equation 3.50 are substantially smoother than those from Figure 3.11, but retain the diagonally oriented process. The time specific displacements $\sqrt{(\mathbf{y}_t - \mathbf{y}_{t-1})'(\mathbf{y}_t - \mathbf{y}_{t-1})}$ are similar in Figures 3.11 and 3.12, but the turning angles are much more consistent (i.e., highly correlated). Thus, integrated time series models (or higher-order VAR models) are good for capturing dynamics of smooth spatio-temporal processes. For example, in the animal movement context, smoothness could be a result of migrational movement (see Chapters 5 and 6).

3.2.2 Implementation

To fit VAR models, we borrow some of the procedures from the preceding section on univariate time series. For example, recognizing that the VAR(1) specification (3.46) can be written as a multivariate Gaussian, where $\mathbf{y}_t \sim \text{N}(\mathbf{A}\mathbf{y}_{t-1}, \boldsymbol{\Sigma})$, the likelihood

* Though we did not mention it earlier, the same is true for the univariate AR(2) model.

becomes

$$L(\mathbf{A}, \boldsymbol{\Sigma}) = \prod_{t=2}^{T} [\mathbf{y}_t | \mathbf{y}_{t-1}, \mathbf{A}, \boldsymbol{\Sigma}]$$
$$= \prod_{t=2}^{T} \mathrm{N}(\mathbf{y}_t | \mathbf{A}\mathbf{y}_{t-1}, \boldsymbol{\Sigma}). \tag{3.51}$$

Then, we maximize Equation 3.51 with respect to $\mathbf{A}$ and $\boldsymbol{\Sigma}$ to obtain the MLEs for model parameters.

From the Bayesian perspective, we need priors for the parameter matrices $\mathbf{A}$ and $\boldsymbol{\Sigma}$. A possible prior for the covariance matrix (depending on its parameterization) is an inverse Wishart (or Wishart for the inverse covariance, or precision, matrix) such that $\boldsymbol{\Sigma}^{-1} \sim \text{Wish}((\mathbf{V}\boldsymbol{\nu})^{-1}, \boldsymbol{\nu})$, where $\text{E}(\boldsymbol{\Sigma}^{-1}) = \mathbf{V}^{-1}$. An appropriate prior for the autoregressive coefficients in $\mathbf{A}$ is not quite as obvious. We could specify independent priors for the individual elements (e.g., $\alpha_{j\tilde{j}} \sim N(0, \sigma^2_{j\tilde{j}})$), but this does not provide a means to correlate them *a priori*. One potential way to generalize the prior for $\mathbf{A}$ is to use the "vec" operator* on $\mathbf{A}$ and the multivariate Gaussian distribution $\text{vec}(\mathbf{A}) \sim \mathrm{N}(\boldsymbol{\mu}_{\mathbf{A}}, \boldsymbol{\Sigma}_{\mathbf{A}})$.

To fit the Bayesian VAR(1) model, we seek the posterior distribution

$$[\mathbf{A}, \boldsymbol{\Sigma} | \mathbf{Y}] \propto \prod_{t=2}^{T} [\mathbf{y}_t | \mathbf{y}_{t-1}, \mathbf{A}, \boldsymbol{\Sigma}][\mathbf{A}][\boldsymbol{\Sigma}], \tag{3.52}$$

where $\mathbf{Y} = (\mathbf{y}_1, \ldots, \mathbf{y}_T)$ is a $J \times T$ matrix containing all of the data. In constructing an MCMC algorithm for this model, as with any other Bayesian model, we first find the full-conditional distributions. In this case, we are fortunate because the full-conditionals for $\text{vec}(\mathbf{A})$ and $\boldsymbol{\Sigma}$ are conjugate (specifically, Gaussian and inverse Wishart), and thus, trivial to sample from sequentially in the algorithm.

3.3 HIERARCHICAL TIME SERIES MODELS

Hierarchical statistical models have been a fundamentally important development in all scientific fields, but in the study of animal movement specifically. Hierarchical models allow us to model many levels of the process under study.† In ecology, hierarchical models are most often used to explicitly couple a measurement error process with an underlying mechanistic process representing the system under study. However, they can also be used to represent a hierarchical mechanistic process as well. For example, in the discrete-time movement models we present in Chapter 5, hierarchical specifications are used to model both the dynamics of movement and an underlying behavioral process of the individual animal. Hierarchical models can also be used to

---

* The "vec" operator converts a $J \times J$ matrix into a $J^2 \times 1$ vector by stacking the columns.

† Hierarchical models are also referred to as multilevel models, state-space models, mixed models, and random effects models in the associated literature.

scale up the inference from individuals to the population. We present model formulations for time series data that will be helpful in each of these settings in the following sections.

Hierarchical models need not be Bayesian, but the Bayesian framework provides a straightforward way to fit hierarchical models. In the Bayesian context, Berliner (1996) provided the first clear description of the structure of a hierarchical model, a structure that we often take for granted now. The hierarchical structure allows a complicated problem to be broken up into several simpler problems (i.e., conditional probability distributions for random variables). Thus, Berliner (1996) formulated a general hierarchical Bayesian model for time series as a sequence of conditional distributions:

$$\text{Stage 1: [data|process, parameters]}, \tag{3.53}$$

$$\text{Stage 2: [process|parameters]}, \tag{3.54}$$

$$\text{Stage 3: [parameters]}, \tag{3.55}$$

where each stage is conditioned on the stages below it in the model. This sequence of distributions appears simple, but provides an incredibly powerful tool for building complicated statistical models. In Stage 1 of a hierarchical framework, we typically find the "data model," which accounts the uncertainty associated with the actual measurements. Stage 2 is composed of the "process model." The term "process" arises from the mechanistic underpinnings associated with our understanding of how the system under study actually works.* The final component is the "parameter model," often referred to as a prior in Bayesian models. This final component is necessary for finding the posterior distribution that is used for Bayesian inference. While helpful in many cases, the parameter model is not necessary for non-Bayesian models.†

3.3.1 Measurement Error

Assume we have a process model for a time series as described in the previous sections, but we are unable to measure the process directly.‡ In these situations, we obtain noisy versions (y_t) of the true underlying process (z_t).§ Then, a generic hierarchical Bayesian model to account for measurement error associated with a first-order

* In the year 1996, Mark Berliner was focused on modeling atmospheric and oceanic processes for which very detailed mathematical models involving the physics of fluid dynamics are available. For this reason, he still prefers we use the term "physical process model" rather than "mechanistic model" for Stage 2. In our presentation here, we have shortened it to "process model."

† Random effects, in the classical sense, are more akin to process models than parameter models. Parameter models are for the bottom-level parameters. Random effects depend on unknown parameters; therefore, they are not at the bottom of the hierarchical structure.

‡ Some would argue that we are never able to directly measure the components of a process we often desire inference for. In which case, hierarchical models are essential.

§ It is also common to see the variable y used to represent the process and z used to represent the data; for example, Cressie and Wikle (2011).

autoregressive process is

$$y_t \sim [y_t|z_t, \boldsymbol{\theta}], \tag{3.56}$$

$$z_t \sim [z_t|z_{t-1}, \boldsymbol{\alpha}], \tag{3.57}$$

$$\boldsymbol{\alpha} \sim [\boldsymbol{\alpha}], \tag{3.58}$$

for $t = 1, \ldots, T$. If the process model is a Gaussian AR(1) and the measurements arise from a Gaussian process centered on the truth, a hierarchical model specification is

$$y_t \sim \text{N}(z_t, \sigma_y^2), \tag{3.59}$$

$$z_t \sim \text{N}(\alpha z_{t-1}, \sigma_z^2), \tag{3.60}$$

$$\alpha \sim \text{N}(0, \sigma_\alpha^2), \tag{3.61}$$

$$\sigma_y^2 \sim \text{IG}(\gamma_1, \gamma_2), \tag{3.62}$$

$$\sigma_z^2 \sim \text{IG}(\gamma_1, \gamma_2), \tag{3.63}$$

where the priors are only necessary if the model is Bayesian. In this case, we specify a normal prior for the autocorrelation parameter (α) and inverse gamma distributions for the two variance components (σ_y^2 and σ_z^2). In this particular model, it can sometimes be difficult to identify both variance components without strong prior information for one of them. Identifiability is a topic we return to in later chapters.

Figure 3.13 shows a simulated time series from a Gaussian hierarchical model with autoregressive parameter $\alpha = 0.95$ and variance components $\sigma_y^2 = \sigma_z^2 = 1$. As the measurement error (σ_y^2) increases, the temporal pattern evident in the latent process (z_t) will be less visible in the observed time series (y_t).

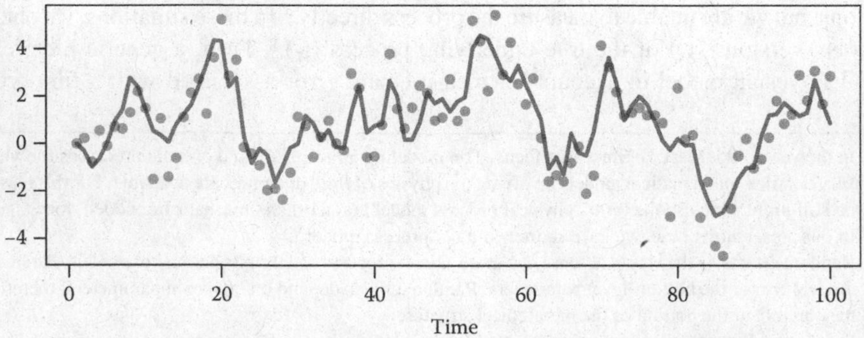

FIGURE 3.13 Simulated time series (y_t, gray points) from hierarchical model with dynamic latent process (z_t, dark line).

We are not obligated to use a normal distribution for the measurement error (although it does yield substantial computational advantages when appropriate). Suppose the measured response variable is a count at each time t.* Then we might choose to model the data as $y_t \sim \text{Pois}(e^{z_t})$, where e^{z_t} represents the underlying intensity process for the behavior of interest. The log of this intensity process is modeled as Equation 3.48 to account for smoothness in behavior over time.

Thus, the options for modeling error are limitless and will explicitly depend on the type of data collected and study design. In animal movement modeling, we often observe positions of individuals as 2-D measurements arising from telemetry data. The ability to account for multivariate measurements is essential, and the hierarchical modeling approach makes it easy to do that.

3.3.2 Hidden Markov Models

Another form of hierarchical model is the hidden Markov model (HMM). The term "Markov," in this sense, is the same as used in the autoregressive time series we have already discussed. Thus, the hierarchical model presented in the previous section is also technically an HMM. However, when the term "HMM" is used, it is typically meant to describe a process model that is discrete (or categorical) and dynamic. In our time series examples thus far, we have focused on processes with continuous support. However, in animal movement modeling, it is common to specify discrete latent processes. For example, suppose we use a hierarchical structure to cluster a process in two different groups. If the process lingers in each group for an amount of time before switching, we could use the model

$$y_t \sim \begin{cases} \text{N}(\mu_0, \sigma_0^2), & w_t = 0 \\ \text{N}(\mu_1, \sigma_1^2), & w_t = 1 \end{cases}, \tag{3.64}$$

$$w_t \sim \begin{cases} \text{Bern}(1-p), & w_{t-1} = 0 \\ \text{Bern}(p), & w_{t-1} = 1 \end{cases}, \tag{3.65}$$

$$\mu_0 \sim \text{N}(\mu_{0,0}, \sigma_{0,0}^2), \tag{3.66}$$

$$\mu_1 \sim \text{N}(\mu_{1,0}, \sigma_{1,0}^2), \tag{3.67}$$

$$\sigma_0^2 \sim \text{IG}(\gamma_1, \gamma_2), \tag{3.68}$$

$$\sigma_1^2 \sim \text{IG}(\gamma_1, \gamma_2), \tag{3.69}$$

$$p \sim \text{Beta}(\alpha, \beta), \tag{3.70}$$

where the two clusters are shaped by Gaussian distributions with potentially different locations and spreads. The key to this HMM is that the cluster probability is a

* For example, when a certain discrete behavior (e.g., forays from a nest) is observed repeatedly during Δt.

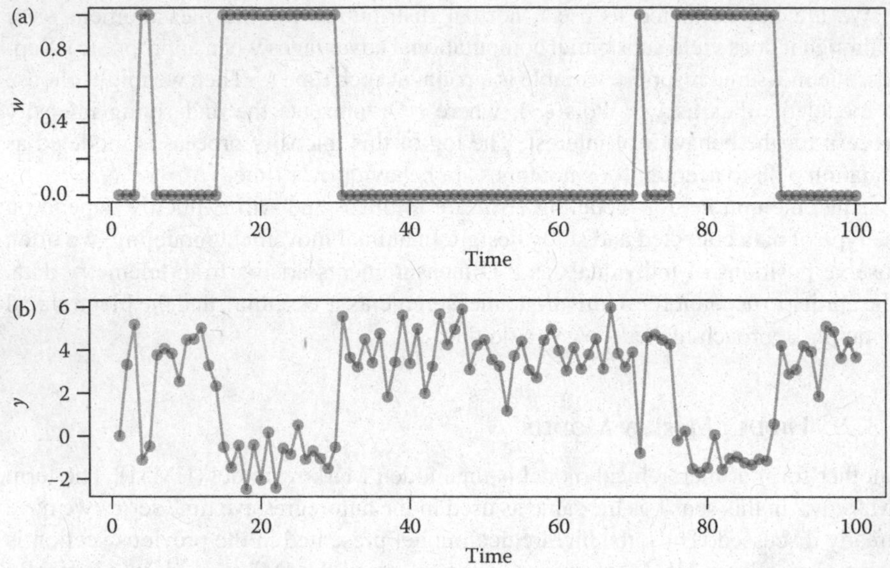

FIGURE 3.14 Simulated time series and dynamic binary process arising from an HMM. Panel (a) shows the time series for the latent process w_t and panel (b) shows the time series for the positions y_t.

Markov autoregressive process, but it is also binary. As p approaches 0 or 1, the process will stay in each cluster longer before shifting to the other cluster. In Chapter 5, we use similar discrete latent variables to represent animal movement behavior and they provide smoothness to the behavioral switching process.

Figure 3.14 shows a simulated time series arising from a latent HMM, as described above. We used $\mu_0 = 4$ and $\mu_1 = -1$ for cluster means and $\sigma_0^2 = \sigma_1^2 = 1$ for variance components in the data model. We specified the probability parameter $p = 0.9$, which imparts a strong smoothness to the latent binary process w_t. The resulting time series y_t, in Figure 3.14, exhibits an uncorrelated random walk within cluster, but abruptly shifts to the other cluster as w_t switches its state. This HMM time series model could be useful for describing the movement of a fish darting between upstream and downstream habitat patches.

3.3.3 Upscaling

Another common usage of hierarchical models in time series is to avoid pseudoreplication by scaling up the inference to the appropriate level. For example, in the animal movement context, we commonly obtain telemetry data for a subsample of individuals from a larger population. Population-level inference is often of interest in many studies, but we need to construct individual-level models to properly represent the movement dynamics. Upscaling can also be useful to help separate measurement uncertainty from process uncertainty.

To demonstrate a model for population-level inference, suppose we have a separate process model for each of J individuals and we wish to estimate a population-level autoregressive parameter α. If we use $y_{j,t} \sim \text{N}(\alpha y_{j,t-1}, \sigma^2_{y,j})$ for $j = 1, \ldots, J$ and $t = 1, \ldots, T$, to estimate α, the model would use all telemetry data for all individuals directly. In reality, each individual probably responds to environmental cues differently and has different physical characteristics; thus, we could let the autoregressive parameter α_j vary by individual. If we substitute the individual-level parameter into each individual model and estimate them all separately, it will not acknowledge any consistent behavior among individuals in the population. Thus, we set up a hierarchical model to allow for structure at the population level:

$$y_{j,t} \sim \text{N}(\alpha_j y_{j,t-1}, \sigma^2_{y,j}), \tag{3.71}$$

$$\alpha_j \sim \text{N}(\mu_\alpha, \sigma^2_\alpha), \tag{3.72}$$

$$\mu_\alpha \sim \text{N}(0, \sigma^2_{\alpha,0}), \tag{3.73}$$

$$\sigma^2_\alpha \sim \text{IG}(\gamma_{\alpha,1}, \gamma_{\alpha,2}), \tag{3.74}$$

$$\sigma^2_{y,j} \sim \text{IG}(\gamma_{y,1}, \gamma_{y,2}), \tag{3.75}$$

where the individual-level parameters arise stochastically from a population-level distribution. For inference, it is most common to focus on the population-level mean μ_α and its uncertainty. However, it can also be useful to interpret population-level variance σ^2_α because it tells us about the spread of individuals. If the spread is relatively small, it implies the individuals are behaving consistently in the population. Thus, population-level influences on the parameter (α) must be stronger than individual-level influences.

Figure 3.15 shows two sets of simulated time series from the hierarchical model with the autocorrelation parameter (α_j) as a random effect. While both panels in Figure 3.15 contain time series that are stationary around zero, the individual time series exhibit similar, but not identical, dynamics. In Figure 3.15a, the time series have positive autocorrelation ranging from $\alpha = 0.55$ to $\alpha = 0.82$. In Figure 3.15b, the time series have negative autocorrelation ranging from $\alpha = -0.75$ to $\alpha = -0.44$. Thus, the individual time series share similar properties because their dynamics arise from a common distribution. Note that the actual time series values need not look similar to have similar dynamics.

Using a similar hierarchical model specification, we can ameliorate the issues with identifiability of measurement variance and process variance. Suppose we have J repeated measurements, $y_{j,t}$, of the underlying process, z_t. Each of the observations is an imperfect measurement of the underlying process, but now with replication, we can properly separate σ^2_z and σ^2_y with the model

$$y_{j,t} \sim \text{N}(z_t, \sigma^2_y), \tag{3.76}$$

$$z_t \sim \text{N}(\alpha z_{t-1}, \sigma^2_z), \tag{3.77}$$

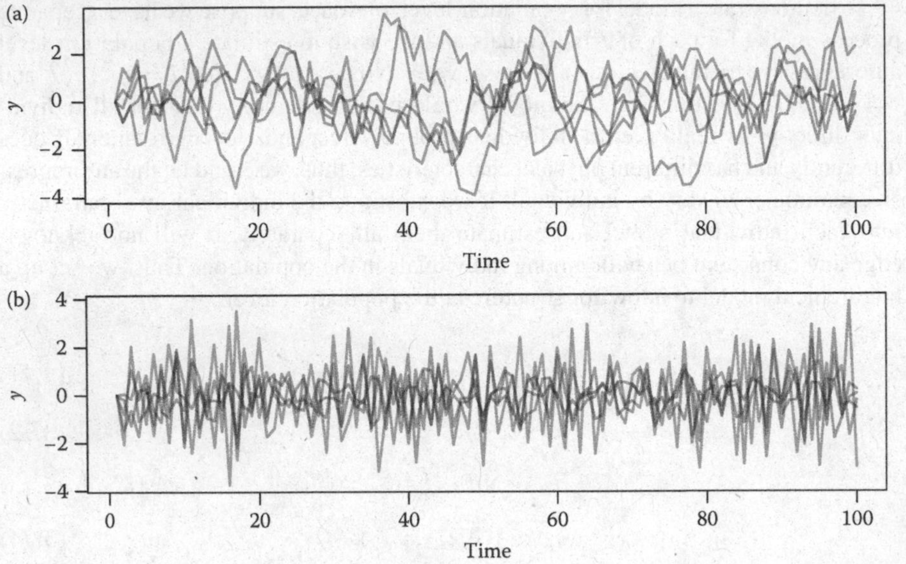

FIGURE 3.15 Five ($J = 5$) simulated time series ($y_{j,t}$) from two different hierarchical models. In panel (a), $\mu_\alpha = 0.7$, and in panel (b), $\mu_\alpha = -0.7$. In both panels, $\sigma^2_{y,j} = 1, \forall j$ and $\sigma^2_\alpha = 0.05$.

$$\alpha \sim \text{N}(0, \sigma^2_\alpha), \tag{3.78}$$

$$\sigma^2_y \sim \text{IG}(\gamma_1, \gamma_2), \tag{3.79}$$

$$\sigma^2_z \sim \text{IG}(\gamma_1, \gamma_2), \tag{3.80}$$

which is very similar to the original hierarchical measurement error model, except that the replication at the data level provides enough information about σ^2_y to separate it from σ^2_z, especially as J increases.

3.3.3.1 Implementation: Kalman Approaches

The implementation of hierarchical time series models from a non-Bayesian perspective usually involves an integrated likelihood approach (integrating the process z_t out of the model). For the nonreplicated measurement error model we discussed previously, one would integrate the process out of the joint model as

$$[\mathbf{y}|\alpha, \sigma^2_y, \sigma^2_z] = \int [\mathbf{y}|\mathbf{z}, \sigma^2_y][\mathbf{z}|\alpha, \sigma^2_z]d\mathbf{z} \tag{3.81}$$

$$= \text{N}(\mathbf{0}, \sigma^2_y\mathbf{I} + \boldsymbol{\Sigma}_z). \tag{3.82}$$

The process model in Equation 3.82 is written jointly as we would write a CAR model in spatial statistics. This specification allows us to write the dynamic structure in terms of covariance, where the matrix $\boldsymbol{\Sigma}_z$ is a function of the parameters α and σ_z^2, such that $\boldsymbol{\Sigma}_z \equiv \sigma_z^2(\text{diag}(\mathbf{W1}) - \alpha\mathbf{W})^{-1}$ and $\mathbf{W}$ is a binary proximity matrix indicating which times are neighbors of each other and diag($\mathbf{W1}$) is a diagonal matrix with row sums of $\mathbf{W}$ along the diagonal. This type of integration is often referred to as "Rao-Blackwellization."

The main drawback of using the integrated likelihood approach is that one cannot simultaneously obtain inference for the latent process. The latent process is one of the key features of interest in most animal ecological studies. Thus, a non-Bayesian alternative to the integrated likelihood approach for estimating the process in hierarchical time series models involves Kalman methods.

Kalman methods allow for the estimation and prediction of latent linear temporal processes such as those described in our hierarchical time series example for measurement error (Kalman 1960). Kalman methods have been extremely popular for signal processing because they are fast to implement and can naturally update inference in real time as new data are obtained.

Consider the simple non-Bayesian hierarchical time series model

$$y_t \sim \text{N}(z_t, \sigma_y^2), \tag{3.83}$$

$$z_t \sim \text{N}(\alpha z_{t-1}, \sigma_z^2). \tag{3.84}$$

To set up basic Kalman terminology, there are three main types of procedures for estimation and prediction. If we are interested in inference about z_t, given data $\mathbf{y}_\tau \equiv (y_1, \ldots, y_\tau)'$, then our problem is prediction if $t > \tau$, it is filtering* if $t = \tau$, and it is smoothing if $t < \tau$.

Thus, to estimate the process sequentially for $t = \tau$, we can use the Kalman filtering algorithm (e.g., Cressie and Wikle 2011):

1. Choose initial values for the prediction mean $E(z_0|\mathbf{y}_0)$ and variance $E((z_0 - E(z_0|\mathbf{y}_0))^2|\mathbf{y}_0)$.
2. Let $t = 1$.
3. Calculate the prediction mean: $E(z_t|\mathbf{y}_{t-1}) = \alpha E(z_{t-1}|\mathbf{y}_{t-1})$.
4. Calculate the prediction variance: $\text{Var}(z_t|\mathbf{y}_{t-1}) = \sigma_z^2 + \alpha^2 \text{Var}(z_{t-1}|\mathbf{y}_{t-1})$.
5. Calculate the Kalman gain† using the prediction variance: $g_t = \text{Var}(z_t|\mathbf{y}_{t-1})(\text{Var}(z_t|\mathbf{y}_{t-1}) + \sigma_y^2)^{-1}$.
6. Calculate the filter distribution mean using the prediction mean and Kalman gain: $E(z_t|\mathbf{y}_t) = E(z_t|\mathbf{y}_{t-1}) + g_t \cdot (y_t - E(z_t|\mathbf{y}_{t-1}))$.
7. Calculate the filter distribution variance using the prediction variance and Kalman gain: $\text{Var}(z_t|\mathbf{y}_t) = (1 - g_t)E((z_t - E(z_t|\mathbf{y}_{t-1}))^2|\mathbf{y}_{t-1})$.
8. Stop if $t = T$, else let $t = t + 1$ and go to step 3.

---

* The term "filtering" is used because it removes unwanted noise from a signal. In this sense, smoothing is also a type of filtering, but one using all the data.

† The "gain" is a multiplier that updates the information from the previous time to provide the expectation at the current time.

This iterative algorithm will result in the correct filter distribution mean and variance for all times. The smoother distribution mean and variance can be obtained using a similar algorithm (see Cressie and Wikle 2011 for details). Furthermore, these algorithms are also easily extended to the multivariate setting. While they are incredibly fast, the drawback to Kalman algorithms is that they do not directly estimate model parameters (i.e., α, σ_y^2, and σ_z^2). Thus, Kalman methods must be paired with parameter estimation algorithms such as the expectation–maximization algorithm or maximum likelihood to provide full model fitting results. See Shumway and Stoffer (2006) for additional details on Kalman methods.

3.3.3.2 Implementation: Bayesian Approaches

In a Bayesian treatment of the hierarchical time series model

$$y_t \sim \text{N}(z_t, \sigma_y^2), \tag{3.85}$$

$$z_t \sim \text{N}(\alpha z_{t-1}, \sigma_z^2), \tag{3.86}$$

$$\alpha \sim \text{N}(0, \sigma_\alpha^2), \tag{3.87}$$

$$\sigma_y^2 \sim \text{IG}(\gamma_1, \gamma_2), \tag{3.88}$$

$$\sigma_z^2 \sim \text{IG}(\gamma_1, \gamma_2), \tag{3.89}$$

we seek the posterior distribution of the latent state variables (z_t) and parameters α, σ_y^2, and σ_z^2:

$$[\mathbf{z}, \alpha, \sigma_y^2, \sigma_z^2 | \mathbf{y}] \propto \prod_{t=1}^{T} [y_t | z_t, \sigma_y^2][z_t | z_{t-1}, \alpha, \sigma_z^2][\alpha][\sigma_y^2][\sigma_z^2]. \tag{3.90}$$

The joint posterior is not analytically tractable, but we can use MCMC to fit the model. For our simple hierarchical time series model, the full-conditional distributions are tractable because we used conjugate prior distributions.* Thus, we construct an MCMC algorithm by sampling from the following distributions sequentially:

$$[\alpha|\cdot] = \text{N}\left(\left(\frac{\sum_t z_{t-1}^2}{\sigma_z^2} + \frac{1}{\sigma_\alpha^2} \right)^{-1} \left(\frac{\sum_t z_t z_{t-1}}{\sigma_z^2} \right), \left(\frac{\sum_t z_{t-1}^2}{\sigma_z^2} + \frac{1}{\sigma_\alpha^2} \right)^{-1} \right), \tag{3.91}$$

$$[\sigma_y^2|\cdot] = \text{IG}\left(\frac{T}{2} + \gamma_1, \frac{\sum_t (y_t - z_t)^2}{2} + \gamma_2 \right), \tag{3.92}$$

$$[\sigma_z^2|\cdot] = \text{IG}\left(\frac{T}{2} + \gamma_1, \frac{\sum_t (z_t - \alpha z_{t-1})^2}{2} + \gamma_2 \right), \tag{3.93}$$

* Recall that conjugacy implies that the form of the full-conditional matches that of the prior.

$$[z_t|\cdot] = \text{N}\left(\left(\frac{1}{\sigma_y^2}+\frac{2}{\sigma_z^2}\right)^{-1}\left(\frac{y_t}{\sigma_y^2}+\frac{z_{t+1}+z_{t-1}}{\sigma_z^2}\right), \left(\frac{1}{\sigma_y^2}+\frac{2}{\sigma_z^2}\right)^{-1}\right),$$

$$\text{for} \quad t = 1, \ldots, T-1, \tag{3.94}$$

$$[z_T|\cdot] = \text{N}\left(\left(\frac{1}{\sigma_y^2}+\frac{1}{\sigma_z^2}\right)^{-1}\left(\frac{y_t}{\sigma_y^2}+\frac{z_{t-1}}{\sigma_z^2}\right), \left(\frac{1}{\sigma_y^2}+\frac{1}{\sigma_z^2}\right)^{-1}\right), \tag{3.95}$$

given an initial value for z_0. As discussed in the previous chapters, after a large number of MCMC samples have been collected, we can obtain inference in the form of posterior means, variances, and credible intervals using Monte Carlo integration. For more details on Bayesian methods and MCMC, see Hobbs and Hooten (2015).

Using the simulated data set, based on $\sigma_y^2 = 0.1$, $\sigma_z^2 = 1$, and $\alpha = 0.95$, we estimated the latent temporal process z_t using maximum likelihood (with Kalman filtering) and the Bayesian hierarchical model (with MCMC). Figure 3.16a shows the time series with Kalman smoother mean and 95% confidence interval while

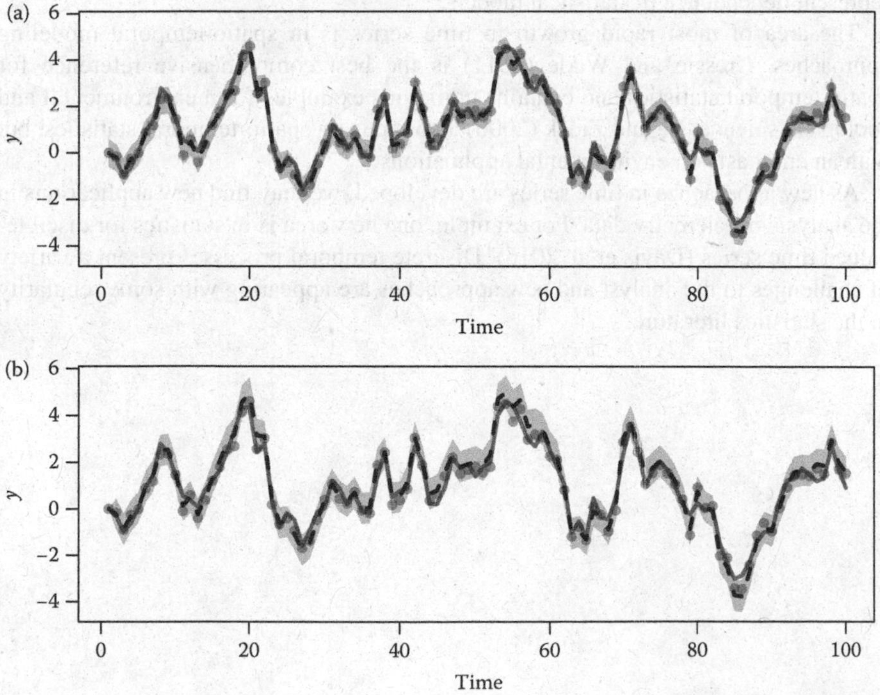

FIGURE 3.16 Estimated latent process for z_t based on simulated data y_t (points). Panel (a) shows the Kalman smoother mean (dashed line) and 95% confidence interval (gray region). Panel (b) shows the Bayesian posterior mean (dashed line) and 95% credible interval (gray region).

Figure 3.16b shows the same time series with the Bayesian posterior mean and 95% credible interval. The confidence interval for the Kalman smoother (Figure 3.16a) is narrower than that of the Bayesian credible interval (Figure 3.16b). While both statistical estimates are obtained via smoothing, the Bayesian credible interval is slightly wider because it accommodates the uncertainty associated with the unknown parameters.

3.4 ADDITIONAL READING

Classical references on time series analysis include Brockwell and Davis (2013) and Shumway and Stoffer (2006); however, the literature for time series is massive because of its importance in econometrics and other fields concerned with the analysis of long term data sets. Chapter 3 of Cressie and Wikle (2011) covers all the basics of time series analysis, from dynamical systems and chaos, to random walks and autoregressive models. Cressie and Wikle (2011) also cover spectral representations of time series, a topic that we only briefly touch upon in this book. The spectral perspective of temporal processes is critical to help understand basis function specifications of spatial and time series models. Hefley et al. (2016a) provide a gentle introduction to basis function concepts for ecologists and how they can be used to represent dependence in statistical models.

The area of most rapid growth in time series is in spatio-temporal modeling approaches. Cressie and Wikle (2011) is the best comprehensive reference for spatio-temporal statistics and contains numerous examples from environmental and ecological science. Le and Zidek (2006) also focus on spatio-temporal statistics, but with an emphasis on environmental applications.

As new approaches in time series are developed, we may find new applications in the analysis of telemetry data. For example, one new area is in statistics for discrete-valued time series (Davis et al. 2016). Discrete temporal processes present a variety of challenges to the analyst and new approaches are appearing with some regularity in the statistics literature.

4 Point Process Models

There may not be another topic in quantitative ecology that is as mystifying and misunderstood as the study of space use and resource selection. Our goal in this chapter is to describe the topic and various approaches for inference, while making both historical and contemporary connections between methods. We begin by describing the concept of space use in the context of spatial point processes and then build on that with the concept of resource selection. This perspective is somewhat new in ecology, as many of the approaches seem to have been developed in different fields over time, but as you will see, it provides a fully rigorous approach for modeling certain types of telemetry data.

4.1 SPACE USE

Space use is the result of an animal movement process. Thus, most space use studies seek to better understand where an individual (or individuals) spent their time. Movement is an inherent trait of all animals, and the moving individuals within populations can be thought of as a dynamic system. The dynamics of individual movement can be high-dimensional but are most often considered from a 2-D perspective.* The true individual locations $\boldsymbol{\mu}_i$ (or observed locations, without measurement error), for a finite set of times t_i $(i = 1, \ldots, n)$, are often considered to represent a spatial point process. Under this interpretation, the $\boldsymbol{\mu}_i$ are considered to be random vectors before they are observed, but fixed and known quantities after they are observed and are then treated as response data. In space use studies, we seek to characterize the distribution from which the individual locations arose. That is, we assume that some multivariate probability distribution exists, and gives rise to $\boldsymbol{\mu}_i$. In the 2-D case, this can be thought of as a PDF in space, $f(\boldsymbol{\mu}_i)$ (or $[\boldsymbol{\mu}_i]$). The animal ecology literature refers to this spatial density function as the "utilization distribution," or "UD" for short.

Our goal is to use the individual locations ($\boldsymbol{\mu}_i$) to learn about the spatial probability distribution that gave rise to them. As we discussed in Section 2.1, one type of nonparametric approach for learning about the spatial probability distribution is called kernel density estimation (KDE). Conventional approaches to KDE use a density estimator of the form given in Equation 2.4, which we reformulate as a function of $\boldsymbol{\mu}$ rather than $\mathbf{s}$:

$$\hat{f}(\mathbf{c}) = \frac{\sum_{i=1}^{n} k((c_1 - \mu_{1,i})/b_1)k((c_2 - \mu_{2,i})/b_2)}{nb_1b_2}. \tag{4.1}$$

* Even though terrestrial animals live on a spheroid that is clearly not 2-D. For small spatial extents, the 2-D assumptions are often sufficient, but keep in mind that we may not be able to reduce the dimensionality of space down to two and still retain the important ecological characteristics for animals that swim or fly.

The true density function $f(\mathbf{c})$ can then be estimated for any location $\mathbf{c}$ given the true individual locations $\boldsymbol{\mu}_i$ for $i = 1, \ldots, n$ and choice of kernel function $k(\cdot)$. Additional quantities in this estimator are the bandwidth parameters b_1 and b_2. These bandwidth parameters control the smoothness of the estimated density surface. Many approaches exist for setting or estimating b_1 and b_2. Most commonly, a default bandwidth is calculated for each margin (i.e., latitude and longitude) as 0.9 times the minimum of the sample standard deviation and the interquartile range divided by 1.34 times the sample size to the negative one-fifth power (e.g., Silverman 1986; Scott 1992). There are many alternative methods for setting an appropriate bandwidth (e.g., cross-validation), but the method described above works well for Gaussian kernels and many data sets. Although the UD is often not estimated in a parametric framework, the estimated density function serves as a basis from which to calculate many important space use metrics.

The GPS telemetry data and estimated UD, based on KDE, for an individual mountain lion (*Puma concolor*) in Colorado, USA, is shown in Figure 4.1. The data in Figure 4.1 were used in an example by Hooten et al. (2013b), and represent 91 positions observed every 3 h over a period of approximately 11 days for an adult mountain lion. The estimated UD in this example indicates that the individual mountain lion likely uses space differentially, with at least two main regions of higher-intensity use

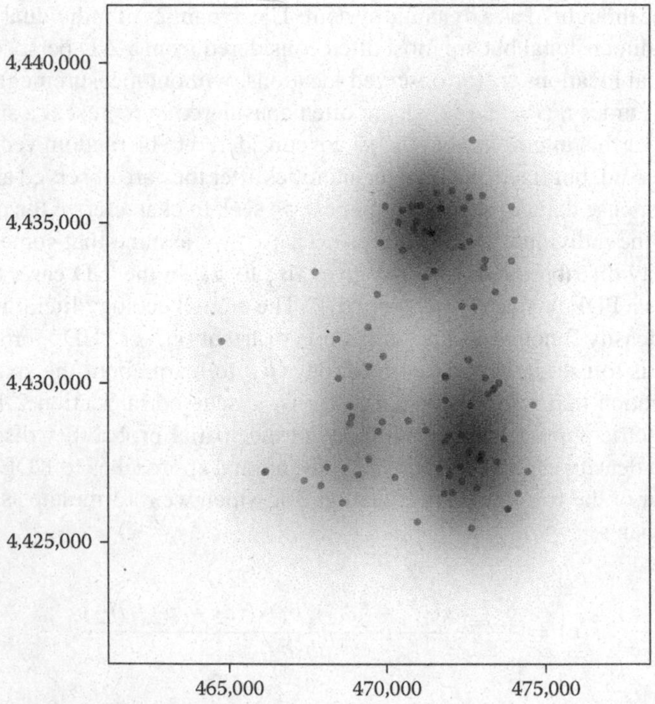

FIGURE 4.1 Mountain lion telemetry locations (points) and utilization distribution estimated using KDE (darker gray shading indicates greater utilization).

in the study area (Figure 4.1). Furthermore, there appears to be at least one telemetry observation that is distant from the regions of highest-intensity use (leftmost point in Figure 4.1), perhaps due to a foray into a neighboring individual's territory.

4.1.1 Home Range

The conventional definition of a home range was put forth by Burt (1943) as the "area traversed by an individual in its normal activities of food gathering, mating, and caring for young. Occasional sallies outside the area, perhaps exploratory in nature, should not be considered part of the home range." Powell and Mitchell (2012) point out that, like many other ecological concepts, the home range is difficult to characterize because it is a function of many interacting endogenous and exogenous factors. Nonetheless, many researchers still wish to estimate the home range despite it being somewhat abstract. Mathematically, we can describe the home range, under Burt's definition, as a nonlinear feature in multidimensional space that serves as a semipermeable boundary to movement. Inside the boundary, the space use pattern (i.e., the UD) may represent an elaborate cognitive map of the environment perceived by the individual (e.g., Borger et al. 2008). Hence, a suite of nonparametric tools have been used to learn about the home range. For example, the concept of an individual home range is often quantified as the 95% isopleth (or density contour containing 95% of the mass) of the estimated UD $\hat{f}$. Alternatively, convex hull (or minimum convex polygon) approaches have been used as a mathematical object that bounds a set of telemetry locations. Development of home range estimation methods has expanded rapidly in recent years (e.g., Getz et al. 2007; Laver and Kelly 2008; Lyons et al. 2013). While home range estimation methods have become popular in animal space use studies, there has been no consensus on which approaches are best.

The term "home range" has come under substantial scrutiny in recent years. At one point, the phrase "home range model" was used as a catchall for animal movement models in general. The home range is an emergent feature of a complicated set of animal movement outcomes and rarely a strict geographic perimeter that the individual delineates.* For our purposes, we consider the home range as a subset of geographic space where it is most likely to find a particular individual animal. Fundamentally, the home range is an individual-based spatial topological feature. It can be estimated using many different approaches, each carrying their own assumptions about the individual's life history and its interaction with the environment. A feature of the home range commonly of interest is its size, often measured in area. All home range estimation methods allow for the estimation of the area enclosed by the boundary.

While home range estimation is not the focus of this book, it can be useful for characterizing the spatial support of animal movement process. The spatial support is critical for most point process modeling approaches. As previously mentioned, two of the most commonly used techniques for estimating the home range are (1) a large isopleth of a KDE and (2) a convex hull of the telemetry data. An isopleth of the

* With the possible exception of true physical constraints such as fenced regions or a very strong territorial effect in a confined space.

UD (or KDE of the UD) is essentially a contour line, or more formally, a line drawn through all of the points on a surface that have the same density value. For example, by convention, the 95% isopleth of a KDE delineates the region that contains 95% of the total density. A convex hull* is the smallest polygon containing all of the telemetry points by connecting the "outside" points while having no acute interior angles (i.e., no interior angles less than 90°).

Returning to the mountain lion example in the previous section, Figure 4.2 demonstrates the similarities and differences among home range estimation methods. For example, the KDE isopleth increases in size as the percentage of the isopleth increases. The 95% KDE isopleth in Figure 4.2a is sufficiently small that

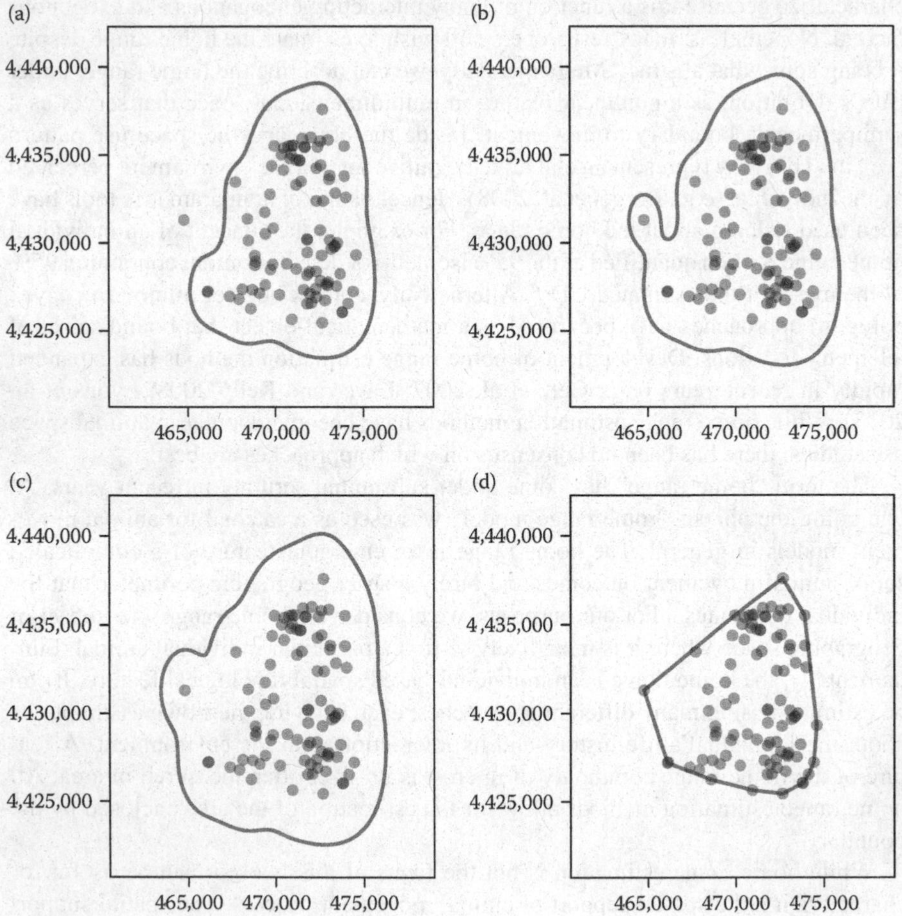

FIGURE 4.2 Mountain lion telemetry locations (points) and estimated home range delineation using (a) 95% KDE isopleth, (b) 97.5% KDE isopleth, (c) 99% KDE isopleth, and (d) convex hull.

* A convex hull is often referred to as a minimum convex polygon (MCP) in the animal ecology literature.

the dominant region of space in the estimated home range does not include one of the telemetry observations. In fact, at the 95% level, there are two distinct regions of space, but we only plot the dominant one (i.e., the one with larger area) for illustration here. Figure 4.2d illustrates that, in this case, the convex hull estimate of the mountain lion home range is substantially smaller than the KDE isopleth estimates and captures all of the telemetry data. The researcher must decide which type of home range estimator to use if inference for the home range is desired. The convex hull method is less subjective, but some would argue that it is also less realistic. Signer et al. (2015) argue that the relative differences in home range size among individuals is most important, and thus, the estimation method may not impact the desired inference.

4.1.2 Core Areas

Within an individual's home range, there may exist regions that are used more intensively. These regions of higher-intensity use are often referred to as "core areas," and are commonly estimated as the region contained within the 50% isopleth of the KDE UD (Laver and Kelly 2008). Core areas need not be contiguous or equally sized or shaped. For temporally independent telemetry data, statistical evidence for core areas can be obtained by examining the telemetry data as a point process using the methods described in Chapter 2. Specifically, the Ripley's K and L functions can be used to assess clustering and regularity in the telemetry data. If evidence of clustering exists within the home range, it suggests the presence of core areas.

In the mountain lion example from the previous sections, we used a home range delineation based on the 99% KDE isopleth and estimated the L functions for a set of distances spanning half the range of the data (Figure 4.3). The mountain lion data indicate the presence of clustering within all distances (due to $\hat{L}$ falling above the CSR simulation interval). These results suggest that differential space use is likely for the mountain lion individual and that there may be a core area or areas of higher-intensity use within the home range.

Wilson et al. (2010) proposed a parametric statistical model to estimate the core areas within an individual's home range. The basic approach considers a discrete set of regions in the home range, each with a distinct intensity of use. The goal is to estimate the number of subregions, their associated intensity, and their shape, given a set of telemetry data. Thus, Wilson et al. (2010) assumed that a KDE isopleth can serve as a constraint to cut the home range into subregions (like a cookie cutter).* That is, after a UD has been estimated using KDE methods and conditioned on, a single chosen isopleth ϕ will yield a number of subregions of the home range and their shape. Thus, for a situation where two different levels of density ($f_{\mathcal{C}}$ and $f_{\tilde{\mathcal{C}}}$) are associated with core and noncore areas, we only need to find the optimal† isopleth ϕ while estimating the two densities.‡

---

* Some type of constraint (e.g., a KDE isopleth) is needed to adhere to contiguous regions of space for more powerful estimation.

† By "optimal," we mean the isopleth that best splits the home range into subregions.

‡ Recall that the density and intensity are proportional to each other, with the intensity carrying more information: the expected number of points in a region. Because there may be several core areas of

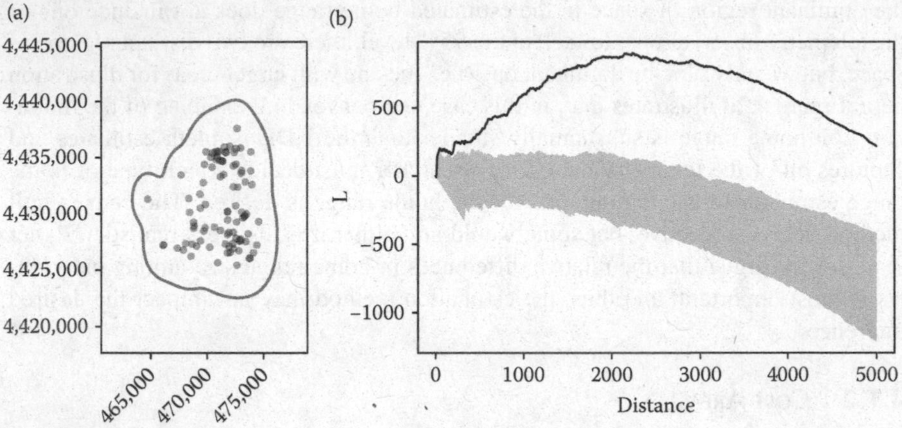

FIGURE 4.3 (a) Mountain lion telemetry locations (points) and estimated home range delineation using a 99% KDE isopleth and (b) the corresponding estimated L function (black line) and Monte Carlo interval based on 1000 CSR simulations of the point process within the home range (gray region).

The model proposed by Wilson et al. (2010) partitions the home range into a large number of m grid cells for computational reasons. The telemetry data are then converted to counts within each grid cell.* Grid cells not containing telemetry observations receive a zero count. To begin, assume that the home range $\mathcal{H}$ can be partitioned into two subsets $\mathcal{H} = \mathcal{C} \cup \tilde{\mathcal{C}}$, where $\mathcal{C}$ and $\tilde{\mathcal{C}}$ represent the nonoverlapping core and noncore areas (i.e., their intersection is zero).† The core area $\mathcal{C}$ may be composed of several distinct subregions itself in cases where the UD is multimodal. If the KDE isopleth ϕ is known, then both $\mathcal{C}$ and $\tilde{\mathcal{C}}$ are known and there are $m_{\mathcal{C}}$ and $m_{\tilde{\mathcal{C}}}$ grid cells that fall within each subregion. Thus, Wilson et al. (2010) used a multinomial framework to model the grid cell counts

$$\mathbf{y}_{\mathcal{C}} \sim \text{MN}(\mathbf{y}_{\mathcal{C}} | n_{\mathcal{C}}, \mathbf{p}_{\mathcal{C}}, \phi), \tag{4.2}$$

$$\mathbf{y}_{\tilde{\mathcal{C}}} \sim \text{MN}(\mathbf{y}_{\tilde{\mathcal{C}}} | n_{\tilde{\mathcal{C}}}, \mathbf{p}_{\tilde{\mathcal{C}}}, \phi), \tag{4.3}$$

where $\mathbf{y}_{\mathcal{C}}$ (an $m_{\mathcal{C}} \times 1$ vector) are the cell counts in the core areas and $\mathbf{y}_{\tilde{\mathcal{C}}}$ (an $m_{\tilde{\mathcal{C}}} \times 1$ vector) are the cell counts in the noncore areas. The total numbers of telemetry observations in core and noncore areas are $n_{\mathcal{C}}$ and $n_{\tilde{\mathcal{C}}}$, and the grid cell probabilities for core and noncore areas are $\mathbf{p}_{\mathcal{C}} \equiv (1/m_{\mathcal{C}}, \ldots, 1/m_{\mathcal{C}})'$ and $\mathbf{p}_{\tilde{\mathcal{C}}} \equiv (1/m_{\tilde{\mathcal{C}}}, \ldots, 1/m_{\tilde{\mathcal{C}}})'$. The multinomial specification and equal grid cell probabilities imply that the total

different sizes within a home range, we use the term "density" here rather than "intensity." While core areas all have density $f_{\mathcal{C}}$ in our model, the intensities will vary with size; larger core areas will have higher expected numbers of telemetry observations even though they all have the same density.

* This is similar to the implementation of resource selection function models, and thus, it serves a good segue to the next section.

† Although the core area will be completely surrounded by noncore area given our model assumptions.

core area intensity is $a_{\mathcal{C}} n_{\mathcal{C}} / m_{\mathcal{C}}$ and noncore area intensity is $a_{\bar{\mathcal{C}}} n_{\bar{\mathcal{C}}} / m_{\bar{\mathcal{C}}}$, where the total core size is $a_{\mathcal{C}}$ and noncore size is $a_{\bar{\mathcal{C}}}$. The important point is that both intensities are known when the isopleth is ϕ is known. Thus, the total likelihood for the model can be written as

$$[\mathbf{y}|\phi] = \text{MN}(\mathbf{y}_{\mathcal{C}}|n_{\mathcal{C}}, \mathbf{p}_{\mathcal{C}}, \phi) \times \text{MN}(\mathbf{y}_{\bar{\mathcal{C}}}|n_{\bar{\mathcal{C}}}, \mathbf{p}_{\bar{\mathcal{C}}}, \phi), \tag{4.4}$$

where $\mathbf{y}$ is the $m \times 1$ vector of all cell counts. The likelihood in Equation 4.4 can be maximized to find the MLE for ϕ or a prior could be specified for ϕ and Bayesian inference can be obtained. Wilson et al. (2010) obtain a Bayesian estimate for ϕ using the likelihood (4.4) and the prior $\phi \sim \text{Beta}(1.1, 1.1)$.*

The core area model described thus far is appropriate when a single isopleth partitions the area into core and noncore areas. However, it is possible that there may be multiple levels of core areas at increasingly higher levels of intensity. In these cases, we can easily generalize the model by allowing for several isopleths (in the vector $\boldsymbol{\phi} \equiv (\phi_1, \ldots, \phi_J)'$) such that they are ordered from small to large. Wilson et al. (2010) use a Dirichlet prior for the isopleth vector because each of the ϕ_j isopleths are bounded by zero and one, and sum to one ($\sum_{j=1}^{J} \phi_j = 1$). In this generalized setting, the likelihood is now a product over all subregions of the home range

$$[\mathbf{y}|\boldsymbol{\phi}] = \prod_{j=1}^{J} \text{MN}(\mathbf{y}_{\mathcal{C}_j}|n_{\mathcal{C}_j}, \mathbf{p}_{\mathcal{C}_j}, \phi_j), \tag{4.5}$$

where the home range is partitioned as $\mathcal{H} = \cup_{j=1}^{J} \mathcal{C}_j$ and $n_{\mathcal{C}_j}$ is number of telemetry observations in the jth subregion.

Wilson et al. (2010) recommended a general procedure for implementation when the number of core areas is unknown:

1. Check for clustering in the observed set of telemetry data using the Ripley's L function and associated Monte Carlo hypothesis tests.
2. If clustering exists, partition the domain into a large number of grid cells (as big as computationally feasible) and fit the core area model assuming only two levels of intensity.
3. Use the posterior mean isopleth $E(\phi|\mathbf{y})$ to split the telemetry observations into two sets, one for the core and one for the noncore.
4. Check for additional clustering in each set separately using the methods in step 1.
5. If checks reveal no further evidence of clustering, stop and obtain the desired inference (e.g., core area size).
6. If additional clustering exists, fit the core area model using three levels of intensity.
7. Check each of these three subregions for further clustering.

* The hyperparameters of this prior were chosen deliberately to keep ϕ away from the unreasonable values of zero and one while still being only weakly informative.

8. If no further clustering is evident, stop and obtain inference.
9. If additional clustering exists, continue to fit core area models with increasingly more levels of intensity until there is no further evidence of clustering within subregions.

Figure 4.4 shows the estimated core area for the mountain lion data in our example from the previous sections. The posterior mean isopleth occurred at 49% and is composed of two core area regions shown as a dashed line in Figure 4.4. The estimated core area itself encompasses approximately one-third of the total home range.

Figure 4.5 shows the core and noncore areas as well as their estimated L functions. The simulation envelopes based on 1000 CSR point processes fully encompass the estimated L functions for the core and noncore areas (Figure 4.5) indicating a lack of clustering or regularity in each of the partitions of the home range. Thus, following the guidance of Wilson et al. (2010), we conclude that the estimated region in Figure 4.4 is sufficient for delineating the core area of space use for our example mountain lion. Had there been evidence of significant clustering in either the core or noncore areas, we would fit the core area model using two partitions, which would result in three areas of distinct space use intensity.

The advantages to this sequential approach of model fitting and model checking is that the assumptions of the model can be verified during the procedure. The drawback

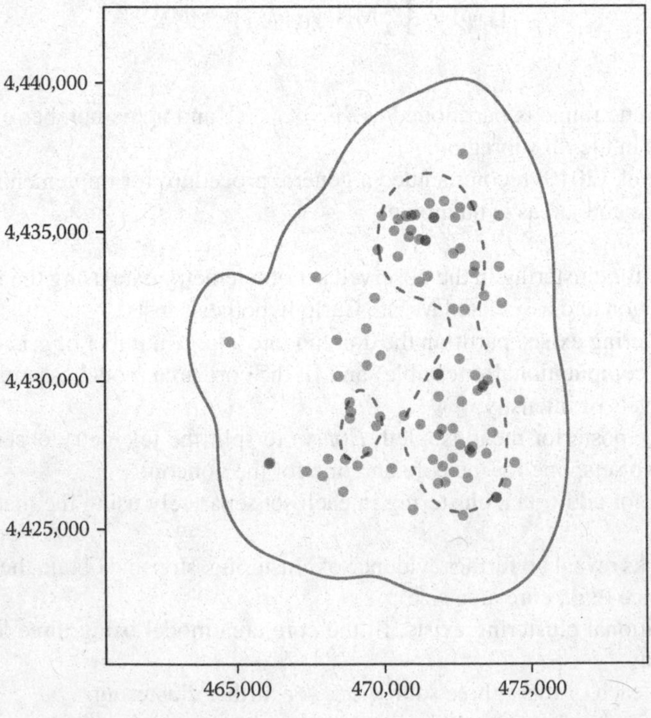

FIGURE 4.4 Mountain lion telemetry locations (points), home range (dark line), and estimated core area delineation (dashed line).

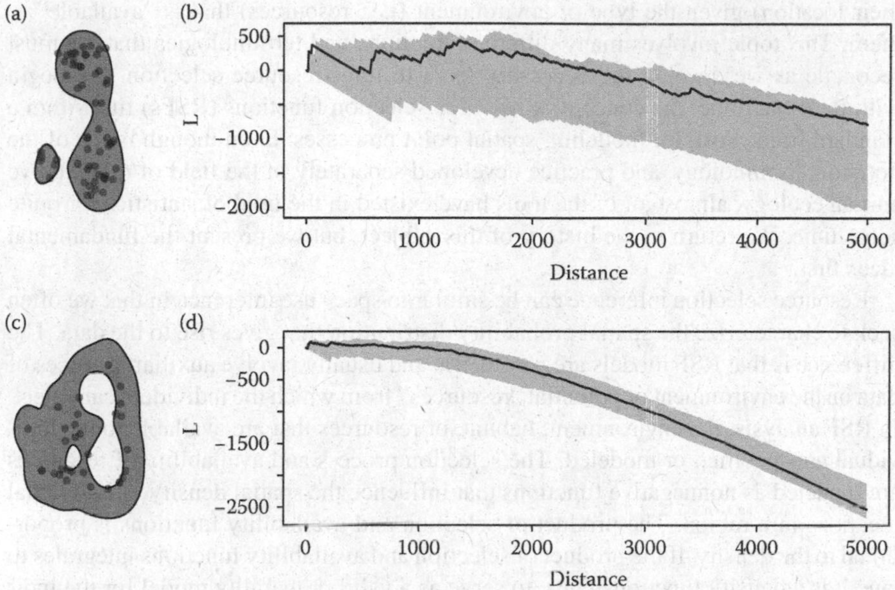

FIGURE 4.5 Mountain lion telemetry locations (points) and home range (dark line). The estimated (a) core area and (c) noncore area are shown as gray regions. The estimated L functions (dark line) and simulation envelopes (gray regions) are shown for the (b) core and (d) noncore areas.

of the approach is that it requires supervision.* However, future extensions of these methods might include more automatic procedures in which the optimal number of levels of core areas could be estimated simultaneously with the isopleths and associated intensities. Another potential drawback is that this type of core area model was designed for use with temporally independent telemetry data. The model would need to be generalized further to properly account for temporal autocorrelation due to the movement process itself. See the latter sections in this chapter for details on how to account for temporal dependence in point process models.

The core area estimation approach described in this section provides a way to optimally partition the home range into discrete regions depending on their intensity of use. However, differential use of space within a home range may depend on various ecological and environmental characteristics. Resource selection models allow us to generalize the core area estimation concept to include potential mechanistic variables that affect space use.

4.2 RESOURCE SELECTION FUNCTIONS

Studies focused on resource selection differ from those interested in space use in that they seek inference concerning the choices that individuals make (as evidenced by

* As with any responsible application of statistics.

their location) given the type of environment (i.e., resources) that is "available" to them. This topic involves many different notations and terminologies that we must reconcile as we develop the necessary tools to infer resource selection. We begin with the punch line: the concept of resource selection functions (RSFs) fits within a standard framework for modeling spatial point processes. Even though much of the notation, terminology, and practice developed separately in the field of quantitative animal ecology, almost all of the tools have existed in the field of statistics for quite some time. We return to the history of this subject, but we present the fundamental ideas first.

Resource selection inference can be similar to space use inference in that we often seek to characterize the spatial probability distribution that gives rise to the data. The difference is that RSF models are parametric and usually involve auxiliary sources of data on the environment or potential "resources" from which the individual can select. In RSF analysis, the environment, habitat, or resources that are available to the individual are specified or modeled. The selection process and availability of resources are modeled as nonnegative functions that influence the spatial density of individual locations in a region. The product of selection and availability functions is proportional to the density. If the product of selection and availability functions integrates to one, it is a density function. Thus, to serve as a valid probability model for the individual locations as a point process, the product of selection and availability functions must be normalized so that it is a proper density function over space.

We describe the RSF model from a somewhat unconventional perspective in wildlife ecology so as to remain consistent with the standard statistical view of a point process model. In doing so, we treat the spatial location $\boldsymbol{\mu}_i$ as the random quantity of interest for which we specify a PDF. The traditional approach in the wildlife ecological literature treats the environment or resources (i.e., $\mathbf{x}(\boldsymbol{\mu}_i)$) as the modeled quantity. Both perspectives are correct in that they are designed to model a point process. In the recent literature, you will see both formulations. We treat the spatial location $\boldsymbol{\mu}_i$ as the point, whereas some other descriptions will treat the set of environmental conditions $\mathbf{x}(\boldsymbol{\mu}_i)$ as the point. We model the spatial location directly because it allows us to generalize the model to accommodate more complicated situations.

Consider the weighted distribution formulation of a point process model for independent individual locations $\boldsymbol{\mu}_i \sim [\boldsymbol{\mu}_i|\boldsymbol{\beta},\boldsymbol{\theta}]$ such that

$$[\boldsymbol{\mu}_i|\boldsymbol{\beta},\boldsymbol{\theta}] \equiv \frac{g(\mathbf{x}(\boldsymbol{\mu}_i),\boldsymbol{\beta})f(\boldsymbol{\mu}_i,\boldsymbol{\theta})}{\int g(\mathbf{x}(\boldsymbol{\mu}),\boldsymbol{\beta})f(\boldsymbol{\mu},\boldsymbol{\theta})d\boldsymbol{\mu}}, \tag{4.6}$$

where the selection function g depends on $\boldsymbol{\beta}$, the selection coefficients. The availability (i.e., f) depends on $\boldsymbol{\theta}$, the availability coefficients. Furthermore, the denominator in Equation 4.6 is necessary so that the entire PDF $[\boldsymbol{\mu}_i|\boldsymbol{\beta},\boldsymbol{\theta}]$ integrates to one over the support of the point process. The RSF model in Equation 4.6 provides a useful example of how we can construct PDFs from scratch for nearly any type of data or process.

In principle, any positive functions can be used for availability (f) and selection (g). However, in basic resource selection studies, the availability function is taken to be the uniform PDF on the support of the point process ($\mathcal{M}$). For such uniform availability specifications, the interpretation is that the individual can occur anywhere in the support $\mathcal{M}$ with equal probability, and thus, the availability coefficients, $\boldsymbol{\theta}$, disappear from the model and the focus shifts toward the selection coefficients $\boldsymbol{\beta}$. Johnson (1980) introduced a natural ordering of four scales for resource selection inference that ecologists may be interested in:

1. First-order selection: The extent of the species distribution.
2. Second-order selection: The home range of an individual or natural group of individuals.
3. Third-order selection: Sub-home-range features (e.g., habitat types within a home range).
4. Fourth-order selection: Micromovement and behavior (e.g., acquisition of food, mating, nest building).

The concept for scales of selection inference proposed by Johnson (1980) are commonly referred to and allow the researcher to define the support $\mathcal{M}$ based on their goals for inference.

The selection function g can assume any positive form; however, two forms are most popular: the exponential and logistic functions. The exponential selection function can be expressed as $g(\mathbf{x}(\boldsymbol{\mu}_i), \boldsymbol{\beta}) \equiv \exp(\mathbf{x}'(\boldsymbol{\mu}_i)\boldsymbol{\beta})$, whereas the logistic selection function takes the form of a probability

$$g(\mathbf{x}(\boldsymbol{\mu}_i), \boldsymbol{\beta}) \equiv \frac{\exp(\mathbf{x}'(\boldsymbol{\mu}_i)\boldsymbol{\beta})}{1 + \exp(\mathbf{x}'(\boldsymbol{\mu}_i)\boldsymbol{\beta})}. \tag{4.7}$$

The $\mathbf{x}'(\boldsymbol{\mu}_i)\boldsymbol{\beta}$ term in Equation 4.7 resembles the mean function in linear regression and the forms of selection as link functions commonly used in generalized linear modeling with Poisson and Bernoulli likelihoods.* In most GLMs, the value of one is included as the first covariate in $\mathbf{x}$ so that the first element of $\boldsymbol{\beta}$ (i.e., β_0) acts as an intercept in the model. However, if we use an intercept in the exponential selection function, it will cancel in the numerator and denominator of Equation 4.6. Thus, an intercept is not included in RSF models that rely on Equation 4.6 directly when the selection function is exponential.

The main difference in the resulting inference from the two common forms of selection functions is that the logit form† (4.7) allows for inference directly on the probability of selection, whereas the exponential form limits inference to the relative intensity of selection. However, even in this case, inference concerning the direction and magnitude of environmental effects on selection can still be obtained directly by learning about $\boldsymbol{\beta}$. Thus, despite this apparent shortcoming, most resource selection studies still rely on the exponential form for the model because of tradition and ease of implementation. Under uniform availability, the RSF with exponential selection

* Recall the Bernoulli distribution is a binomial distribution with one trial.

† Sometimes referred to as a resource selection probability function (RSPF) (Lele and Keim 2006).

function is

$$[\boldsymbol{\mu}_i|\boldsymbol{\beta}] \equiv \frac{\exp(\mathbf{x}'(\boldsymbol{\mu}_i)\boldsymbol{\beta})}{\int \exp(\mathbf{x}'(\boldsymbol{\mu})\boldsymbol{\beta})d\boldsymbol{\mu}}. \tag{4.8}$$

Notice the similarity of this resource selection model (4.8) with that of the heterogeneous spatial point process model (2.8) described in Section 2.1. Thus, to form a likelihood under the assumption of conditional independence* for the points (i.e., $\boldsymbol{\mu}_i$, for $i = 1, \ldots, n$), we take the product of Equation 4.8 over the n individual locations

$$\prod_{i=1}^{n} \frac{\exp(\mathbf{x}'(\boldsymbol{\mu}_i)\boldsymbol{\beta})}{\int \exp(\mathbf{x}'(\boldsymbol{\mu})\boldsymbol{\beta})d\boldsymbol{\mu}}. \tag{4.9}$$

An MLE for $\boldsymbol{\beta}$ is obtained by maximizing Equation 4.9 with respect to $\boldsymbol{\beta}$. Likewise, in a Bayesian setting, we specify a prior for $\boldsymbol{\beta}$ and find the posterior distribution $[\boldsymbol{\beta}|\boldsymbol{\mu}_1, \ldots, \boldsymbol{\mu}_n]$. In either case, we need to evaluate the integral in the denominator of Equation 4.9 explicitly because it involves $\boldsymbol{\beta}$. This integral is typically multivariate† and not analytically tractable. Thus, unlike most parametric statistical models used in maximum likelihood or Bayesian analyses, we cannot actually evaluate the likelihood directly and a numerical approach is necessary.‡

4.2.1 Implementation of RSF Models

Mystery and misunderstanding surrounds the proper implementation of RSF models. There are at least three main methods for fitting RSF models (Warton and Shepherd 2010; Aarts et al. 2012). Perhaps the most common implementation of RSF models takes the form of logistic regression. This is followed closely by Poisson regression and then resource utilization function (RUF) approaches (which we discuss in the next section).

We begin by describing the logistic regression approach to fitting RSF models. The main idea in the logistic regression approach is to convert the individual spatial locations into a binary data set where the observed locations are represented by ones and the available locations are represented by zeros. That is, a background sample is typically taken from the availability distribution f in such a way that it represents a large but finite set of possible locations the individual could have occupied. The environmental covariates ($\mathbf{x}_i$) at the individual locations ($\boldsymbol{\mu}_i$) are associated with each of the ones for $i = 1, \ldots, n$, and similarly, the covariates are recorded for the background sample of $m - n$ locations. The response variable is then specified as $\mathbf{y} \equiv (1, \ldots, 1, 0, \ldots, 0)'$ and used in a standard logistic regression with the complete set of covariates. That is, the model becomes $y_i \sim \text{Bern}(p_i)$, where $\text{logit}(p_i) = \beta_0 + \mathbf{x}_i'\boldsymbol{\beta}$, for $i = 1, \ldots, m$ total binary observations.

---

* Independent, given other features of the model (e.g., $\boldsymbol{\beta}$).

† The integral has the same dimension as the points (i.e., $\boldsymbol{\mu}$). In most cases, the dimension of $\boldsymbol{\mu}$ is two.

‡ Compare with standard PMFs available in R such as "dpois()" and "dbinom()."

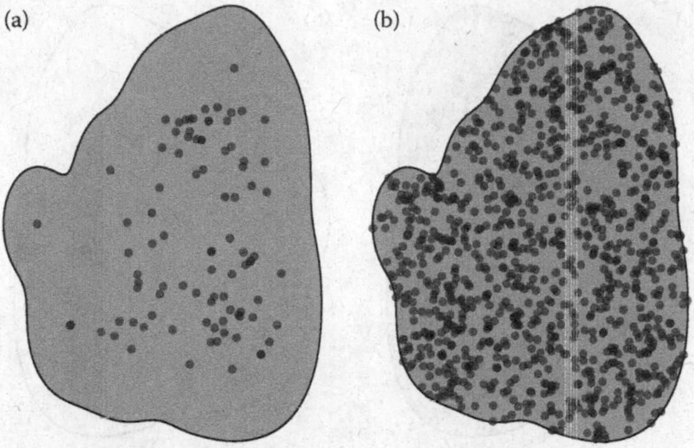

FIGURE 4.6 (a) Mountain lion telemetry locations (points) and home range (gray region). (b) Background sample based on 1000 samples from a CSR point process (points) and home range (gray region).

Figure 4.6a shows the original point process for the mountain lion data and estimated home range based on a 99% KDE isopleth. Figure 4.6b shows a background sample of size 1000 based on a CSR point process within the estimated home range.

In the Poisson regression approach to fitting RSF models, the support of the individual locations $\mathcal{M}$ is gridded up into L areal units (i.e., grid cells or pixels). The covariates ($\mathbf{x}_l$ for $l = 1, \ldots, L$, for large L) are associated with each grid cell and the individual locations ($\boldsymbol{\mu}$) are counted in each grid cell and recorded in an $L \times 1$ vector of cell frequencies $\mathbf{z}$. The model becomes $z_l \sim \text{Pois}(\lambda_l)$, where $\log(\lambda_l) = \beta_0 + \mathbf{x}_l'\boldsymbol{\beta}$ for all grid cell counts $l = 1, \ldots, L$. The intercept β_0 is related to grid cell size,* but usually ignored, and only the estimates of $\boldsymbol{\beta}$ are used for resource selection inference.

Figure 4.7a shows the original point process for the mountain lion data and estimated home range based on a 99% KDE isopleth. Figure 4.7b shows the gridded point process with counts within 1 km grid cells in the estimated home range.

The striking result is that both the logistic and Poisson regression approaches to implementing the RSF model yield the same inference about $\boldsymbol{\beta}$, under conditions we explain in what follows (Warton and Shepherd 2010; Aarts et al. 2012). In the mountain lion example, we fit the spatial point process model using both Poisson and logistic regression to yield inference for the selection coefficients $\boldsymbol{\beta}$ based on the standardized covariates in Figure 4.8. The results indicate that both the logistic and Poisson regression approach to fitting the point process model are very similar (Table 4.1). Furthermore, resource selection inference resulting from the model fits indicates that this mountain lion is selecting for lower elevations and steeper slopes relative to available terrain; there is no evidence of selection for exposure given the other covariates in the model.

The necessary conditions for equivalence in the inference are that the background sample used in the logistic regression and the number of grid cells used in the Poisson

* β_0 increases as grid cell size increases.

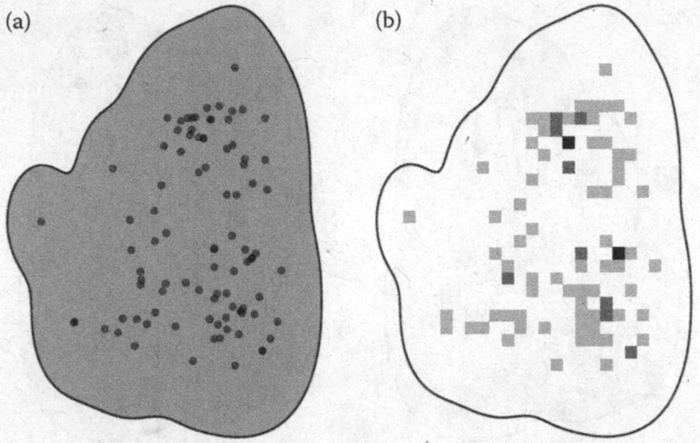

FIGURE 4.7 (a) Mountain lion telemetry locations (points) and home range (gray region). (b) Gridded cell counts of the mountain lion telemetry locations within the home range (light gray = 1, medium gray = 2, and dark gray = 3 points in the pixel).

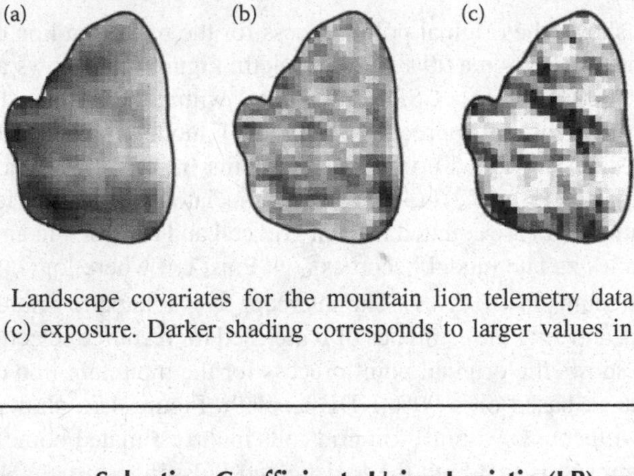

FIGURE 4.8 Landscape covariates for the mountain lion telemetry data: (a) elevation, (b) slope, and (c) exposure. Darker shading corresponds to larger values in the covariates.

TABLE 4.1
Estimated Resource Selection Coefficients Using Logistic (LR) and Poisson Regression (PR) Based on the Covariates in Figure 4.8

	LR		**PR**	
	Estimate	**p-Value**	**Estimate**	**p-Value**
(a) Elevation	−0.311	0.008	−0.295	0.008
(b) Slope	0.244	0.033	0.215	0.035
(c) Exposure	−0.139	0.265	−0.131	0.286

regression often need to be very large, sometimes on the order of tens of thousands (Northrup et al. 2013). Perhaps the most common mistake made in most classical RSF implementations is that a large-enough background sample is not used. The easiest way to check whether a large-enough sample is chosen is to try larger background sample sizes until the inference stabilizes. The reason that a large background sample or number of grid cells is so important is that it is implicitly sidestepping the integral in the denominator of the RSF model (4.8) for us. However, when the covariate information is only available at a certain resolution, then additional grid cells at resolutions smaller than the covariates in the Poisson regression do not improve the approximation.

We could choose to numerically compute the integral and then use maximum likelihood or Bayesian methods, but the easy implementation in statistical software using a GLM seems much more straightforward for most ecologists. There can be some advantages to the explicit integration approach however, and we discuss these in the next section.

4.2.2 Efficient Computation of RSF Integrals

Two main advantages can be achieved by evaluating the point process likelihood (4.9) directly when fitting RSF models:

1. Computational efficiency can be improved.
2. The model formulation can be readily extended.

We discuss the first item here and return to the topic of model extensions in Section 4.4.

One approach for improving computational performance in the evaluation of a point process likelihood (4.9) is to find efficient approximations for the integral in the denominator. The function $\exp(\mathbf{x}'(\boldsymbol{\mu})\boldsymbol{\beta})$ will inevitably be quite complicated;* thus, the integral of the function with respect to $\boldsymbol{\mu}$ will be analytically intractable. However, a fairly simple approximation to the integral $\int \exp(\mathbf{x}'(\boldsymbol{\mu})\boldsymbol{\beta})d\boldsymbol{\mu}$ can be found using numerical quadrature. That is, break up the support of the point process ($\mathcal{M}$) into a large number of equally sized grid cells (as in the Poisson regression procedure described above) and then evaluate $\exp(\mathbf{x}'(\boldsymbol{\mu}_l)\boldsymbol{\beta})$ for each grid cell (assuming that $\boldsymbol{\mu}_l$ is the grid cell center for cell l). We then multiply by the area of a grid cell (a) and sum to obtain the integral approximation. Thus, we approximate the integral with

$$\int \exp(\mathbf{x}'(\boldsymbol{\mu})\boldsymbol{\beta})d\boldsymbol{\mu} \approx a \sum_{l=1}^{L} \exp(\mathbf{x}'(\boldsymbol{\mu}_l)\boldsymbol{\beta}) \tag{4.10}$$

for grid cells $l = 1, \ldots, L$.

For a grid consisting of thousands of cells (e.g., 100 × 100 grid), the quadrature approximation (4.10) will be sufficiently fast for most purposes; however, for grids on the order of 10^6 grid cells (e.g., 1000 × 1000 grid) and larger, the large sums could be a bottleneck for computation. The bottleneck is magnified when such approximations need to be computed in an iterative algorithm like MCMC. Still, for independent individual locations $\boldsymbol{\mu}_i$, the integral only needs to be approximated once

* Owing to the combined complexity of the covariates.

on each MCMC iteration. Therefore, if the integral takes 0.1 s to approximate once and you need 100,000 MCMC iterations, then your algorithm will never take less than about 2.8 h to fit the model (assuming all other required calculations are negligible). If we could speed up the integral approximation by an order of magnitude, we could reduce the total required computational time down to 16 min, which is reasonable for contemporary statistical model fitting.

In what follows, we describe two approaches for approximating the required RSF integral that can be made as accurate as quadrature without increasing computational time. If a minimal amount of additional approximation error is acceptable, these approaches can offer an order of magnitude faster integral approximation.

To simplify our presentation of the integral approximation techniques, we first reduce the complexity of the required integral through orthogonalization and integration. Most statistical algorithms are iterative and we typically need to evaluate the RSF integral when optimizing or sampling the RSF coefficients. Thus, if we can reduce the problem to one where a single coefficient is dealt with at a time, then a few good tricks for reducing computational burden become apparent.

For computing purposes only, we begin by transforming the environmental variables $\mathbf{X}$ to create a new set of orthogonal covariates $\tilde{\mathbf{X}} = \mathbf{XV}$. To perform this transformation, we acquire $\mathbf{V}$ from the singular value decomposition of the matrix of environmental variables $\mathbf{X} = \mathbf{UDV}'$ such that $\mathbf{U}$ and $\mathbf{V}$ are the left and right singular vectors, respectively, and $\mathbf{D}$ is a diagonal matrix with the singular values on the diagonal. The orthogonalization allows us to perform a type of principal components regression because we can always express the selection function as $\exp(\mathbf{X}\boldsymbol{\beta}) = \exp(\tilde{\mathbf{X}}\tilde{\boldsymbol{\beta}})$, where $\tilde{\boldsymbol{\beta}}$ are a set of selection coefficients associated with the transformed covariates. These new covariates are linear combinations of the original covariates but can be difficult to interpret, although they can be easily visualized as spatial maps. For example, the principal component scores resulting from the orthogonalization of the original mountain lion covariates are shown in Figure 4.9.

The advantages of using this orthogonal covariate transformation are manifold. It would appear that we lose the ability to interpret the selection coefficients; however, we can always recover them with the inverse transformation $\boldsymbol{\beta} = \mathbf{V}\tilde{\boldsymbol{\beta}}$. Moreover, the orthogonalization results in a much more stable computational algorithm if the original covariates were correlated (i.e., multicollinear). Finally, in situations where

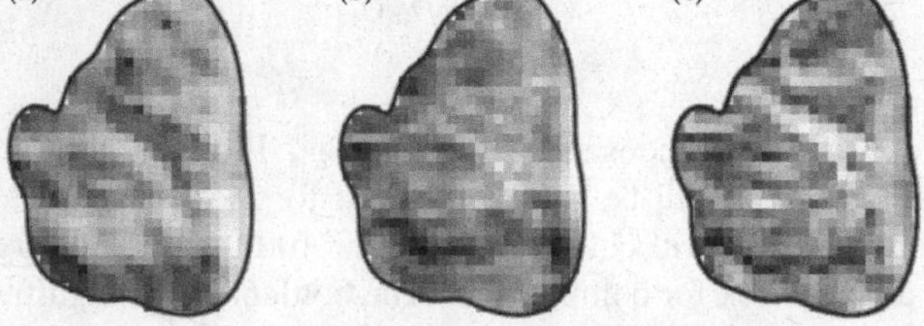

FIGURE 4.9 Principal components of landscape covariates for the mountain lion telemetry data: (a) component 1 (41% of variation), (b) component 2 (37% of variation), and (c) component 3 (22% of variation).

there are many original environmental variables, we can reduce the dimension of the orthogonal covariate set by retaining only the first q columns of $\mathbf{V}$ when calculating $\tilde{\mathbf{X}}$. This approximation is a common technique used in spatial statistics, but will only be worthwhile for large sets of covariates (i.e., more than 10).

The orthogonalization affords us one final benefit. It allows us to construct a conditional algorithm where we only sample (or optimize) a single coefficient $\tilde{\beta}_j$ at a time. Normally, in an optimization (i.e., Newton–Raphson) or sampling algorithm (e.g., MCMC), we would prefer to handle a set of regression coefficients jointly; however, in this case, it will pay off to deal with them one at a time.

In an effort to make this point as clearly as possible, we focus on a Bayesian RSF model using an MCMC algorithm. In this case, we seek the posterior distribution of $[\tilde{\boldsymbol{\beta}}|\boldsymbol{\mu}_1,\ldots,\boldsymbol{\mu}_n]$. Using MCMC, we can sample from full-conditional distributions for each of the coefficients sequentially to fit the model. That is, we need to be able to efficiently sample each coefficient $\tilde{\beta}_j$ from the full-conditional distribution

$$[\tilde{\beta}_j|\cdot] \propto [\tilde{\beta}_j]\prod_{i=1}^{n}\frac{\exp(\tilde{\mathbf{x}}'(\boldsymbol{\mu}_i)\tilde{\boldsymbol{\beta}})}{\int \exp(\tilde{\mathbf{x}}'(\boldsymbol{\mu})\tilde{\boldsymbol{\beta}}d\boldsymbol{\mu})}, \tag{4.11}$$

where the initial term $[\tilde{\beta}_j]$ is the marginal prior distribution* for $\tilde{\beta}_j$ and the integral in the denominator needs to be approximated (because it involves $\tilde{\beta}_j$). In working with Equation 4.11, we can expand the exponential such that

$$\begin{aligned}\exp(\tilde{\mathbf{x}}'(\boldsymbol{\mu}_i)\tilde{\boldsymbol{\beta}}) &= \exp\left(\sum_{\forall j}\tilde{x}_j(\boldsymbol{\mu}_i)\tilde{\beta}_j\right)\\ &= \exp(\tilde{x}_j(\boldsymbol{\mu}_i)\tilde{\beta}_j)\exp\left(\sum_{\forall l\neq j}\tilde{x}_l(\boldsymbol{\mu}_i)\tilde{\beta}_l\right).\end{aligned} \tag{4.12}$$

This nicely isolates the jth effect $\tilde{x}_j(\boldsymbol{\mu}_i)\tilde{\beta}_j$ from the rest, allowing us to expand terms in the full-conditional distribution from Equation 4.11 to

$$\begin{aligned}[\tilde{\beta}_j|\cdot] &\propto [\tilde{\beta}_j]\prod_{i=1}^{n}\frac{\exp(\tilde{\mathbf{x}}'(\boldsymbol{\mu}_i)\tilde{\boldsymbol{\beta}})}{\int \exp(\tilde{\mathbf{x}}'(\boldsymbol{\mu})\tilde{\boldsymbol{\beta}})d\boldsymbol{\mu}}\\ &\propto \frac{[\tilde{\beta}_j]\prod_{i=1}^{n}\exp(\tilde{\mathbf{x}}'(\boldsymbol{\mu}_i)\tilde{\boldsymbol{\beta}})}{\left(\int \exp(\tilde{\mathbf{x}}'(\boldsymbol{\mu})\tilde{\boldsymbol{\beta}})d\boldsymbol{\mu}\right)^n},\end{aligned}$$

* For example, $[\tilde{\beta}_j] \equiv \mathrm{N}(0,\sigma_\beta^2)$.

$$\propto \frac{[\tilde{\beta}_j]\prod_{i=1}^n \exp(\tilde{x}_j(\boldsymbol{\mu}_i)\tilde{\beta}_j)\prod_{i=1}^n \exp\left(\sum_{\forall l\neq j}\tilde{x}_l(\boldsymbol{\mu}_i)\tilde{\beta}_l\right)}{\left(\int\prod_{\forall j}\exp\left(\tilde{x}_j(\boldsymbol{\mu})\tilde{\beta}_j\right)d\boldsymbol{\mu}\right)^n}$$

$$\propto \frac{[\tilde{\beta}_j]\prod_{i=1}^n \exp(\tilde{x}_j(\boldsymbol{\mu}_i)\tilde{\beta}_j)}{\left(\int\prod_{\forall j}\exp\left(\tilde{x}_j(\boldsymbol{\mu})\tilde{\beta}_j\right)d\boldsymbol{\mu}\right)^n}, \tag{4.13}$$

where the last product in the numerator drops out in the proportionality in Equation 4.13 because it does not involve $\tilde{\beta}_j$. To simplify the integral in the denominator of Equation 4.13, we change the integral so that it integrates over the covariates ($\tilde{x}_j$) rather than the spatial locations ($\boldsymbol{\mu}$).* The full-conditional distribution for $\tilde{\beta}_j$ then becomes

$$[\tilde{\beta}_j|\cdot] \propto \frac{[\tilde{\beta}_j]\prod_{i=1}^n \exp(\tilde{x}_j(\boldsymbol{\mu}_i)\tilde{\beta}_j)}{\left(\int\prod_{\forall j}\exp\left(\tilde{x}_j(\boldsymbol{\mu})\tilde{\beta}_j\right)d\boldsymbol{\mu}\right)^n}$$

$$\propto \frac{[\tilde{\beta}_j]\prod_{i=1}^n \exp(\tilde{x}_j(\boldsymbol{\mu}_i)\tilde{\beta}_j)}{\left(\prod_{\forall j}\int\exp\left(\tilde{x}_j\tilde{\beta}_j\right)[\tilde{x}_j]d\tilde{x}_j\right)^n}$$

$$\propto \frac{[\tilde{\beta}_j]\prod_{i=1}^n \exp(\tilde{x}_j(\boldsymbol{\mu}_i)\tilde{\beta}_j)}{\left(\int\exp\left(\tilde{x}_j\tilde{\beta}_j\right)[\tilde{x}_j]d\tilde{x}_j\prod_{\forall l\neq j}\int\exp\left(\tilde{x}_l\tilde{\beta}_l\right)[\tilde{x}_l]d\tilde{x}_l\right)^n}$$

$$\propto \frac{[\tilde{\beta}_j]\prod_{i=1}^n \exp(\tilde{x}_j(\boldsymbol{\mu}_i)\tilde{\beta}_j)}{\left(\int\exp\left(\tilde{x}_j\tilde{\beta}_j\right)[\tilde{x}_j]d\tilde{x}_j\right)^n}, \tag{4.14}$$

where the product over the integrals in the denominator is valid when the covariates are independent. The covariates $\tilde{x}_j$ should be independent if they are normally distributed because they have been orthogonalized.† This new parameterization does not immediately seem to help. However, because we can approximate the integral with a sum (i.e., the quadrature concept discussed earlier), in certain circumstances, components of the sum can be precalculated, which increases computational efficiency. To illustrate this idea, consider a summation approximation of the required integral

$$\int\exp\left(\tilde{x}_j\tilde{\beta}_j\right)[\tilde{x}_j]d\tilde{x}_j \approx a\sum_{l=1}^{L}\exp\left(\tilde{x}_j^{(l)}\tilde{\beta}_j\right)[\tilde{x}_j^{(l)}], \tag{4.15}$$

* Recall that $\int\exp\left(\tilde{x}(\boldsymbol{\mu})\tilde{\beta}\right)d\boldsymbol{\mu} = \int\exp\left(\tilde{x}\tilde{\beta}\right)[\tilde{x}]d\tilde{x}$, where $[\tilde{x}]$ is the distribution of the covariate implied by a uniform distribution on $\boldsymbol{\mu}$. Also, keep in mind that our initial integral over $\boldsymbol{\mu}$ is multivariate; we just use the single integral notation to simplify things.

† Independence and orthogonality are equivalent for normally distributed random variables, but that is not true for all random variables. Therefore, this technique requires potentially strong assumptions.

where "a" is the area of quadrature grid cell as before and L is the number of total cells. Now suppose that a discretization can be found for the variable $\tilde{x}_j$ such that it falls in one of C classes. Also suppose that the loss in precision due to the discretization can be decreased with a larger number of classes. Then, we replace the quadrature sum in Equation 4.15 with a different sum involving the classes as

$$\int \exp\left(\tilde{x}_j\tilde{\beta}_j\right)[\tilde{x}_j]d\tilde{x}_j \approx a\sum_{l=1}^{L}\exp\left(\tilde{x}_j^{(l)}\tilde{\beta}_j\right)[\tilde{x}_j^{(l)}]$$

$$\approx a\sum_{c=1}^{C} n_c \exp\left(\tilde{x}_j^{c}\tilde{\beta}_j\right)[\tilde{x}_j^{c}], \tag{4.16}$$

where n_c corresponds to the number of cells containing that particular class for the covariate. This reduces the sum from a potentially very large L dimension down to a much smaller C dimension because n_c can be precalculated after the optimal discretization into classes is performed.

Many methods exist for finding an optimal (i.e., minimal loss) discretization for the covariates. Perhaps the simplest approach is to cluster each covariate using a K-means approach (or other clustering algorithm) prior to model fitting to determine the classes. In our experience, this type of preclustering can speed up fitting algorithms by an order of magnitude or more depending on the complexity of the covariate.

4.3 RESOURCE UTILIZATION FUNCTIONS

Marzluff et al. (2004) introduced the RUF as an alternative way to study the use of space and resources by individual animals. The essential idea underpinning RUF analysis is to first estimate the density or intensity of space use over the geographic domain of interest (i.e., typically the relevant home range or study area) and second to link that resulting spatial map to a set of spatially explicit covariates in a regression model. The concept is similar to that used in RSF analyses, but rather than modeling the points directly as a response variable, the RUF approach uses a two-stage analysis that ultimately treats the estimated density or intensity as the response variable.

We begin by describing step 1 of the typical RUF analysis. For a set of individual locations $\boldsymbol{\mu}_1, \ldots, \boldsymbol{\mu}_n$, estimate the density (or intensity, if unnormalized) of the point process at a grid of regular locations, $\mathbf{c}_1, \ldots, \mathbf{c}_m$, similar to that described in the previous sections. KDE methods such as those described in Chapter 2 and at the beginning of this chapter could be used to obtain these estimates $\hat{f}(\mathbf{c}_j)$. The second step in an RUF analysis traditionally involves fitting a linear model to the estimated density given a set of covariates on the regular grid such that

$$\hat{f}(\mathbf{c}_j) = \mathbf{x}_j'\boldsymbol{\beta} + \eta_j + \varepsilon_j, \tag{4.17}$$

for $j = 1, \ldots, m$, and where the error terms are normally distributed. The errors, ε_j, are assumed to be independent and identically Gaussian, whereas η_j are allowed to be spatially correlated such that $\boldsymbol{\eta} \sim \text{N}(\mathbf{0}, \boldsymbol{\Sigma})$. The covariance matrix is typically parameterized assuming a continuous spatial process. For example, the elements of the covariance matrix $\boldsymbol{\Sigma}$ are often defined as $\Sigma_{jl} \equiv \exp(-d_{jl}/\phi)$, where d_{jl} is the Euclidean distance between cell locations $\mathbf{c}_j$ and $\mathbf{c}_l$ in the grid, and ϕ controls the range of spatial dependence as described in Chapter 2.

The advantages of the basic RUF concept are that it is intuitive and straightforward to implement. It is intuitive because it attempts to link the estimated density (or the UD) associated with animal relocations to the environment in a regression framework. It is straightforward because the actual implementation only requires two lines of computer code: one to estimate $\hat{\mathbf{f}}$ and another to fit the regression model. At the time of the development of the RUF concept (i.e., early 2000s), it was especially attractive as compared to RSF fitting procedures, and seems like it should yield similar inference. At present, it is now widely known that RSFs can be implemented with only one line of computer code (after the initial preprocessing of the data). That is, to fit an RSF model, only software for fitting a GLM is required, either Bernoulli (i.e., using the background sample approach) or Poisson regression (i.e., the count modeling approach). Thus, any computational advantages to the RUF may be moot at this point.

However, the traditional RUF approach highlights some important issues that we describe further. Hooten et al. (2013b) compared and contrasted the RUF and RSF procedures in an attempt to reconcile the inference they provide. Among other things, they found two key differences that we describe in what follows: the "support" for the response variable in the model and the relationship between first- and second-order (i.e., mean and variance) components of the RUF model.

Recall that when we use the word "support" we are talking about the values that a certain variable can assume. In the case of the RUF model, the approach described by Marzluff et al. (2004), and later Millspaugh et al. (2006), links the estimated density ($\hat{\mathbf{f}}$ or UD) directly to the covariates without transformation. Thus, the RUF model implies an identity link function (i.e., no transformation). On the other hand, if we consider the Poisson regression approach to fitting the RSF model, it is customary to use the log link function such that $\log(\lambda(\mathbf{c}_j)) = \mathbf{x}_j'\boldsymbol{\beta}$. As the grid cell area approaches zero, the intensity function λ is proportional to the density function f. Thus, the log of λ plus a constant is equivalent to the log of f, which implies that we may want to use the log of the estimated density surface (or UD) as the response variable in the RUF model such that

$$\log(\hat{f}(\mathbf{c}_j)) = \mathbf{x}_j'\boldsymbol{\beta} + \eta_j + \varepsilon_j. \tag{4.18}$$

We refer to this new model as a modified RUF model. The modified RUF (4.18) more closely mimics the RSF and provides more similar inference.

The second issue concerning the RUF pertains to the second-order spatial structure of the model ($\boldsymbol{\eta}$). In the early development of the RUF concept, Marzluff et al. (2004) noticed that the first-stage density estimation procedure induced a form of spatial autocorrelation that was exogenous to the resource selection process. Therefore, a model checking effort for the standard regression form of RUF (i.e., without spatially autocorrelated random effects) indicates spatial dependence in the residuals. In fact, based on our own experience with these models, an almost "textbook" empirical

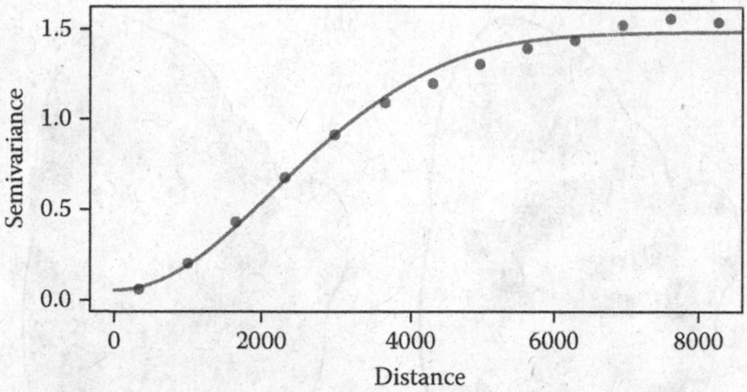

FIGURE 4.10 Empirical (points) and fitted (line) semivariogram based on the residuals of regressing the log estimated UD on the exposure covariate for the mountain lion data. The fitted semivariogram is based on a Gaussian model for covariance.

variogram for the residuals often results. For example, Figure 4.10 shows the empirical and fitted semivariogram for the residuals when regressing the log estimated UD on the exposure covariate for the mountain lion data. The smoothly increasing semivariance with distance is due to the UD estimation using KDE. Therefore, because it is *ad hoc* to use a model for inference when evidence for a lack of fit is present.* Marzluff et al. (2004) suggested adding a correlated random effect to the model, as is typically done in spatial statistics. The result is a model with the same basic form as the right-hand side of Equation 4.18. This is an excellent model for spatial prediction, but prediction is not the goal of RUF analysis. Instead, we seek to learn about the regression coefficients $\boldsymbol{\beta}$ as surrogates for selection coefficients in a point process RSF model. Inference concerning $\boldsymbol{\beta}$ requires the covariates $\mathbf{X}$ to be linearly independent of the random effect $\boldsymbol{\eta}$. Evaluating the collinearity assumption is not trivial, and thus, it is not often checked.

A number of recent studies have shown that the inference for $\boldsymbol{\beta}$ can be affected if collinearity exists among the fixed and random effects (e.g., Hodges and Reich 2010; Hughes and Haran 2013; Hanks et al. 2015b). For example, in the mountain lion application, the exposure covariate is negatively correlated ($\rho = -0.29$) with the eleventh eigenvector of the estimated covariance matrix (Figure 4.11). Recall the discussion of spatial confounding in the context of general spatial statistics in Chapter 2. Spatial confounding can also occur in RUF models because, when fitting RUF models with and without the spatial random effect, one arrives at potentially different inference. In an attempt to help alleviate the problem, Hooten et al. (2013b) evaluated restricted spatial regression (RSR) for RUF models and found that it can yield improved inference in some cases (e.g., less biased estimates of coefficients). As a reminder, the idea with RSR is to force the random effect to be orthogonal to the fixed effects. The use of RSR is only warranted when the first-order effects (i.e., $\mathbf{x}_j'\boldsymbol{\beta}$) take precedence over the second-order effects (i.e., η_j). Using RSR for inference can

* In this case, due to lack of residual independence.

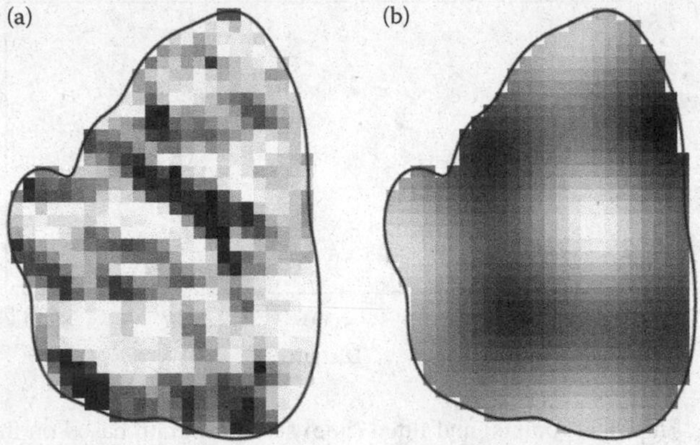

FIGURE 4.11 Exposure covariate (a) and eleventh eigenvector (b) of $\hat{\boldsymbol{\Sigma}}$ based on the fitted semivariogram using a Gaussian model for covariance. The correlation for the covariate and eigenvector was approximately 0.3.

be detrimental in cases when the covariates are collinear with a true additive random effect in reality (Hanks et al. 2015b). Thus, caution must be exercised in specifying and fitting RUF models.

A final potential issue with using RUF inference in lieu of RSF inference relates to model misspecification (i.e., an incorrect formulation of the model for the desired type of inference). To illustrate this issue, consider the following simplified modeling scenario. Imagine a very basic linear regression model, where a spatially indexed response variable $\mathbf{y}$ is regressed on a single covariate $\mathbf{x}$ using the model

$$\mathbf{y} = \beta_0 + \beta_1\mathbf{x} + \boldsymbol{\varepsilon}. \tag{4.19}$$

Suppose that the response variable $\mathbf{y}$ is smoothed using an $n \times n$ linear smoother matrix $\mathbf{M}$ to yield a new response variable $\mathbf{My}$. The correct model for the new smooth response variable is then

$$\mathbf{My} = \beta_0\mathbf{M1} + \beta_1\mathbf{Mx} + \mathbf{M}\boldsymbol{\varepsilon}. \tag{4.20}$$

If the rows of $\mathbf{M}$ each sum to one and the errors $\boldsymbol{\varepsilon}$ are normal and independent with homogeneous variance σ^2, then this new model for the smoothed data can also be written as

$$\mathbf{My} \sim \text{N}(\beta_0 + \beta_1\mathbf{Mx}, \sigma^2\mathbf{MM}'). \tag{4.21}$$

Notice that the new model (4.21) is very similar to the original, but with two important differences. The first difference is that the original covariate is replaced with a new smoothed version of it. The model formulation in Equation 4.21 suggests that, if you have data that are smoothed after the process of interest occurs, you should use a model containing the same type of smoothing on the covariate as on the response. Intuitively, this seems sensible, but might have gone unnoticed if we had not written the transformed model explicitly. The second difference is that the original errors were uncorrelated but the smoothing induces a specific type of correlation in the new model via the covariance matrix $\sigma^2\mathbf{MM}'$ in Equation 4.21. In fact, the type of correlation induced is the same type used in kernel convolution approaches for fitting spatially explicit models (Higdon 1998).

The bottom line is that the simple regression example in Equations 4.19 through 4.21 illustrates that the inference could be affected when using standard RUF approaches. Because the original density of the point process is being estimated without the use of finer-scale underlying covariates (i.e., using KDE alone), it will likely be smoothed in a similar fashion as in the simple case outlined above, thus affecting inference if the covariates are not smoothed appropriately first. Hooten et al. (2013b) empirically demonstrated that it was possible to obtain better inference using an RUF with smoothed covariates and a spatially correlated error structure (similar to that proposed by Marzluff et al. 2004). In this case, the phrase "better inference" pertains to inference closer to that arising from a Poisson regression implementation of the RSF model. Thus, while it is possible to fix up the RUF model, it may no longer be worthwhile because the RSF model is simpler to fit. Having said that, there may still be some uses for two-stage procedures like that used in RUF models. For example, in cases where complicated model extensions are required or the amount of data becomes too large, some form of multiple imputation may be necessary.* The two-stage aspect of multiple imputation is similar to that used in the RUF procedure. We return to this idea in Chapter 7.

4.4 AUTOCORRELATION

As we have seen in the preceding sections, obtaining inference for resource selection using animal telemetry data can be tricky. In addition to the spatial autocorrelation issues that we discussed in the previous section, further consideration of the temporal form of autocorrelation in the analysis of telemetry data is critical.

The fundamental issue with temporal autocorrelation arises because the point process models used to obtain RSF inference often assume that each point (i.e., observed animal position) arises independently of the others. When the telemetry fixes are obtained close together in time, the points will naturally be closer together due to the physics involved in movement (e.g., animals have limited speed when moving). If short time gaps between telemetry fixes creates a form of dependence in the observations that cannot be accounted for by the standard RSF model, then the model assumptions will not be valid and we cannot rely on the resulting statistical inference.

For these reasons, building on the work of Dunn and Gipson (1977) and Schoener (1981), Swihart and Slade (1985) developed a method for assessing temporal dependence in telemetry data. A function of distance moved and distance from activity center serves as the basis for assessing dependence. For a given time lag l,

* Multiple imputation is a two-stage procedure where an imputation distribution is first estimated and then realizations from it are used as data in secondary models. It can be useful in situations with missing data.

Swihart and Slade (1985) relied on the statistic

$$\frac{\sum_{i=l+1}^{n}(\mu_{1,i}-\mu_{1,i-l})^2+(\mu_{2,i}-\mu_{2,i-l})^2}{\sum_{i=1}^{n}(\mu_{1,i}-\bar{\mu}_1)^2+(\mu_{2,i}-\bar{\mu}_2)^2}\cdot\frac{n}{n-l}, \tag{4.22}$$

assuming that the positions $\boldsymbol{\mu}_i \equiv (\mu_{1,i}, \mu_{2,i})'$ for $i = 1, \ldots, n$ are observed directly without measurement error. Thus, the autocorrelation statistic (4.22) is essentially a multivariate Durbin–Watson statistic* that accounts for the home range (Durbin and Watson 1950). By calculating Equation 4.22 for a set of time lags ranging from small to large, one could look for a temporal lag at which the autocorrelation levels off. This leveling off suggests a time lag beyond which pairs of telemetry observations can be considered independent. For large-enough data sets, the original set of telemetry observations could be thinned such that no two points occur within the determined time lag and the usual RSF model then can be fit to the subsampled data set.

The papers by Swihart and Slade (1985) and Swihart and Slade (1997) are important contributions to the animal movement literature because they remind us to check the assumptions of our models. The downside is that we leave out data if we are interested in using standard approaches for analyzing telemetry data. A similar dilemma occurred early in the development of spatial statistics. Before modern methods for model-based geostatistics existed, researchers finding evidence for residual spatial autocorrelation would resort to subsampling data at spatial lags beyond which the errors were considered to be independent.

Numerous authors have challenged the claim that autocorrelation can affect animal space use inference (e.g., Rooney et al. 1998; deSolla et al. 1999; Otis and White 1999; Fieberg 2007). However, most of those studies were specifically focused on home range estimation rather than resource selection inference. Despite the different focus, Otis and White (1999) issue an important reminder to always consider the temporal extent of the study when collecting and analyzing telemetry data. While Otis and White (1999) opt for design-based approaches (i.e., those that rely on random sampling for frequentist inference) that minimize the effects of temporal autocorrelation for the estimation of quantities they were interested in, it is generally important to obtain a representative sample of the process under study. Fieberg et al. (2010) provide an excellent overview of different approaches for dealing with autocorrelation in resource selection inference, ranging from the subsampling approach we just described to hybrid models containing both movement and selection components.

We agree with Fieberg et al. (2010) that newer sources of telemetry data collected at fine temporal resolutions present both a challenge and opportunity for new modeling and inference pertaining to animal movement. We return to some of these approaches discussed by Fieberg et al. (2010) in what follows.

In terms of model-based methods to properly account for temporal autocorrelation, we would normally turn to those approaches used in time series (e.g., Chapter 3). That is, for temporally indexed data, y_t, we could model it in terms of mean effects

* As discussed in Chapter 3.

(μ_t, i.e., a trend) and temporally correlated random effects (η_t) as

$$y_t = \mu_t + \eta_t + \varepsilon_t,$$

$$\eta_t = \eta_{t-1} + \nu_t.$$

However, this type of linear model structure does not neatly fit into the point process framework, nor will it play nicely with typical telemetry data. The time gaps between telemetry fixes are almost always irregular in practice, despite intentional regularity in the duty cycling. Also, latency in the time required to obtain a fix is a random quantity that is difficult to control. In the collection of most time series data, we often assume that stochasticity associated with observation time is inconsequentially small relative to the desired inference, and thus, most fixes are mapped to a set of regular time intervals. Missing data between fixes is still an issue however, but statistical methods have been developed for dealing with that issue, as we will see in Chapters 5 and 6.

Fleming et al. (2015) present a generalization of the KDE isopleth approach for estimating animal home ranges when telemetry data are autocorrelated; they use an alternate form of bandwidth in the KDE to properly adjust for autocorrelation in the data (Fleming et al. 2015), but their method is for home range estimation without explicitly considering movement constraints or resource selection. To generalize the point process model such that it explicitly accommodates temporal variation and autocorrelation veers toward the broader concepts in mechanistic animal movement modeling. Thus, we return to this in the upcoming section on spatio-temporal point process (STPP) models.

4.5 POPULATION-LEVEL INFERENCE

Fieberg et al. (2010) discussed one additional theme, somewhat unrelated to temporal autocorrelation. Along with others, Fieberg et al. (2010) offered a substantial discussion of population-level inference. They contended that the individual animal should be the "sample unit" in studies of animal populations. That is, each individual should exhibit its own response to the environment, but there should also be a more general response by the population of that species as a whole. This same concept has become quite common in multispecies occupancy modeling (e.g., Kery and Royle 2008), where models are specified so that there is a "borrowing of strength" at some level among individual-level or species-level parameters.

The main concept of borrowing strength is inherent in hierarchical modeling and has been discussed extensively in the statistics literature (e.g., Gelman and Hill 2006; Hobbs and Hooten 2015). It is easiest to first demonstrate how group-level inference works in a simpler setting, then we can move to the RSF context next. Thus, consider a linear model framework in which the response variables $y_{i,j}$ represent a sequence of measurements collected for each individual j (for $j = 1, \ldots, J$ and $i = 1, \ldots, n_j$). A multilevel (i.e., hierarchical) Gaussian regression model can be written as

$$y_{i,j} \sim \text{N}(\mathbf{x}'_{i,j}\boldsymbol{\beta}_j, \sigma^2),$$

$$\boldsymbol{\beta}_j \sim \text{N}(\boldsymbol{\mu}_\beta, \boldsymbol{\Sigma}_\beta),$$

where each individual has its own set of coefficients ($\boldsymbol{\beta}_j$ and hence, response to the environmental conditions), but shared error variance (σ^2). The individual-level effects are then assumed to arise from a population-level distribution with mean $\boldsymbol{\mu}_\beta$ and covariance ($\boldsymbol{\Sigma}_\beta$). On average, we expect the individuals to respond to the environment like $\boldsymbol{\mu}_\beta$, but with variation corresponding to $\boldsymbol{\Sigma}_\beta$. This concept is often referred to as "shrinkage" because, as the diagonal elements of $\boldsymbol{\Sigma}_\beta$ get small, all of the individual sets of coefficients become more like the population-level mean $\boldsymbol{\mu}_\beta$. Another descriptor for this framework is a "random effects model." Despite the ongoing debate about the phrase "random effects" (especially in Bayesian statistics), it is often used to describe animal movement models because at least some subset of the $\boldsymbol{\beta}_j$ can be thought of as arising from a distribution with unknown parameters (i.e., $\boldsymbol{\mu}_\beta$ and $\boldsymbol{\Sigma}_\beta$).

It is important to note that this form of random effects model is more general than what is commonly used in ecology. It is much more common to let an intercept be the random effect and let the remaining regression parameters be the fixed effects. A model set up this way can be written as

$$y_{i,j} \sim \text{N}(\beta_{0,j} + \mathbf{x}'_{i,j}\boldsymbol{\beta}, \sigma^2),$$

$$\beta_{0,j} \sim \text{N}(\mu_0, \sigma_0^2).$$

Notice that this simpler type of random effect model can only shrink the individual-level intercepts ($\beta_{0,j}$) back to a population-level intercept. Therefore, it is much less flexible than the case where all coefficients (i.e., $\boldsymbol{\beta}_j$) are allowed to random effects.

Regardless of how many parameters are considered as random effects, the advantages of the hierarchical model in this setting are that we can obtain rigorous statistical population-level inference by building the population mechanism into the model directly, effectively providing more power to estimate model parameters because we can borrow strength among individuals. The population-level inference is often obtained by estimating the population-level mean $\boldsymbol{\mu}_\beta$ and its associated uncertainty. For example, if one of the coefficients in $\boldsymbol{\mu}_\beta$ corresponding to a particular type of covariate is substantially larger than zero, it would imply that the population is responding positively to that covariate on the whole. This type of inference could occur even if some individuals are responding negatively to the covariate.

The hierarchical model appropriately weights unbalanced data sets and allows us to properly scale the inference to the correct level so that the individual, rather than each observation, is the sample unit. To visualize the effect this can have on inference, consider an alternative model where each observation is the sample unit: $y_{i,j} \sim \text{N}(\mathbf{x}'_{i,j}\boldsymbol{\mu}_\beta, \sigma^2)$. In this simplified nonhierarchical model, there are essentially $J\sum_{j=1}^{J} n_j$ total observations to estimate q coefficients (given there are $q-1$ covariates). However, in the original hierarchical version of the model, even at best (i.e., very small σ^2), there are only Jq effective observations to estimate the q population-level coefficients. This reduction in effective sample size is a result of the goals for inference in the study. While it might seem like a bad thing, it keeps us from being too optimistic about population-level effects by appropriately increasing the uncertainty associated with the estimator for $\boldsymbol{\mu}_\beta$.

How can the random-effect concept be used for inferring population-level resource selection? As it turns out, the population-level RSF model can easily be formulated by indexing the selection coefficients by individual and specifying a distribution for them. The individual-level coefficients are essentially means, like those in the simple regression model; thus, we use the same multivariate Gaussian distribution for them as random effects

$$\boldsymbol{\mu}_{i,j} \sim \prod_{i=1}^{n_j} \frac{\exp(\mathbf{x}'(\boldsymbol{\mu}_{i,j})\boldsymbol{\beta}_j)}{\int \exp(\mathbf{x}'(\boldsymbol{\mu})\boldsymbol{\beta}_j)d\boldsymbol{\mu}},$$

$$\boldsymbol{\beta}_j \sim \text{N}(\boldsymbol{\mu}_\beta, \boldsymbol{\Sigma}_\beta).$$

In practice, the implied spatial point process model could still be fit using either logistic or Poisson regression after properly transforming the data as described in the earlier sections. After fitting the model, population-level inference for resource selection can be obtained by assessing the estimate for $\boldsymbol{\mu}_\beta$.

To demonstrate the benefit of using a hierarchical RSF model for population-level inference, we simulated point processes arising from 10 individuals (Figure 4.12). Each simulated individual in Figure 4.12 has a positive response to the exposure covariate (from the previously analyzed mountain lion data), but the selection for exposure is stronger for some individuals. Ultimately, inference is desired for resource selection at the population level, but we analyzed the individuals separately

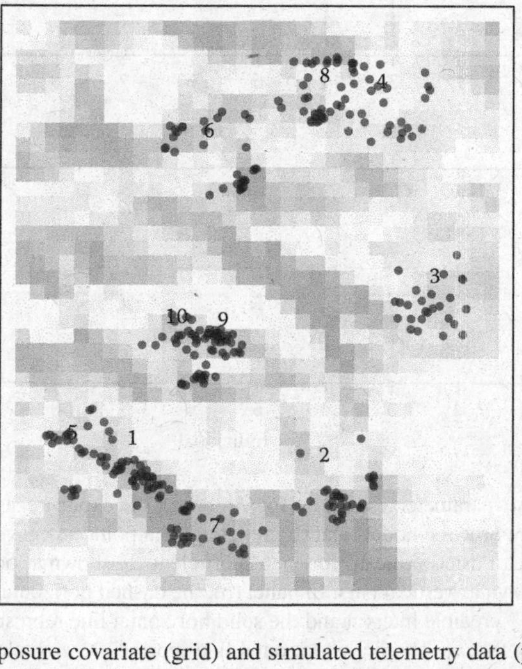

FIGURE 4.12 Exposure covariate (grid) and simulated telemetry data (points) for 10 individuals. The individuals are denoted by number at each home range centroid.

first based on point process models of the form

$$\mu_{i,j} \sim \prod_{i=1}^{n_j} \frac{\exp(\beta_{j,1}x(\mu_{i,j}))}{\int \exp(\beta_{j,1}x(\mu))d\mu} \tag{4.23}$$

that are implemented using Bayesian Poisson GLMs, where $y_{j,l} \sim \text{Pois}(e^{\beta_{j,0}+\beta_{j,1}x_{j,l}})$ for cell counts $y_{j,l}$, at grid cells c_l ($l = 1, \ldots, L$). Gaussian priors were specified for the coefficients such that $\beta_{j,k} \sim \text{N}(0, 16)$ for $k = 0, 1$ and $j = 1, \ldots, J$. Figure 4.13a shows the point estimates and 95% credible intervals for each individual. Thus, as expected, most of the selection coefficients are estimated to be positive, indicating a preference for more exposed terrain, while a few (i.e., individual 3 and 4) do not appear to be significant. Do we have sufficient evidence to conclude that the simulated population of individuals is positively selecting for exposure at the population level?

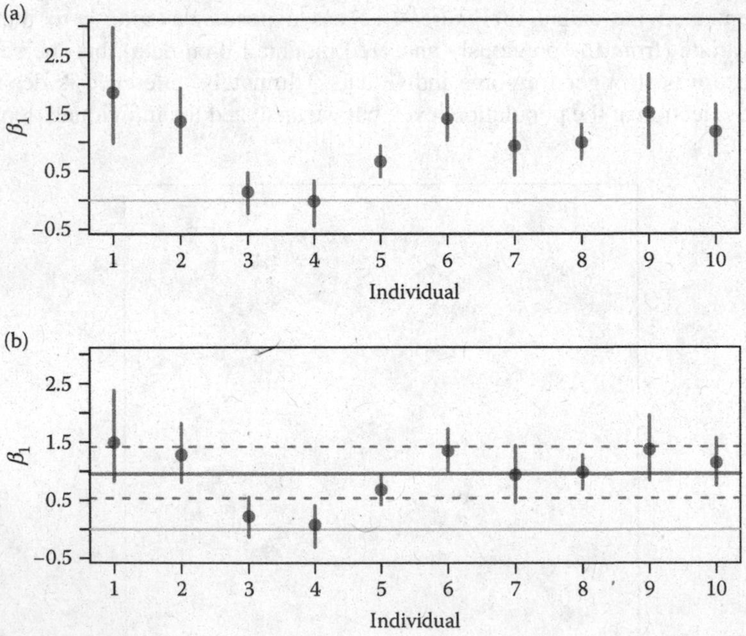

FIGURE 4.13 RSF parameter estimates for β_1 based on exposure as a covariate using (a) independent point process models and (b) a hierarchical point process model with pooling at the individual level. Posterior means for each coefficient are shown as points and 95% credible intervals are shown as vertical bars. In panel (b), the dashed horizontal lines represent the population-level 95% credible interval and the solid horizontal line represents the population-level posterior mean for μ_β. The gray horizontal line represents zero selection and is shown for reference only.

A hierarchical point process can be used to obtain inference for population-level selection. A simple hierarchical model for one resource covariate (x) is

$$\mu_{i,j} \sim \frac{\exp(x(\mu_{i,j})\beta_{1,j})}{\int \exp(x(\mu)\beta_{1,j})d\mu},$$

$$\beta_{1,j} \sim \text{N}(\mu_\beta, \sigma_\beta^2),$$

$$\mu_\beta \sim \text{N}(0, 100),$$

$$\sigma_\beta \sim \text{Unif}(0, 100),$$

where $i = 1, \ldots, n_J, j = 1, \ldots, J$, and μ_β represents the population-level mean selection. The individual-level coefficients $\beta_{1,j}$ are often referred to as random effects in this type of hierarchical model because they arise from a distribution with unknown parameters. Figure 4.13b shows the estimates for $\beta_{1,j}$ and μ_β. As compared with Figure 4.13a, the estimated individual coefficients appear to be reduced (i.e., shrunk) toward the population-level mean (black horizontal line) in the hierarchical model (Figure 4.13b). Population-level inference is obtained by assessing the posterior distribution of μ_β. Based on the simulated data in Figure 4.12, the 95% credible interval for μ_β, in the hierarchical model, does not include zero, suggesting a positive selection for exposure for the study population.

Overall, hierarchical models provide a natural way to obtain population-level inference treating the individual as the sample unit. However, when the number of observations per individual is large and the measurement error is small, very little population-level shrinkage will occur. Thus, in those cases, it has been argued that we could fit the individual-level models independently and use a secondary statistical model to obtain population-level inference.

4.6 MEASUREMENT ERROR

Thus far, we have assumed that the individual positions of the animals are measured without error. However, all telemetry data are subject to some amount of measurement error. As we discussed in Chapter 1, in VHF data, the measurement error can be associated with numerous intrinsic and extrinsic features. For example, the largest source of telemetry error probably arises from observer experience and ability. Variation due to different observers can cause systematic errors throughout a data set that may or may not be correctable after the data are collected. Aside from observer error, there can be environmental and instrumental differences in the ability to collect data, both in the telemetry device itself and the antenna array system. In the new era of satellite telemetry data collection, there are clear differences between the two primary methods: Argos and GPS. Both forms of data are affected by the location of the telemetry device on the globe in relation to the arrangement of overhead satellites. Usually, a greater number of overhead satellites (or satellite passes) lowers the associated measurement error, but observation quality can also be affected by weather, topography, land cover, and even animal behavior. Some of the quality is accounted for in the "dilution of precision" (i.e., DOP) metadata that often accompanies GPS telemetry data. DOP calculations are made based on the geometry of the positions of the satellites and the telemetry device. Small DOP values imply high-quality measurements and large values imply low quality. GPS measurement error distributions are often assumed to be multivariate Gaussian, but can vary both spatially and temporally.

Let $\mathbf{s}_i$ for $i = 1, \ldots, n$ represent the measured telemetry locations, then the simplest parametric model for the error conditioned on the true but unknown location $\boldsymbol{\mu}_i$ is $\mathbf{s}_i \sim \text{N}(\boldsymbol{\mu}_i, \sigma^2\mathbf{I})$. In this case, the error variance (σ^2) is assumed to be homogeneous, but it could be generalized such that it is a function of the provided DOP information for each measurement ($\sigma_i^2 = g(\text{DOP}_i)$). A simple link function relating the DOP to the error variance is the logarithm. In this case, we might choose to model the error standard deviation as a linear function of DOP such that $\log(\sigma_i) = \alpha_0 + \alpha_1 \text{DOP}_i$. When fitting this model, we expect the slope coefficient α_1 to be positive because DOP increases as error variance increases. The multivariate Gaussian model for error, in this case, provides circular error isopleths, implying that there is symmetry and no directional bias in the telemetry errors. Clearly, these assumptions may not always hold, but the basic framework for modeling the error structure we present is capable of being extended for more complicated situations. For example, if we expected the errors to be greater in the longitudinal direction than the latitudinal direction, we could replace the error covariance matrix (i.e., $\sigma^2\mathbf{I}$) with one that is still diagonal, but with two variance components as the diagonal elements (i.e., $\text{diag}(\sigma_1^2, \sigma_2^2)$). An example of independent Gaussian errors, with covariance $\sigma^2\mathbf{I}$, is shown in Figure 4.14a.

In contrast to GPS data, Argos telemetry data are subject to an entirely different type of measurement error due to the polar orbiting nature of the associated satellites. Some of the same environmental and behavioral features that affect GPS error can also influence Argos error, but the actual mechanics of the instrumentation often cause the largest errors. In particular, Argos telemetry errors often assume an X-pattern due to the polar orbit of the satellites and which side of the individual they pass on (e.g., Costa et al. 2010; Douglas et al. 2012). Fortunately, Argos provides auxiliary information associated with the error class for each fix. For data prior to the

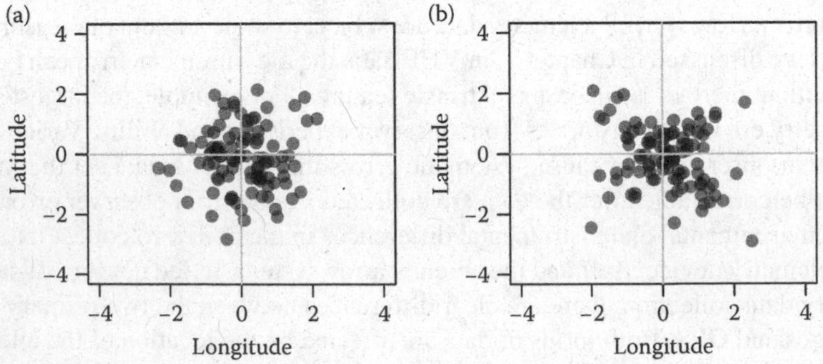

FIGURE 4.14 Two examples of telemetry position errors (i.e., $\mathbf{s}_i - \boldsymbol{\mu}_i$, $i = 1, \ldots, n$). (a) Independent Gaussian with single variance parameter $\sigma^2 = 1$ and (b) mixture Gaussian with nondiagonal covariance (4.24) with mixture probability $p = 0.5$, variance $\sigma^2 = 1$, and covariance parameters $\rho = 0.8$ and $a = 1$.

year 2007, Argos used categorical error classes that are ordinal, taking on the values 3, 2, 1, 0, A, B, Z, with 3 corresponding to the smallest error and Z the largest.

For recently collected Argos data (i.e., since 2007), a new algorithm has been created for providing more detailed information about the type of error distribution (e.g., Boyd and Brightsmith 2013). This new algorithm allows for elliptical-shaped distributions such as the multivariate Gaussian (McClintock et al. 2015).* In the absence of further modeling, these newer techniques for processing raw Argos data can be useful in providing a better understanding of the error associated with the observed locations. However, newer processing methods rely on Kalman methods that imply linear dynamics in the associated underlying movement process (Silva et al. 2014). Thus, researchers should be careful in how they interpret Argos error information in conjunction with ongoing modeling efforts that may or may not share similar dynamic properties.

Given the clear X-shaped pattern in the distribution of most Argos telemetry errors, Brost et al. (2015) and Buderman et al. (2016) suggested accounting for the measurement distribution in a hierarchical framework that can contain any modeled movement process one chooses. We return to these specific movement models in later sections and chapters, but for now, we just describe a measurement model, assuming that there is an underlying model for the true positions $\boldsymbol{\mu}_i$.

The method for accommodating Argos telemetry error presented by Brost et al. (2015) and Buderman et al. (2016) allows the error to arise from a mixture of two elliptically shaped distributions. The use of two distributions accounts for the X-pattern that arises from the direction that the satellite passes overhead. The multivariate Gaussian is incredibly useful for this type of model and can serve a starting point. In our proposed measurement model for the GPS data, we suggested using a multivariate Gaussian that is potentially elliptical in the cardinal directions only. We seek a more flexible specification that can account for an elliptical shape on a diagonal axis. Thus, if we know which side of the telemetry device the satellite passes over, we can use a fully parameterized multivariate Gaussian measurement model: $\mathbf{s}_i \sim \text{N}(\boldsymbol{\mu}_i, \boldsymbol{\Sigma})$, where the covariance matrix $\boldsymbol{\Sigma}$ is completely unknown and need not be diagonal. For example, the covariance matrix

$$\boldsymbol{\Sigma} \equiv \sigma^2 \begin{bmatrix} 1 & \rho\sqrt{a} \\ \rho\sqrt{a} & a \end{bmatrix} \tag{4.24}$$

is quite flexible. In this case, some combination of the three covariance parameters can provide an appropriate amount of eccentricity and tilt for the error ellipses. This measurement model is very similar to that used by McClintock et al. (2015), which relies on information from Argos about the direction of tilt in the ellipse. In older data sets, where such information is not available, we need a mixture model to account for tilt in either direction. Thus, consider a generalization of the measurement model

$$\mathbf{s}_i \sim p \cdot \text{N}(\boldsymbol{\mu}_i, \boldsymbol{\Sigma}) + (1-p) \cdot \text{N}(\boldsymbol{\mu}_i, \boldsymbol{\Psi}\boldsymbol{\Sigma}\boldsymbol{\Psi}'), \tag{4.25}$$

* Elliptical-shaped distributions are also now integrated into the "crawl" R package. For details on "crawl," see Johnson et al. (2008a) and Chapter 6.

where p represents a mixture probability* and the matrix $\mathbf{\Psi}$ rotates the first distribution to provide an X-shape to overall mixture distribution. The rotation can be achieved by specifying $\mathbf{\Psi}$ as

$$\mathbf{\Psi} \equiv \begin{bmatrix} 1 & 0 \\ 0 & -1 \end{bmatrix}. \tag{4.26}$$

Figure 4.14b shows telemetry position errors (i.e., $\mathbf{s}_i - \boldsymbol{\mu}_i$, $i = 1, \ldots, n$) associated with the Gaussian mixture model (4.25).

Mixture models can be represented in many ways. The model presented in Equation 4.25 is one of the most common forms for mixture models, but there can be value in using a hierarchical structure with auxiliary variables to specify the mixture model. For example, Buderman et al. (2016) used the form

$$\mathbf{s}_i \sim \begin{cases} \mathrm{N}(\boldsymbol{\mu}_i, \mathbf{\Sigma}) & \text{if } z_i = 1 \\ \mathrm{N}(\boldsymbol{\mu}_i, \mathbf{\Psi}\mathbf{\Sigma}\mathbf{\Psi}') & \text{if } z_i = 0 \end{cases}, \tag{4.27}$$

where the latent binary process is modeled as $z_i \sim \mathrm{Bern}(p)$ and acts like a switch, turning on and off each distribution as needed. Perhaps surprisingly, this new mixture specification (4.27) yields exactly the same inference as the previous one (4.25) and has other benefits in terms of implementation. In the simple RSF context we have described in this chapter, consider a fully specified model that accounts for Argos telemetry error and uses the RSF point process model for the underlying true observations:

$$\mathbf{s}_i \sim p \cdot \mathrm{N}(\boldsymbol{\mu}_i, \mathbf{\Sigma}) + (1-p) \cdot \mathrm{N}(\boldsymbol{\mu}_i, \mathbf{\Psi}\mathbf{\Sigma}\mathbf{\Psi}'),$$
$$\boldsymbol{\mu}_i \sim \frac{\exp(\mathbf{x}'(\boldsymbol{\mu}_i)\boldsymbol{\beta})}{\int \exp(\mathbf{x}'(\boldsymbol{\mu})\boldsymbol{\beta}) d\boldsymbol{\mu}}. \tag{4.28}$$

Unfortunately, standard statistical software for implementing this type of model has not been developed yet. Therefore, like many of the contemporary movement models, a custom algorithm needs to be developed to either maximize the implied likelihood for this model or calculate the Bayesian posterior distribution.

Another potential issue with the model presented in Equation 4.28 is that it may be difficult to simultaneously identify the variation arising from the process and the uncertainty in the measurements without some form of replication or strong prior information about the measurement error. Identifiability is a potential issue in nearly all measurement error models. Heuristically, identifiability can be a problem in the point process model because the model may not have enough information to position $\boldsymbol{\mu}_i$ in the correct place without some other type of constraint on the true positions. It is possible that enough structure could be provided if the RSF signal is strong enough (i.e., individuals are responding strongly to environmental covariates) and the measurement error is small compared to the scale of the covariates.

* Probability that the measurement arises from the first distribution.

A second form of structure arises in the support for the point process, that is, the spatial domain where the points are restricted to occur. Brost et al. (2015) demonstrates the effect of barriers to movement on the ability to estimate the true underlying point process and resource selection. In the case of a marine species, the shoreline can serve as an adequate boundary and allow the model to separate measurement error from process-based variation (Brost et al. 2015). Finally, natural temporal autocorrelation in the process can also provide enough structure in some cases to separate measurement error from process-based variation (e.g., Brost et al. 2015). This concept is fundamental to the dynamic movement models we describe in the next section and later chapters.

Regardless of the type of telemetry device used, it is important to understand the potential influence of measurement error on the desired inference as well as how to properly account for it. The power of model-based approaches for animal movement inference is that one can generalize the model structures as needed to accommodate intricacies of the data and type of movement behavior.

4.7 SPATIO-TEMPORAL POINT PROCESS MODELS

Returning to the point process model (4.6) for temporally independent telemetry data,

$$[\boldsymbol{\mu}_i|\boldsymbol{\beta},\boldsymbol{\theta}] \equiv \frac{g(\mathbf{x}(\boldsymbol{\mu}_i),\boldsymbol{\beta})f(\boldsymbol{\mu}_i,\boldsymbol{\theta})}{\int g(\mathbf{x}(\boldsymbol{\mu}),\boldsymbol{\beta})f(\boldsymbol{\mu},\boldsymbol{\theta})d\boldsymbol{\mu}}, \tag{4.29}$$

we can generalize it for situations where the time steps between telemetry observations are small. When the time steps are small, we would expect to see a movement signal in the data themselves. Such a signal arises from the physical limitations of the movement process. That is, there is some reasonable finite upper bound to the distance an animal can travel, or is willing to travel, in a fixed amount of time. Heuristically, constraints provide smoothness to the individual's path based on its true positions at each time. Conditioning on the position at the previous time step ($\boldsymbol{\mu}_{i-1}$), envision a spatial map corresponding to the probability the animal will occur at the next time in the absence of other environmental information. For example, the maps in Figure 4.15 indicate that locations near the previous position ($\boldsymbol{\mu}_{i-1}$) would be more likely to host the next position ($\boldsymbol{\mu}_i$). As the distance increases from the previous position, we would be less likely to find the next position. The position labeled $\boldsymbol{\mu}_i$ in Figure 4.15 is more likely under the availability in panel (b) than panel (a). Furthermore, as the time between positions (Δ_i) increases, we would expect the map to be flatter, indicating the animal could be farther away. With increasing Δ_i, we would expect a completely flat surface over the support of the point process ($\mathcal{M}$) such that the effective distribution for that particular position ($\boldsymbol{\mu}_i$) is uniform (or CSR, using the jargon from the point process literature). The surface we are describing corresponds to the availability surface ($f(\boldsymbol{\mu}_i,\boldsymbol{\theta})$) for each particular time t_i and will change over time depending on $\boldsymbol{\mu}_{i-1}$ and Δ_i. Moorcroft and Barnett (2008) refer to this time-varying availability distribution as a "redistribution kernel."

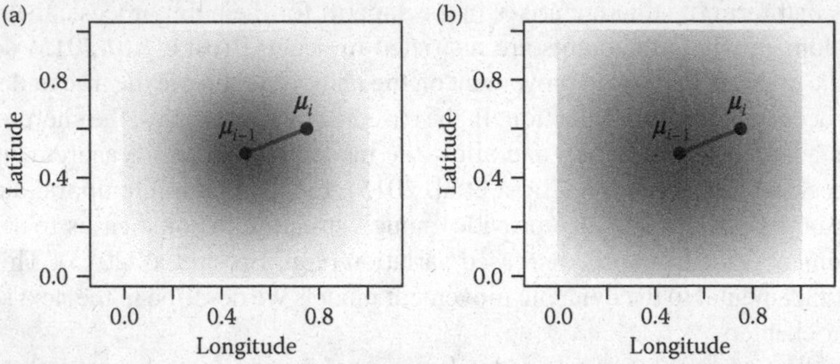

FIGURE 4.15 Examples of two different availability functions $f(\boldsymbol{\mu}_i, \boldsymbol{\theta})$ (shaded surface with darker corresponding to greater availability). (a) Less diffuse availability and (b) more diffuse availability. Two consecutive positions (i.e., $\boldsymbol{\mu}_{i-1}$ and $\boldsymbol{\mu}_i$) are shown for reference.

To translate the concept of time-varying availability into the point process model itself, we need to allow for dependence in the availability distribution such that

$$[\boldsymbol{\mu}_i|\boldsymbol{\mu}_{i-1}, \boldsymbol{\beta}, \boldsymbol{\theta}] \equiv \frac{g(\mathbf{x}(\boldsymbol{\mu}_i), \boldsymbol{\beta})f(\boldsymbol{\mu}_i|\boldsymbol{\mu}_{i-1}, \Delta_i, \boldsymbol{\theta})}{\int g(\mathbf{x}(\boldsymbol{\mu}), \boldsymbol{\beta})f(\boldsymbol{\mu}|\boldsymbol{\mu}_{i-1}, \Delta_i, \boldsymbol{\theta})d\boldsymbol{\mu}}. \tag{4.30}$$

The new model in Equation 4.30 has the same basic form as the original point process model in Equation 4.6, but contains an explicit dependence in time through the availability function $f(\boldsymbol{\mu}_i|\boldsymbol{\mu}_{i-1}, \Delta_i, \boldsymbol{\theta})$. Christ et al. (2008) and Johnson et al. (2008b) presented this STPP model as part of a general framework for accounting for both animal movement and resource selection simultaneously. Later, Forester et al. (2009), Potts et al. (2014a), and Brost et al. (2015) used similar approaches to model telemetry data from elk (*Cervus canadensis*), caribou (*Rangifer tarandus*), and harbor seals (*Phoca vitulina*), respectively.

4.7.1 General Spatio-temporal Point Processes

It is worth taking a step back to examine general STPP models to show how we arrive at the spatio-temporal model in Equation 4.30. We provide an overview of STPP models, but additional detail appears in Johnson et al. (2013) and Schoenberg et al. (2002). As one might expect, a general STPP turns out to be a direct combination of the spatial point process in Chapter 2 and the temporal point process of Chapter 3. The intensity function of an STPP is a function of locations, $\boldsymbol{\mu}$, and time t. However, unlike spatial point processes, the STPP intensity also depends on the history $\mathcal{H}_t$ of the process up time t, like the temporal point process (e.g., $\lambda(\boldsymbol{\mu}, t|\mathcal{H}_t, \boldsymbol{\theta})$).*

If the STPP is orderly (as defined in Section 3.1.6), then the intensity function can be interpreted as the approximate probability of an event occurring in a small space

* We switched to using $\boldsymbol{\mu}$ for a spatial location instead of $\mathbf{s}$ as was used in Chapters 2 and 3 for point process description.

around $\boldsymbol{\mu}$, in a small time interval near t. If $\mathcal{B} = (\boldsymbol{\mu} + \Delta\boldsymbol{\mu}) \times (t + \Delta t)$ is a small cube in space and time, then, under certain conditions,

$$P(n(\mathcal{B}) = 1|\mathcal{H}_t) \approx \lambda(\boldsymbol{\mu}, t|\mathcal{H}_t, \boldsymbol{\theta})|\mathcal{B}|, \tag{4.31}$$

where $n(\mathcal{B})$ is the number of events in $\mathcal{B}$ and $|\mathcal{B}|$ is the volume of the cube. If $\lambda(\boldsymbol{\mu}, t|\mathcal{H}_t) = \lambda(\boldsymbol{\mu}, t)$, that is, it does not depend on the history up to time t, then it is a spatio-temporal Poisson process with the properties given in Chapters 2 and 3 (with respect to spatial and temporal Poisson processes). Following the derivations from each of the two previously discussed processes,* we arrive at the likelihood for the STPP as

$$[(\boldsymbol{\mu}_1, t_1), \ldots, (\boldsymbol{\mu}_n, t_n)|\boldsymbol{\theta}] = \left(\prod_{i=1}^{n} \lambda(\boldsymbol{\mu}_i, t_i|\mathcal{H}_{t_i}, \boldsymbol{\theta})\right) \exp\left(-\int_{\mathcal{T}}\int_{\mathcal{M}} \lambda(\mathbf{u}, v|\mathcal{H}_v, \boldsymbol{\theta})\, d\mathbf{u}\, dv\right), \tag{4.32}$$

where $(\boldsymbol{\mu}_i, t_i)$ are the locations and times of observed events, $\mathcal{M}$ is the spatial study area, and $\mathcal{T}$ is the time window of the study.

Notice that the model in Equation 4.32 does not look like Equation 4.30 yet. Thus, we investigate further, providing more details and one additional result. First, the intensity function is usually decomposed as

$$\lambda(\mathbf{u}, v|\mathcal{H}_v, \boldsymbol{\theta}) = g(\boldsymbol{\mu}|\boldsymbol{\theta})h(t|\boldsymbol{\theta})f(\boldsymbol{\mu}, t|\mathcal{H}_v, \boldsymbol{\theta}), \tag{4.33}$$

where $g(\boldsymbol{\mu}|\boldsymbol{\theta})$ represents the purely spatial component, $h(t|\boldsymbol{\theta})$ is a deterministic baseline temporal intensity, and $f(\boldsymbol{\mu}, t|\mathcal{H}_v, \boldsymbol{\theta})$ is a spatio-temporal interaction effect. Second, if one is not interested in the actual times of events, rather, just the effect event times have on the spatial intensity, then we can condition on the observed times to obtain the conditional likelihood (Diggle et al. 2010b; Johnson et al. 2013),

$$[\boldsymbol{\mu}_1, \ldots, \boldsymbol{\mu}_n|t_1, \ldots, t_n, \boldsymbol{\theta}] = \prod_{i=1}^{n} \frac{\lambda(\boldsymbol{\mu}_i, t_i|\mathcal{H}_{t_i}, \boldsymbol{\theta})}{\int_{\mathcal{M}} \lambda(\mathbf{u}, t_i|\mathcal{H}_{t_i}, \boldsymbol{\theta})d\mathbf{u}}. \tag{4.34}$$

The likelihood resembles the form of a weighted distribution. If we substitute the decomposed STPP intensity from Equation 4.33 into Equation 4.34, we obtain

$$[\boldsymbol{\mu}_1, \ldots, \boldsymbol{\mu}_n|t_1, \ldots, t_n, \boldsymbol{\theta}] = \prod_{i=1}^{n} \frac{g(\boldsymbol{\mu}_i)f(\boldsymbol{\mu}_i, t_i|\mathcal{H}_{t_i}, \boldsymbol{\theta})}{\int_{\mathcal{M}} g(\boldsymbol{\mu})f(\boldsymbol{\mu}, t_i|\mathcal{H}_{t_i}, \boldsymbol{\theta})d\boldsymbol{\mu}}, \tag{4.35}$$

where the temporal baseline intensity $h(t|\boldsymbol{\theta})$ does not appear because it cancels in the numerator and denominator. If the intensity changes depending only on the last

* The derivations to arrive at the STPP likelihood are similar to what was presented in Chapters 2 and 3; thus, we omit it here.

observed event location and time interval since the last event, the resulting conditional distribution of event locations is

$$[\boldsymbol{\mu}_1, \ldots, \boldsymbol{\mu}_1 | t_1, \ldots, t_n, \boldsymbol{\theta}] = \prod_{i=1}^{n} \frac{g(\boldsymbol{\mu}_i) f(\boldsymbol{\mu}_i | \boldsymbol{\mu}_{i-1}, \Delta_i, \boldsymbol{\theta})}{\int_{\mathcal{M}} g(\boldsymbol{\mu}) f(\boldsymbol{\mu} | \boldsymbol{\mu}_{i-1}, \Delta_i, \boldsymbol{\theta}) d\boldsymbol{\mu}}$$

$$= \prod_{i=1}^{n} [\boldsymbol{\mu}_i | \boldsymbol{\mu}_{i-1}, \Delta_i, \boldsymbol{\theta}], \quad (4.36)$$

and we arrive at the full likelihood for the model given by the transitions in Equations 4.30. In the references provided at the beginning of the section (i.e., Christ et al. 2008; Johnson et al. 2008b; Forester et al. 2009; Potts et al. 2014a; Brost et al. 2015), the conditional STPP model was developed under the weighted distribution paradigm (i.e., using expressions like Equation 4.29; Patil and Rao 1977). Those papers developed weighted distributions by specifying movement models for $\boldsymbol{\mu}_i$ given $\boldsymbol{\mu}_{i-1}$ and weighting the spatial distribution by the spatial effects in $g(\boldsymbol{\mu})$. The integral in the denominator of Equation 4.36 results out of necessity to normalize the PDF to integrate to one over the spatial domain $\mathcal{M}$. Johnson et al. (2013) arrived at the same result using STPP concepts directly.

4.7.2 Conditional STPP Models for Telemetry Data

The vast majority of STPP models for animal telemetry data have been developed using the conditional (weighted distribution) approach as an extension to the temporally static resource selection models of Section 4.2. Thus, we begin by investigating weighted distribution specifications. The early developments of this style of STPP model were presented by Arthur et al. (1996) and later generalized by Rhodes et al. (2005). Arthur et al. (1996) presented the basic idea that availability could change as a function of the individual's position and time. They suggested the use of a circular availability function

$$f(\boldsymbol{\mu}_i | \boldsymbol{\mu}_{i-1}, r) = \begin{cases} \frac{1}{\pi r^2} & \text{if } ||\boldsymbol{\mu}_i - \boldsymbol{\mu}_{i-1}|| \leq r \\ 0 & \text{if } ||\boldsymbol{\mu}_i - \boldsymbol{\mu}_{i-1}|| > r \end{cases}, \quad (4.37)$$

where $||\boldsymbol{\mu}_i - \boldsymbol{\mu}_{i-1}||$ is the Euclidean distance between the two positions ($\boldsymbol{\mu}_i$ and $\boldsymbol{\mu}_{i-1}$) and r is the radius of the circular availability area. This early work led to a suite of similar methods known as "step selection functions" (Boyce et al. 2003; Fortin et al. 2005; Potts et al. 2014a; Avgar et al. 2016). The classical step selection function approach defines the availability circle using the empirical step lengths associated with the telemetry data. A background sample of availability locations is selected within the associated circle for each telemetry observation. Then a conditional logistic regression approach is used to associate the covariates at the background sample locations with each telemetry location. Similar methods were developed for use in medical statistics to account for variation in patients that have similar backgrounds to control for potentially confounding factors in life history (Rahman et al. 2003). Fortin et al. (2005) claimed that the remaining temporal dependence in these models will not affect inference on selection coefficients; however, it has been shown that there are exceptions (e.g., Fieberg and Ditmer 2012; Hooten et al. 2013b). For example, when the covariates influencing selection are smoothly varying, there is an increased risk of temporal confounding.

An alternative availability model where the availability range is estimated simultaneously with the other parameters was proposed by Christ et al. (2008) and generalized to uneven times of location by Johnson et al. (2008b) is

$$f(\boldsymbol{\mu}_i|\boldsymbol{\mu}_{i-1},\boldsymbol{\theta}) \propto \exp(-(\boldsymbol{\mu}_i - \tilde{\boldsymbol{\mu}}_i)'\mathbf{Q}_i^{-1}(\boldsymbol{\mu}_i - \tilde{\boldsymbol{\mu}}_i)/2), \tag{4.38}$$

such that $\tilde{\boldsymbol{\mu}}_i = \bar{\boldsymbol{\mu}} + \mathbf{B}_i(\boldsymbol{\mu}_{i-1} - \bar{\boldsymbol{\mu}})$ and $\bar{\boldsymbol{\mu}}$ is a central place of attraction. The components controlling the dispersion of the availability distribution are $\mathbf{B}_i \equiv \exp(-(t_i - t_{i-1})/\phi)\mathbf{I}$, where $\mathbf{Q}_i = \mathbf{Q} - \mathbf{B}_i\mathbf{Q}\mathbf{B}_i$. Johnson et al. (2008b) arrived at this specific form for availability because they were assuming a stochastic process for animal movement called the Ornstein–Uhlenbeck (OU) model (e.g., Dunn and Gipson 1977; Blackwell 1997). The parameter ϕ controls the range of availability as the r parameter does in the "step selection" models. However, in the OU model, the availability limit is soft, meaning the availability function never drops all the way to zero for any distance from the current location, but the function decreases and approaches zero for very large distances. The early step selection models had a hard availability limit (i.e., there is no availability of locations for distances larger than r). Additionally, the OU-based model allows for a central attraction point (or multiple attraction points, e.g., Johnson et al. 2008b). We provide additional details of OU processes for modeling animal movement in Chapter 6. Similar to Johnson et al. (2008b), Moorcroft and Barnett (2008) also described a unification of resource selection models and what they call "mechanistic home range" models. The mechanistic home range models essentially model the movement process in terms of partial differential equations (Moorcroft et al. 2006). Moorcroft and Barnett (2008) also point out that the model in Equation 4.30 rigorously accommodates autocorrelation if it exists.

Potts et al. (2014a) discussed the same framework presented by Johnson et al. (2008b), but referred to Equation 4.30 as the "master equation." Potts et al. (2014a) parameterized the time-varying availability function $f(\boldsymbol{\mu}_i|\boldsymbol{\mu}_{i-1},\theta)$ in terms of bearing θ so that $\boldsymbol{\mu}_i$ and $\boldsymbol{\mu}_{i-1}$ are related by

$$\boldsymbol{\mu}_i = \boldsymbol{\mu}_{i-1} + \begin{pmatrix} \cos(\theta+\pi)\sqrt{(\boldsymbol{\mu}_i - \boldsymbol{\mu}_{i-1})'(\boldsymbol{\mu}_i - \boldsymbol{\mu}_{i-1})} \\ \sin(\theta+\pi)\sqrt{(\boldsymbol{\mu}_i - \boldsymbol{\mu}_{i-1})'(\boldsymbol{\mu}_i - \boldsymbol{\mu}_{i-1})} \end{pmatrix}, \tag{4.39}$$

where $\sqrt{(\boldsymbol{\mu}_i - \boldsymbol{\mu}_{i-1})'(\boldsymbol{\mu}_i - \boldsymbol{\mu}_{i-1})}$ is the Euclidean distance between $\boldsymbol{\mu}_i$ and $\boldsymbol{\mu}_{i-1}$. Additionally, Potts et al. (2014a) were interested in discrete habitat types, and thus, they modified the traditional RSF $g(\mathbf{x}(\boldsymbol{\mu}_i),\boldsymbol{\beta})$ to be the proportion line segment from $\boldsymbol{\mu}_{i-1}$ to $\boldsymbol{\mu}_i$ of habitat $\mathbf{x}$, for example. Potts et al. (2014a) ultimately decomposed the availability function into a finite sum of habitat-specific components. The habitat-specific components involved a product of turning angle and step length distributions (e.g., Weibull and von Mises distributions). Rather than maximize the likelihood based on Equation 4.30 directly, Potts et al. (2014a) used an approximate conditional logistic regression procedure similar to that described in Section 4.2 on RSFs to estimate parameters.

Most (if not all) STPP analysis of telemetry data assumes that the locations are observed without error. If the locations are observed with a significant amount of error, then that must be taken into account. We present an example analysis of the harbor seal telemetry data found in Brost et al. (2015) that uses a hierarchical framework to accommodate complicated telemetry error distributions.

Rather than rely on a specific stochastic process as a model for animal movement, Brost et al. (2015) specified an availability distribution directly based on a particular form of smoothness

$$f(\boldsymbol{\mu}_i|\boldsymbol{\mu}_{i-1},\boldsymbol{\theta}) \propto \exp\left(-\frac{||\boldsymbol{\mu}_i-\boldsymbol{\mu}_{i-1}||}{\Delta_i\phi}\right), \tag{4.40}$$

where $||\boldsymbol{\mu}_i-\boldsymbol{\mu}_{i-1}||$ is a distance measure between true positions $\boldsymbol{\mu}_i$ and $\boldsymbol{\mu}_{i-1}$, Δ_i is the elapsed time between positions, and ϕ acts as a smoothing parameter. The availability distribution in Equation 4.40 is very similar to an exponential model for correlation in a spatial covariance matrix (Chapter 2). In their analysis of harbor seals, Brost et al. (2015) considered the shortest water distance as the distance metric in the availability function. This distance metric allowed them to appropriately accommodate the shoreline as a hard constraint for movement of harbor seals. While increasing the realism and utility of the model, formally accounting for such a constraint adds a nontrivial amount of complexity to the model implementation. It is worth noting that, although the exponential function was used in the study of harbor seals, many other functional forms are reasonable. Forester et al. (2009) describe several different functional forms and state that exponential family functions are preferable.

Following Brost et al. (2015), we analyzed Argos telemetry data arising from an individual seal in the Gulf of Alaska. The telemetry data in our example (Figure 4.16) occur at irregular temporal intervals, ranging from minutes to hours, with the majority of observations occurring less than 2 h apart. The telemetry data are composed of a range of error classifications with the majority of data in the lower-quality Argos error categories (e.g., 0, A, and B classes), which is why many observed positions occur on land, far from water (Figure 4.16).

We specified a hierarchical STPP model for the harbor seal telemetry data such that

$$\mathbf{s}_i \sim \mathrm{p}\cdot \mathrm{t}(\boldsymbol{\mu}_i, \boldsymbol{\Sigma}_i, \nu_i) + (1-\mathrm{p})\cdot \mathrm{t}(\boldsymbol{\mu}_i, \mathbf{H}\boldsymbol{\Sigma}_i\mathbf{H}', \nu_i),$$

$$\boldsymbol{\mu}_i \sim \frac{\exp(\mathbf{x}'(\boldsymbol{\mu}_i)\boldsymbol{\beta})f(\boldsymbol{\mu}_i|\boldsymbol{\mu}_{i-1},\boldsymbol{\theta})}{\int \exp(\mathbf{x}'(\boldsymbol{\mu})\boldsymbol{\beta})f(\boldsymbol{\mu}|\boldsymbol{\mu}_{i-1},\boldsymbol{\theta})d\boldsymbol{\mu}},$$

$$\boldsymbol{\beta} \sim \mathrm{N}(\mathbf{0}, \sigma^2_\beta \mathbf{I}),$$

where the t-distribution allows for extreme telemetry observations and has heavier tails than the Gaussian distribution as the degrees of freedom parameter ν_i decreases

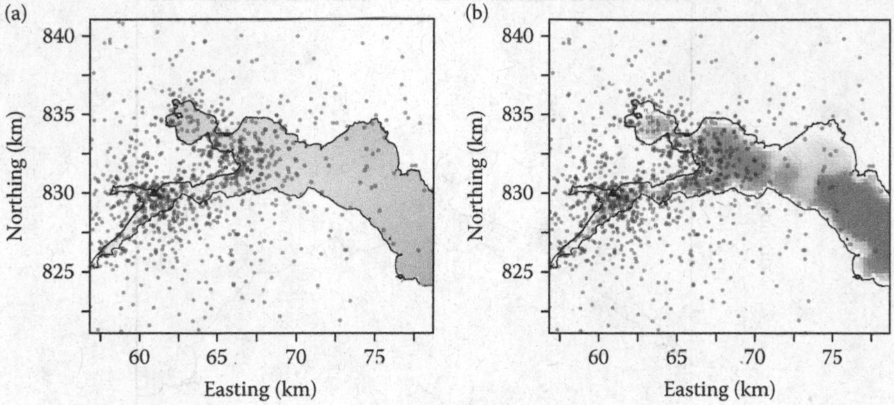

FIGURE 4.16 Argos telemetry data ($\mathbf{s}_i$, for $i = 1, \ldots, n$; shown as points) for an individual harbor seal and two different environmental covariates ($\mathbf{X}$) influencing harbor seal movement: (a) distance from known haul out (i.e., distance from position shown with a dark triangle in the left of each panel) and (b) bathymetry (i.e., ocean depth). Both covariates were standardized and are shown with darker shading as the values of the covariate increase.

and p = 0.5. The measurement scale matrix $\boldsymbol{\Sigma}_i$ was specified as in Equation 4.24 and data model parameters σ_i^2, a_i, ρ_i, and ν_i assume one of six distributions depending on which error class was recorded for that telemetry observation.* We assumed uniform priors on ecologically reasonable ranges of support for the standard deviation σ_i as well as ρ_i and ν_i. The time-varying availability function $f(\boldsymbol{\mu}_i|\boldsymbol{\mu}_{i-1}, \boldsymbol{\theta})$ in the hierarchical model was specified as in Equation 4.40, where the distance metric was the shortest water distance between $\boldsymbol{\mu}_i$ and $\boldsymbol{\mu}_{i-1}$.

Fitting the hierarchical STPP model of Brost et al. (2015) to the harbor seal telemetry data has additional benefits. For example, because the individual is constrained to be in the water and adjacent shorelines only, erroneous telemetry observations over land will naturally be constrained to occur in the correct support (i.e., the water). Furthermore, the constraint itself actually aids in the estimation of measurement error–specific parameters (i.e., σ_i^2, a_i, ρ_i, and ν_i) because the model knows that positions on land are incorrect. To summarize the estimated true individual positions $\boldsymbol{\mu}_i$, we calculated the posterior mean UD for all positions $E(\{\boldsymbol{\mu}_i, \forall i\}|\{\mathbf{s}_i, \forall i\})$ for the entire support considered in the study area (Figure 4.17). The selection coefficients associated with the distance to haul out and bathymetry covariates were both estimated to be negative. Therefore, after controlling for potential autocorrelation due to temporal proximity of telemetry fixes, complicated Argos measurement error, and barriers to movement, the data suggest that this individual harbor seal selects for aquatic environments nearer the haul out and in shallower water. These findings agree with the central place foraging behavior of harbor seals in the North Pacific Ocean.

* If telemetry observation $\mathbf{s}_i$ is measured with error class $c_i = 2$, then the variance parameter for observation i assumes the variance for error class 2: $\sigma_i^2 = \sigma_{c_i=2}^2$. The other parameters are defined similarly. Priors for the parameters of different error classes can be specified such that they contain differing information about the precision of the measurement at that time.

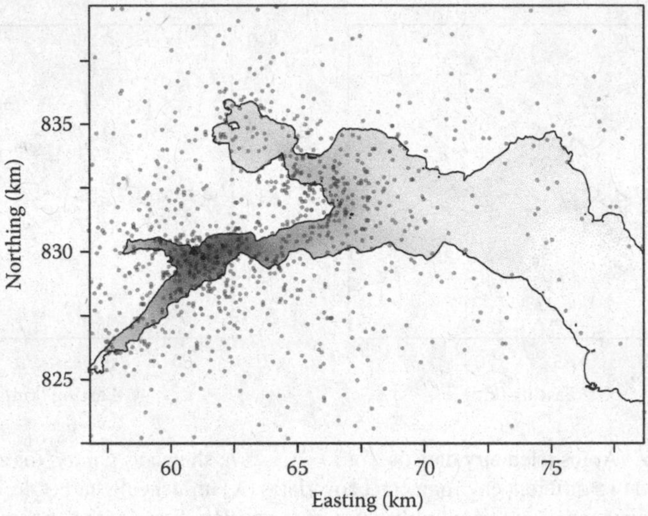

FIGURE 4.17 Argos telemetry data (i.e., $\mathbf{s}_i$, for $i = 1, \ldots, n$; shown as points) for an individual harbor seal and the estimated posterior mean UD (i.e., utilization distribution) (i.e., $E(\{\boldsymbol{\mu}_i, \forall i\}|\{\mathbf{s}_i, \forall i\})$) based on true underlying positions $\boldsymbol{\mu}_i$ and known covariates $\mathbf{X}$.

4.7.3 Full STPP Model for Telemetry Data

We are only aware of a single study that analyzes telemetry data using the full STPP specification, that is, the likelihood in Equation 4.32, rather than the conditional likelihood in Equation 4.36 (Johnson et al. 2013). To demonstrate movement and resource selection analysis using the full unconditional likelihood, we analyze the brown bear (*Ursus arctos*) data previously analyzed by Christ et al. (2008) and Johnson et al. (2008b) using a time-indexed redistribution kernel in the conditional likelihood.

The bear data are composed of $n = 475$ GPS locations of a brown bear in Southeast Alaska. The data were analyzed in Johnson et al. (2008b) using the OU model with two centers of attraction, which the bear moved between at a known time. The model also included the influence of two habitat covariates, distance from nearest stream and vegetation classification. For simplicity in this example, we only use the stream distance covariate. Figure 4.18a shows our brown bear telemetry observations and the distance from nearest stream covariate.

It is obvious when the change in centers of attraction occurs in Figure 4.18, so the known switching time in the previous analysis is not a serious shortcoming, however, we choose a different approach based on a static nonparametric region of attraction. We model the STPP log intensity function as

$$\log \lambda(\boldsymbol{\mu}_i, t_i|\mathcal{H}_{t_i}) = \mathbf{x}'(\boldsymbol{\mu}_i)\boldsymbol{\beta} + \eta(\boldsymbol{\mu}_i, \boldsymbol{\theta}) - \alpha d(\boldsymbol{\mu}_i, \boldsymbol{\mu}_{i-1})^2/\Delta_i, \tag{4.41}$$

where $\mathbf{x}'(\boldsymbol{\mu}_i)$ is a vector containing a 1 (for the intercept) and the stream distance covariate at $\boldsymbol{\mu}_i$, and $\eta(\boldsymbol{\mu}_i, \boldsymbol{\theta})$ is a thin-plate regression spline in 2-D (df = 25, Wood 2003), $d(\boldsymbol{\mu}_i, \boldsymbol{\mu}_{i-1})$ is the Euclidean distance from $\boldsymbol{\mu}_{i-1}$ to $\boldsymbol{\mu}_i$, and $\Delta_i \equiv t_i - t_{i-1}$.

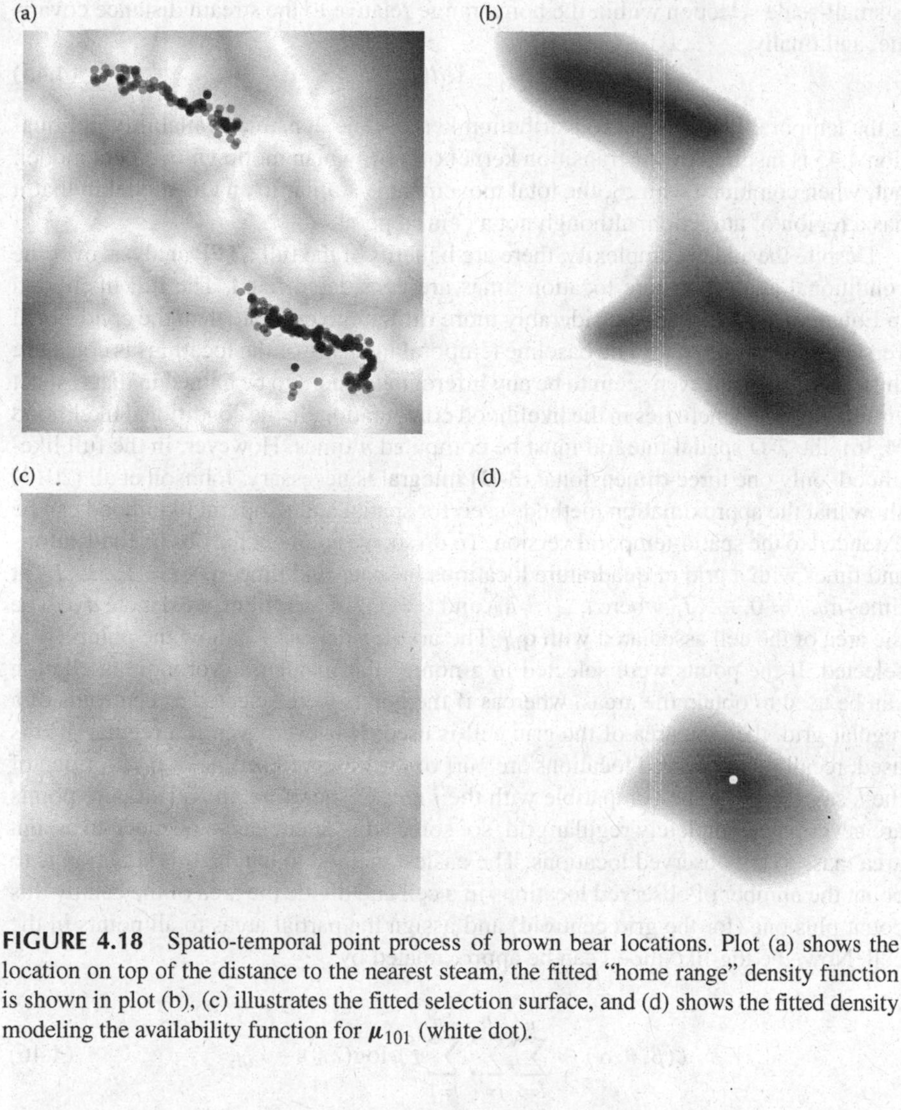

FIGURE 4.18 Spatio-temporal point process of brown bear locations. Plot (a) shows the location on top of the distance to the nearest steam, the fitted "home range" density function is shown in plot (b), (c) illustrates the fitted selection surface, and (d) shows the fitted density modeling the availability function for $\boldsymbol{\mu}_{101}$ (white dot).

To examine how this relates to the other models in this section, we factor the intensity as

$$\lambda(\boldsymbol{\mu}_i, t_i|\mathcal{H}_{t_i}) = g_1(\boldsymbol{\mu}_i)g_2(\boldsymbol{\mu}_i)f(\boldsymbol{\mu}_i|\boldsymbol{\mu}_{i-1}, \Delta_i), \tag{4.42}$$

where

$$g_2(\boldsymbol{\mu}_i) = e^{\eta(\boldsymbol{\mu}_i, \boldsymbol{\theta})} \tag{4.43}$$

might be considered broad-scale selection within the study area, perhaps a home range,

$$g_1(\boldsymbol{\mu}) = e^{\mathbf{x}'(\boldsymbol{\mu}_i)\boldsymbol{\beta}} \tag{4.44}$$

is small-scale selection within the home range relative to the stream distance covariate, and finally,

$$f(\boldsymbol{\mu}_i|\boldsymbol{\mu}_{i-1}, \Delta_i) = \alpha d(\boldsymbol{\mu}_i, \boldsymbol{\mu}_{i-1})^2/\Delta_i \tag{4.45}$$

is the temporally dependent redistribution kernel. The dynamic availability in Equation 4.45 is inspired by the transition kernel of a Brownian motion movement model, but, when combined with g_2, the total movement is similar to an OU model in that it has a region of attraction, although not a central point.

Despite the added complexity, there are benefits of the full STPP analysis over the conditional analysis where location times are considered fixed. The full likelihood in Equation 4.32 appears considerably more difficult to evaluate than the conditional version in Equation 4.36. The baseline temporal intensity of the locations is constant; thus, there does not even seem to be any inferential benefit to be gained in that respect either. The real benefit lies in the likelihood computation. In the conditional likelihood (4.36), the 2-D spatial integral must be computed n times. However, in the full likelihood, only one three-dimensional (3-D) integral is necessary. Johnson et al. (2013) show that the approximation methods used for spatial and temporal likelihood can be extended to the spatio-temporal version. To do so, we augment the observed locations and times with a grid of quadrature locations in space and time, $\mathbf{q}_{ijl}$, $l = 1, \ldots, L_{ij}$ at times $u_{ij}, j = 0, \ldots, J_i$, where $t_{i-1} = u_{i0}$ and $t_i = u_{iJ_i}$. In addition, we denote a_{ijl} to be the area of the cell associated with $\mathbf{q}_{ijl}$. The area a_{ijl} depends on how the points were selected. If the points were selected in a nonregular manner, a Voronoi tessellation can be used to obtain the areas, whereas if the points were selected as centroids of a regular grid, then the area of the grid cell is used. However, even if a regular grid is used, recall that observed locations are part of the observed set; $\boldsymbol{\mu}_i = \mathbf{q}_{iJ_il}$ for one of the l, say $l = L_{ij}$, to be compatible with the j index. Therefore, the quadrature points are never on a completely regular grid, so, some adjustment has to be made to assign area mass to the observed locations. The easiest method to handle this situation is to count the number of observed locations in a cell and divide the area of the cell by this count plus one (for the grid centroid) and assign the partial areas to all points in the cell. Now, the log-likelihood can be approximated by

$$\ell(\boldsymbol{\beta}, \boldsymbol{\theta}, \alpha) \approx \sum_{i=1}^{n}\sum_{j=1}^{J_i}\sum_{l=1}^{L_{ij}} z_{ijl}\log(\lambda_{ijl}) - \lambda_{ijl}, \tag{4.46}$$

where

$$\lambda_{ijl} = \exp(\log(v_{ijl}) + \mathbf{x}'(\mathbf{q}_{ijl})\boldsymbol{\beta} + \eta(\mathbf{q}_{ijl}, \boldsymbol{\theta}) + \alpha d(\mathbf{q}_{ijl}, \boldsymbol{\mu}_{i-1})^2/\Delta_{ij}), \tag{4.47}$$

and $v_{ijl} = a_{ijl}(u_{ij} - u_{i,j-1})$, $\Delta_{ij} = u_{ij} - t_{i-1}$, and $z_{ijl} = 1$ for $j = J_i$ and $l = L_{ij}$ and zero elsewhere. As in Chapters 2 and 3, the z_{ijl} can be thought of as independent Poisson variables and we can fit the model with any GLM fitting software using $\log(v_{ijl})$ as an offset. Fitting a single model may not be much faster using the full likelihood versus the conditional likelihood; however, after the "model data" have been created, that is, z_{ijl}, $\mathbf{q}_{ijl}$, $\mathbf{x}(\mathbf{q}_{ijl})$, and $B_{ijl} = d(\mathbf{q}_{ijl}, \boldsymbol{\mu}_{i-1})^2/\Delta_{ij}$, any number of other submodels or alternate models that use the quantities can be fit using the optimized GLM algorithms in most statistical software. Thus, a full analysis, including model selection or multimodel inference, can proceed quickly after the data are created.

We fit the full STPP model to the brown bear telemetry data using the R package "mgcv" (Wood 2003) to implement the thin-plate spline (Figure 4.18). The larger-scale home range surface, $g_2(\boldsymbol{\mu})$ in Figure 4.18b, shows the bimodal surface found by Johnson et al. (2008b) when using an OU movement model and two centers of attraction. The difference between this analysis and that described by Johnson et al. (2008b) is that we did not have to specify the number of points of attraction or the switching time. The small-scale resource selection surface, $g_1(\boldsymbol{\mu})$, is shown in Figure 4.18c, where one can see that the bear selects for habitat in close proximity to streams with coefficient estimate $\hat{\beta} = -2.41$ and 95% confidence interval $(-2.71, -2.11)$ for the distance from the nearest stream covariate. Finally, the OU-like transition kernel, $g_2(\boldsymbol{\mu}_i)f(\boldsymbol{\mu}_i|\boldsymbol{\mu}_{i-1}, \Delta_i)$, is shown in Figure 4.18d for $i = 101$. Notice that the mass of the transition density is centered on the current location (white point) and decreases to zero as the distance from the current location increases.

4.7.4 STPPs as Spatial Point Processes

In addition to analyzing telemetry data using the full STPP likelihood, Johnson et al. (2013) showed how to use a spatial point process model to implement an STPP for telemetry data. Similar to how we integrated a variable out of a higher-dimensional distribution in the previous chapters, we can marginalize over the time dimension of an STPP to obtain a spatial point process. The method is based on a result given by Illian et al. (2008). If a spatio-temporal Poisson process is defined by the intensity function $\lambda(\boldsymbol{\mu}, t)$, then the observed locations (ignoring the times that they were observed) follow a spatial Poisson process with intensity $\lambda(\boldsymbol{\mu}) = \int_{\mathcal{T}} \lambda(\boldsymbol{\mu}, t)\, dt$. Thus, if we want to consider analyzing animal telemetry with a spatial point process as described by Warton and Shepherd (2010) and Aarts et al. (2012), we can use the marginalization result to create a spatial Poisson process approximation to the marginal process.

For the STPP intensity function in the brown bear analysis in Equation 4.41, the spatial point process intensity function is given by

$$\begin{aligned}\lambda(\boldsymbol{\mu}) &= \int_{\mathcal{T}} \lambda(\boldsymbol{\mu}, t|\mathcal{H}_t)\, dt \\ &= \exp\left(\mathbf{x}'(\boldsymbol{\mu})\boldsymbol{\beta} + \eta(\boldsymbol{\mu}, \boldsymbol{\theta})\right) \int_{\mathcal{T}} \exp\left(-\alpha d(\boldsymbol{\mu}, \boldsymbol{\mu}_{i-1})^2/(t - t_{i-1})\right) dt \\ &= \exp\left(\mathbf{x}'(\boldsymbol{\mu})\boldsymbol{\beta} + \eta(\boldsymbol{\mu}, \boldsymbol{\theta})\right) \sum_{i=1}^{n} \int_0^{\Delta_i} \exp\left(-\alpha d(\boldsymbol{\mu}, \boldsymbol{\mu}_{i-1})^2/u\right) du, \end{aligned} \quad (4.48)$$

where we assume that $\alpha > 0$. The integral on the right-hand side of Equation 4.48 does not exist in a closed form, but symbolically,

$$\gamma_i(\boldsymbol{\mu}) = \int_0^{\Delta_i} \exp\left(-\alpha d(\boldsymbol{\mu}, \boldsymbol{\mu}_{i-1})^2/u\right) du$$
$$= \Delta_i \exp(-\alpha d(\boldsymbol{\mu}, \boldsymbol{\mu}_{i-1})^2/\Delta_i) - \alpha d(\boldsymbol{\mu}, \boldsymbol{\mu}_{i-1})^2 \Gamma(0, \alpha d(\boldsymbol{\mu}, \boldsymbol{\mu}_{i-1})^2/\Delta_i), \tag{4.49}$$

where $\Gamma(\cdot,\cdot)$ is the incomplete gamma function. Although $\Gamma(\cdot,\cdot)$ is not available in closed form, numerical solutions are available in most statistical software. It is hardly apparent what the $\gamma_i(\boldsymbol{\mu})$ function looks like in geographic space, but as Johnson et al. (2013) noted, it is similar in shape to a bivariate normal density centered on $\boldsymbol{\mu}$ (Figure 4.19). Substituting Equation 4.49 into Equation 4.48, we obtain the spatial intensity

$$\lambda(\boldsymbol{\mu}) = \exp\left(\mathbf{x}'(\boldsymbol{\mu})\boldsymbol{\beta} + \eta(\boldsymbol{\mu}, \boldsymbol{\theta})\right) \sum_{i=1}^{n} \gamma_i(\boldsymbol{\mu})$$
$$= \exp\left(\mathbf{x}'(\boldsymbol{\mu})\boldsymbol{\beta} + \eta(\boldsymbol{\mu}, \boldsymbol{\theta}) + u(\boldsymbol{\mu})\right), \tag{4.50}$$

where $u(\boldsymbol{\mu}) = \log \sum_{i=1}^{n} \gamma_i(\boldsymbol{\mu})$ is a log kernel density estimate made by placing the $\gamma_i(\boldsymbol{\mu})$ kernel over every observed location.

Using this spatial marginalization approach, Johnson et al. (2013) showed that standard GLM fitting software can be used with the Berman–Turner quadrature method or Poisson cell counts for parameter estimation. However, some level of approximation is still required beyond the quadrature or Poisson representation of

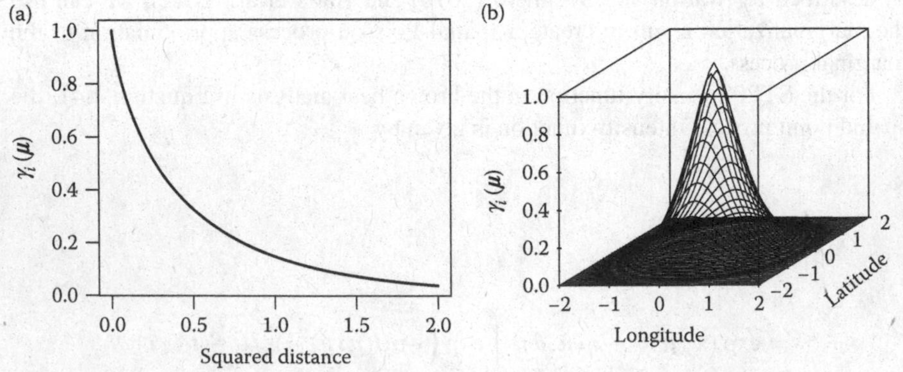

FIGURE 4.19 Illustration of $\gamma_i(\boldsymbol{\mu})$. The plot on the left depicts a 1-D view, where $\gamma_i(\boldsymbol{\mu})$ is a function of squared distance from $\boldsymbol{\mu}$. The plot on the right shows the full 3-D view of the $\gamma_i(\boldsymbol{\mu})$ function illustrating that it assumes the form of a kernel similar to a bivariate normal kernel.

the likelihood. First, note that there is a movement-related parameter, α, that is part of the $\gamma_i(\boldsymbol{\mu})$ calculation. Thus, if one uses GLM fitting routines in an efficient manner, α must be known and fixed at $\hat{\alpha}$ so that we can calculate one kernel density map, $\hat{u}(\boldsymbol{\mu})$, which may be used as a covariate in the GLM model. However, this assumes that the Brownian kernel is correct and uncertainty about α is ignored. Technically, $u(\boldsymbol{\mu})$ is a random spatial field that controls interactions between observed locations, much like the Gibbs spatial point process models of Section 2.1.3, for which model fitting is notoriously difficult. Illian et al. (2012) suggested a log-Gaussian Cox process (Section 2.1.3) approximation, which can be fit numerically using readily available software. The basic premise of the Illian et al. (2012) approach is to create a constructed covariate that captures interaction effects, then add the covariate to a random effects version of the Poisson GLM representation of the Poisson spatial point process. Thus, for the spatial marginalization model in Equation 4.50, this can be accomplished by the following procedure:

1. Partition the region into fine set of grid cells.
2. Choose a reasonable value for $\hat{\alpha}$ and calculate $\hat{u}(\boldsymbol{\mu}_l)$ at the cell centroids, $\boldsymbol{\mu}_l$.
3. Count the number of telemetry locations within each cell, y_l.
4. Fit a Poisson generalized additive model (GAM) to the y_l with log rate

$$\log(\lambda_l) = \mathbf{x}'(\boldsymbol{\mu}_l)\boldsymbol{\beta} + \eta_1(\boldsymbol{\mu}_l, \boldsymbol{\theta}_1) + \eta_2(\hat{u}(\boldsymbol{\mu}_l), \boldsymbol{\theta}_2), \tag{4.51}$$

where η_1 is a 2-D thin-plate regression spline and η_2 is a 1-D thin-plate regression spline or other nonparametric smooth.

Instead of GAM smoothing, Johnson et al. (2013) and Illian et al. (2012) used ICAR models (Section 2.3.2), which provide an acceptable alternative in a Bayesian framework.

To demonstrate the spatial marginalization of STPP models, we reanalyzed the bear data presented in the last section. In their example, Johnson et al. (2013) chose a value for α based on a commonly held belief about the maximum speed of travel for northern fur seals (*Callorhinus ursinus*). We take an empirical approach by setting $\hat{\alpha}$ equal to 1/mean(observed velocity)^2, because, for Brownian motion, the expected displacement in one unit of time is approximately $1/\sqrt{\alpha}$. After selecting $\hat{\alpha}$, we created a heterogeneous kernel UD using Equation 4.49; the kernel is shown in Figure 4.20a. The remaining effects in the model were as described in the previous section and the `R` package "mgcv" was used to fit the model. The estimated resource selection coefficient for stream distance was $\hat{\beta} = -1.78$, with 95% confidence interval $(-2.12, -1.44)$ (fitted selection surface shown in Figure 4.20b). What might be termed the "availability" surface, $\eta_1(\boldsymbol{\mu}, \boldsymbol{\theta}_1) + \eta_2(\hat{u}(\boldsymbol{\mu}), \boldsymbol{\theta}_2)$ (Figure 4.20c) accounts for all the other influences beyond resource selection, that is temporal autocorrelation and home range effects. The availability surface functions as a trade-off of

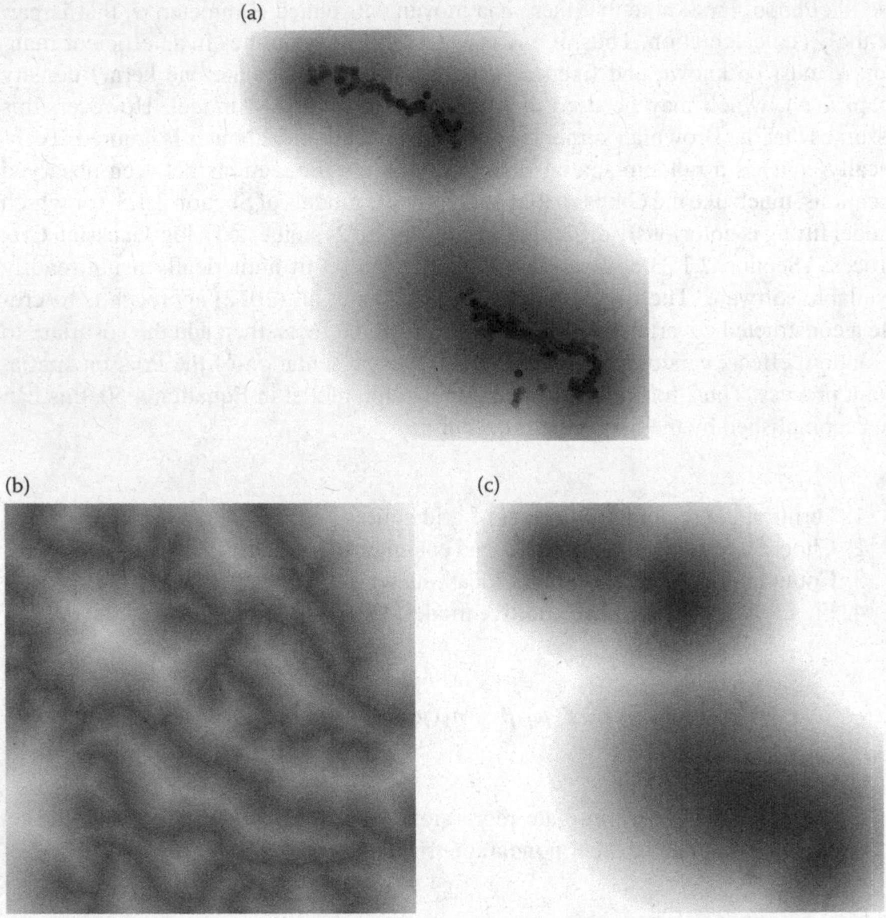

FIGURE 4.20 Spatial point process model fit to spatio-temporal brown bear telemetry data using the temporal marginalization approximation. Plot (a) illustrates the observed data and the $\log \hat{u}(\boldsymbol{\mu})$ surface. The fitted resource selection surface for the stream distance covariate is shown in (b). Plot (c) illustrates the fitted availability surface; $\eta_1(\boldsymbol{\mu}, \boldsymbol{\theta}_1) + \eta_2(\hat{u}(\boldsymbol{\mu}), \boldsymbol{\theta}_2)$. Plots (b) and (c) partition space utilization to the components attributable to known covariates and those components that cannot be assigned to a specific habitat trait.

temporal autocorrelation in a spatio-temporal model for spatial autocorrelation in a spatial model.

Overall, the advantages to using the full STPP or the temporally marginalized spatial method are substantial. One can use available software packages to fit spatio-temporal models to their temporally correlated movement data and it will be tractable for large data sets and populations of telemetered individuals. However, as with any approximate method, the quality of inference for selection coefficients ($\boldsymbol{\beta}$) depends on how well the availability is approximated.

4.8 ADDITIONAL READING

As we discussed in Chapter 2, spatial point process models have a long history in the fields of stochastic processes and statistics. Therefore, there are numerous references. We referenced Illian et al. (2008) in Chapter 2, and find it to be an excellent reference on the subject from the statistics perspective. From the animal ecology perspective, the classical reference for RSFs is Manly et al. (2007). Johnson (1980) is the most well-known reference describing (and naming) the different scales associated with resource selection inference (e.g., global, regional, home range level, sub-home range level).

Nielson and Sawyer (2013) proposed the use of a negative binomial likelihood for modeling aggregations of telemetry data in discrete units of space (instead of a Poisson). A negative binomial is one type of overdispersed Poisson distribution (Ver Hoef and Boveng 2007) and could, in principle, account for overdispersion in the counts. When point processes are clustered at small scales (subpixel scales) as opposed to completely random (CSR), the counts will be overdispersed relative to the Poisson. White and Bennetts (1996) suggest that many types of count data in ecological research may be subject to overdispersion; thus, finding ways to properly account for overdispersion in point process models for telemetry data is an ongoing area of research.

While we are mainly focused on model-based approaches for animal movement inference, there is a large body of literature devoted to density estimation (e.g., Silverman 1986). More recently, there have been several new developments in spatio-temporal density estimation for telemetry data (e.g., Keating and Cherry 2009; Fleming et al. 2015).

There is substantial literature on animal home ranges and methods for estimating them; perhaps, in part, because of the overuse of the phrase "home range" to represent all animal movement models. However, focusing strictly on delineating home range boundaries, the classical overview was written by Worton (1987). Subsequent follow-up reviews can be found in Worton (1989) and Powell (2000). Furthermore, many other nonparametric home range estimators have been developed (e.g., Getz et al. 2007).

Finally, an important area of related research is species distribution modeling (SDM) based on "presence-only" data. The most common methods used to analyze species distribution data are point process models, although they are not commonly referred to as point process models in the SDM literature. Critical recent references in SDM are Warton and Shepherd (2010), Aarts et al. (2012), and Dorazio (2012), who reconciled many related methods, and as we saw in the previous section, much of their results also pertain to the analysis of telemetry data. Nielson et al. (2009) discussed the issue of nonignorable missing telemetry data due to failed position acquisitions correlating with certain environmental or behavioral variables. Dorazio (2012) discussed a similar situation in species distribution analysis based on point process models where there may be imperfect detection of the species. Many of the concepts Dorazio (2012) discussed transfer to the analysis of telemetry data as well. Related concepts have been discussed in the spatial statistics literature, and each field has recommended methods for accommodating preferential sampling (e.g., Diggle et al. 2010a).

5 Discrete-Time Models

5.1 POSITION MODELS

5.1.1 RANDOM WALK

Obtaining resource selection inference using the point process models described in the previous chapter is often straightforward when the temporal component of the process is unimportant or ignored. However, point process models become increasingly sophisticated when the temporal component is accommodated explicitly. If the physical process of movement itself is of interest, it may be useful to take a different modeling perspective. An alternative to the point process perspective considers the data and underlying process in the time domain directly, allowing for explicit forms of temporal dependence. If the process is considered in discrete time, we are firmly back in the realm of time series statistics, as discussed in Chapter 3. We begin our discussion of continuous-space discrete-time movement models by introducing the random walk and then make a sequence of extensions that provide additional insight about the dynamics and behavior of moving animals.

When telemetry measurement error is formally accounted for, the phrase "state-space model" has been commonly used to describe these classes of models (e.g., Jonsen et al. 2005; Patterson et al. 2008). Recall that a state-space model is a hierarchical model in which the data are modeled conditioned on the underlying process and then the underlying process is also modeled (usually dynamically, but not always). Thus, the hierarchical models for point processes (or RSFs) in the previous chapter are also technically state-space models. In that case, the "state" is the underlying point process representing true animal locations. In what follows, we describe models for the position process as the state and then extend them so that the state represents the behavioral mode of the individual.

For now, assume there is no (or very little) measurement error associated with our telemetry data so that we can model the true individual locations $\boldsymbol{\mu}_t$ directly. Next, assume that an appropriate time scale is known in advance. That is, the temporal "grain" of our model can be thought of as Δ_t, the length of time between two successive animal locations. For now, if we assume that Δ_t is constant through time, we can drop the Δ_t from the notation for the nearest time ahead $\boldsymbol{\mu}_{t+\Delta_t}$ and behind $\boldsymbol{\mu}_{t-\Delta_t}$ so that we have $\boldsymbol{\mu}_{t+1}$ and $\boldsymbol{\mu}_{t-1}$, without any loss in generality.*

The key to formulating a random walk is to recall Markovian dynamics from Chapter 3. In the simplest case, we assume the location at time t depends on all of the

* We use the t subscript here instead of i for simplicity and consistency with the time series notation. Also, we can always just linearly rescale the entire temporal extent so that Δ_t is with respect to the units of interest. For example, in that case, if Δ_t is an hour, then the $+1$ and -1 correspond to the hour after and before.

other locations but only through its nearest neighbors in time. That is, if the random walk is of order 1 (e.g., an AR(1) time series model), we can write

$$\boldsymbol{\mu}_t = \boldsymbol{\mu}_{t-1} + \boldsymbol{\varepsilon}_t, \tag{5.1}$$

for $t = 1, \ldots, T$, where the errors are often assumed to be independent and normally distributed such that $\boldsymbol{\varepsilon}_t \sim \text{N}(\mathbf{0}, \boldsymbol{\Sigma})$. In the simplest case, the error covariance matrix could be specified as $\boldsymbol{\Sigma} \equiv \sigma^2\mathbf{I}$, so that the errors are symmetric. In time series statistics, this model is often referred to as a vector autoregressive model (i.e., VAR(1); because $\boldsymbol{\mu}_t$ is multidimensional) of order one. Recall, from Chapter 3, that an alternative way to write the random walk model is using distribution notation such that $\boldsymbol{\mu}_t \sim \text{N}(\boldsymbol{\mu}_{t-1}, \sigma^2\mathbf{I})$. The distribution notation is a theme throughout this book and can be helpful when formulating hierarchical models, especially in a Bayesian framework.

In terms of mechanisms, the VAR(1) model implies that the displacement of the individual during each time step occurs in a random direction with step length governed by a univariate Weibull distribution. In this case, the variance component σ^2 controls the step lengths between successive locations. For example, Figure 5.1 shows both the empirical and theoretical distributions (histogram based on $T = 10{,}000$ time steps and $\sigma^2 = 0.5, 1, 2$) of the step lengths resulting from three simulated 2-D trajectories using Equation 5.1. Notice how both the central tendency and spread in step length distribution increase as the random walk variance parameter (σ^2) increases (Figure 5.1).

The formulation in Equation 5.1 is often referred to as an "intrinsic" conditional autoregressive model (ICAR) because the effect of the location at the previous time step is not attenuated or mixed with another location-based force. ICAR models are nonstationary in the sense that the process is not being shrunk back toward some fixed location in space and there are no other constraints on the process (e.g., that the $\boldsymbol{\mu}_t$ sum to $\mathbf{0}$). There is no assumed center of gravity in the model to keep the individual in one general area; thus, it lacks that mechanism for modeling a central place forager like a pygmy rabbit (*Brachylagus idahoensis*) or a harbor seal (*Phoca vitulina*). However, substantial flexibility can be accommodated in the autoregressive framework, and the VAR(1) specification can serve as a basis from which we can generalize to account for more complicated mechanisms of movement.

Finally, one of the unique aspects of conditional autoregressive models is that it is straightforward to translate the first-order (i.e., mean) dynamics into second-order dependence (covariance). That is, if we vectorize all of the $\boldsymbol{\mu}_t$ and concatenate such that $\boldsymbol{\mu} \equiv (\boldsymbol{\mu}_1', \ldots, \boldsymbol{\mu}_T')'$, then the same properties used in spatial statistics allow us to write the joint distribution for all of the individual locations as $\boldsymbol{\mu} \sim \text{N}(\mathbf{1} \otimes \bar{\boldsymbol{\mu}}, \boldsymbol{\Sigma}_\mu \otimes \mathbf{I})$. This type of formulation can sometimes be advantageous for computational reasons because of the sparsity of $\boldsymbol{\Sigma}_\mu^{-1}$ or various basis function expansions of the covariance structure. We return to this concept of modeling dynamics in the second-order component of the model in Chapter 6.

We consider each of the following generalizations to the simple random walk model in turn:

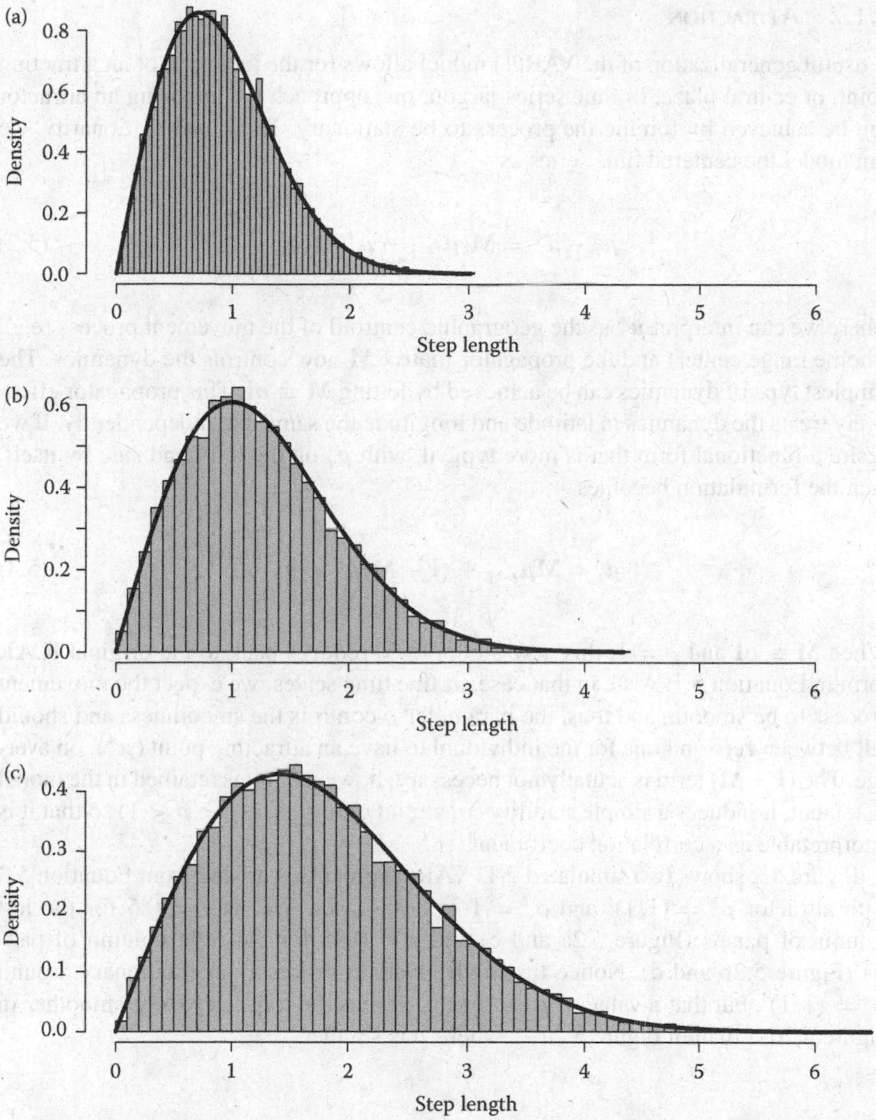

FIGURE 5.1 Empirical (histogram) and theoretical (solid line) step length distributions based on simulated trajectories using Equation 5.1 with $T = 10{,}000$ and (a) $\sigma^2 = 0.5$, (b) $\sigma^2 = 1$, and (c) $\sigma^2 = 2$. The theoretical distribution for step lengths arising from the model in Equation 5.1 is Weibull$(2, \sqrt{2\sigma^2})$.

1. Attraction
2. Measurement error
3. Temporal alignment (i.e., irregular data)
4. Heterogeneous behavior (e.g., covariate-based, change-point-based)

5.1.2 Attraction

A useful generalization of the VAR(1) model allows for the inclusion of an attracting point, or central place. In time series jargon, one approach for imposing an attractor can be achieved by forcing the process to be stationary. To impose stationarity, we can model the centered time series as

$$\boldsymbol{\mu}_t - \boldsymbol{\mu}^* = \mathbf{M}(\boldsymbol{\mu}_{t-1} - \boldsymbol{\mu}^*) + \boldsymbol{\varepsilon}_t, \tag{5.2}$$

where we can interpret $\boldsymbol{\mu}^*$ as the geographic centroid of the movement process (e.g., a home range center) and the propagator matrix $\mathbf{M}$ now controls the dynamics. The simplest type of dynamics can be achieved by letting $\mathbf{M} \equiv \rho\mathbf{I}$. This propagator effectively treats the dynamics in latitude and longitude the same, but independently. If we desire a functional form that is more typical, with $\boldsymbol{\mu}_t$ on the left-hand side by itself, then the formulation becomes

$$\boldsymbol{\mu}_t = \mathbf{M}\boldsymbol{\mu}_{t-1} + (\mathbf{I} - \mathbf{M})\boldsymbol{\mu}^* + \boldsymbol{\varepsilon}_t. \tag{5.3}$$

When $\mathbf{M} \equiv \rho\mathbf{I}$ and $\rho = 1$, this new model (5.3) reduces back to the original ICAR form in Equation 5.1. Also, in that case, at fine time scales, we expect the movement process to be smooth, and thus, the parameter ρ controls the smoothness and should fall between zero and one for the individual to have an attracting point ($\boldsymbol{\mu}^*$), on average. The $(\mathbf{I} - \mathbf{M})$ term is actually not necessary; however, if it is retained in the model statement, it induces a simple stability constraint on ρ (i.e., $-1 < \rho < 1$) so that it is interpretable as a correlation coefficient.

Figure 5.2 shows two simulated 2-D VAR(1) processes arising from Equation 5.3 with attractor $\boldsymbol{\mu}^* = (1,1)'$ and $\sigma^2 = 1$ in both cases. We set $\rho = 0.5$ for the left column of panels (Figure 5.2a and c) and $\rho = 0.95$ for the right column of panels (Figure 5.2b and d). Notice that both bivariate processes are stationary around $\boldsymbol{\mu}^* = (1,1)'$, but that a value of ρ closer to 1 forces the trajectory to be smoother in Figure 5.2d–f than in Figure 5.2a–c, where ρ is smaller.

5.1.3 Measurement Error

To account for measurement error in the random walk model, we add a level of hierarchy to the model structure for the observed data with error (as in Chapters 3 and 4). As before, suppose that the observed telemetry locations are $\mathbf{s}_t$, and they arise as Gaussian random variables with mean $\boldsymbol{\mu}_t$ and error variance σ_s^2. The resulting hierarchical model with a single attracting location $\boldsymbol{\mu}^*$ is

$$\begin{aligned} \mathbf{s}_t &\sim \mathrm{N}(\boldsymbol{\mu}_t, \sigma_s^2\mathbf{I}), \\ \boldsymbol{\mu}_t &\sim \mathrm{N}(\mathbf{M}\boldsymbol{\mu}_{t-1} + (\mathbf{I} - \mathbf{M})\boldsymbol{\mu}^*, \sigma_\mu^2\mathbf{I}). \end{aligned} \tag{5.4}$$

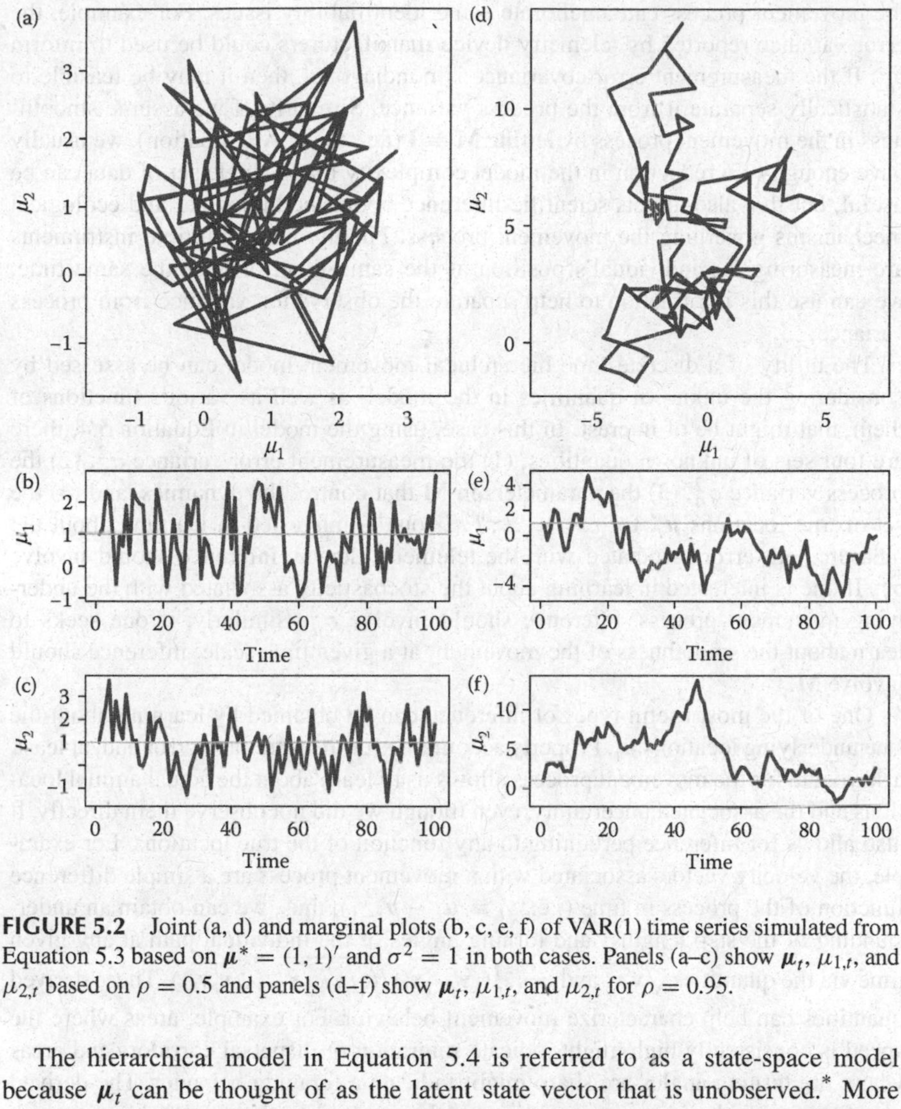

FIGURE 5.2 Joint (a, d) and marginal plots (b, c, e, f) of VAR(1) time series simulated from Equation 5.3 based on $\boldsymbol{\mu}^* = (1, 1)'$ and $\sigma^2 = 1$ in both cases. Panels (a–c) show $\boldsymbol{\mu}_t$, $\mu_{1,t}$, and $\mu_{2,t}$ based on $\rho = 0.5$ and panels (d–f) show $\boldsymbol{\mu}_t$, $\mu_{1,t}$, and $\mu_{2,t}$ for $\rho = 0.95$.

The hierarchical model in Equation 5.4 is referred to as a state-space model because $\boldsymbol{\mu}_t$ can be thought of as the latent state vector that is unobserved.* More complicated error models, such as those described in the previous chapter, could also be used. However, it is important to recognize that the variance components (i.e., σ_s^2 and σ_μ^2) in this hierarchical model may not be identifiable without some form of replication at the data level. That is, there may not be enough information in a single set of telemetry data for an individual movement path to separate the signal from the noise. More information or constraints on either the measurement error or

* Again, we feel that the term "state-space" is a bit too broad to be used to effectively differentiate random walk models for animal movement because any hierarchical model can be thought of as a state-space model. Outside of the animal ecology world, the term "state-space" is often reserved for temporal and spatio-temporal processes.

the movement process can ameliorate some identifiability issues. For example, the error variance reported by telemetry device manufacturers could be used to inform σ_s^2. If the measurement error covariance is nondiagonal, then it may be feasible to statistically separate it from the process variance. Similarly, if we assume smoothness in the movement process by letting $\mathbf{M} \equiv \mathbf{I}$ (i.e., the ICAR situation), we usually have enough of a reduction in the model complexity that a single set of data can be useful, but this also affects scientific inference about the biological and ecological mechanisms governing the movement process. Finally, when multiple instruments are measuring the individual's position at the same time, or near the same time, we can use this information to help separate the observation variance from process variance.*

The utility of a discrete-time hierarchical movement model can be assessed by considering the unknown quantities in the model, as well as various functions of them, that might be of interest. In this case, using the model in Equation 5.4, there are four sets of unknown quantities: (1) the measurement error variance σ_s^2, (2) the process variance σ_μ^2, (3) the parameters in $\mathbf{M}$ that control the dynamics, and (4) the set of true locations $\boldsymbol{\mu}_t$, for $t = 1, \ldots, T$. If one is interested in learning about the measurement error associated with the telemetry device, inference should involve σ_s^2. If one is interested in learning about the stochasticity associated with the underlying movement process, inference should involve σ_μ^2. Similarly, if one seeks to learn about the smoothness of the movement at a given time scale, inference should involve $\mathbf{M}$.

One of the most useful types of inference can be obtained by learning about the true underlying locations $\boldsymbol{\mu}_t$. Properly accounting for measurement error and, at least, a surrogate for the movement process allows us to learn about the actual animal locations and the associated uncertainty, even though we did not observe them directly. It also allows for inference pertaining to any function of the true locations. For example, the velocity vectors associated with a movement process are a simple difference function of the process in time (i.e., $\mathbf{v}_t \equiv \boldsymbol{\mu}_t - \boldsymbol{\mu}_{t-1}$); thus, we can obtain an understanding of the step lengths† and turning angles of the individual path at any given time via the quantities $\sqrt{\mathbf{v}_t'\mathbf{v}_t}$ and $\cos^{-1}((\mathbf{v}_{t-1}'\mathbf{v}_t)/(\sqrt{\mathbf{v}_{t-1}'\mathbf{v}_{t-1}}\sqrt{\mathbf{v}_t'\mathbf{v}_t}))$. These derived quantities can help characterize movement behavior. For example, areas where the speed is consistently high might indicate migration or dispersal corridors and areas where the turning angles are sharp might indicate a foraging behavior. The derived quantities are indexed in time so they can be mapped to the spatial domain (with associated uncertainty) because we have formal inference for the true locations in space (i.e., $\boldsymbol{\mu}_t$). In some sense, this could be viewed as an emergent or derived form of inference.

* This could occur when an individual is telemetered with a GPS and Argos device simultaneously, for example. While it may be a good idea to use multiple telemetry devices for statistical reasons, it may not always be practical. However, telemetry data sets collected with two devices do exist (e.g., Argos and VHF devices; Buderman et al. 2016).

† To obtain speed from step length when the trajectory is temporally irregular, divide $\sqrt{\mathbf{v}_i'\mathbf{v}_i}$ by Δ_i, the difference in time between fixes $\Delta_i = t_i - t_{i-1}$. Then the speed has the same units as Δ_i.

5.1.4 Temporal Alignment (Irregular Data)

We have relied on evenly spaced data and process time steps in our presentation of the continuous-space discrete-time models described thus far. While it simplifies the expressions in the model statement to assume a perfect alignment of data and process time steps, it also sweeps some of the more complicated technical challenges for implementation under the rug.

Given that the movement process can be embedded as a latent component in a hierarchical model, the temporal resolution is user-defined. The actual choice of the time step, Δt, is directly related to the inference one obtains from the model. For example, if $\Delta t = 1$ h, the parameters controlling the movement dynamics are interpretable on the 1-h time scale. Thus, a quantity such as turning angle represents the angle associated with the overall displacement vector over the 1 h period. The formal identification of the most appropriate temporal resolution for modeled movement processes is an important area of ongoing research (e.g., Gurarie and Ovaskainen 2011; Fleming et al. 2014; Hooten et al. 2014; Schlägel and Lewis 2016).

Nonetheless, it is convention to choose the process time scale at the most regular scale the data appear and then develop a measurement error model that meshes with the process scale. The approach used by Jonsen et al. (2005), McClintock et al. (2012), and others is to align the data and process scales by linear interpolation. To include this interpolation in our model specification, we modify our temporal notation such that $\mathbf{s}_i$ represents the observed animal location at time i, for $i = 1, \ldots, n$ measurements (i.e., telemetry fixes). The measurement times can then be indexed as t_i. We link the measurement times with the process times via a weighted average

$$\mathbf{s}_i \sim \text{N}((1 - w_i)\boldsymbol{\mu}_{t-\Delta t} + w_i\boldsymbol{\mu}_t, \sigma_s^2\mathbf{I}), \tag{5.5}$$

where $\boldsymbol{\mu}_{t-\Delta t}$ and $\boldsymbol{\mu}_t$ correspond to the nearest process time before and after t_i, respectively. The weight w_i is a function of the time interval between t_i and t such that

$$w_i = \frac{t - t_i}{\Delta t}. \tag{5.6}$$

This model is general enough that, when t_i co-occurs with t, the data point is exactly associated with the underlying process location. For cases when Δt is small relative to the movement frequency of the animal, this type of linear interpolation model performs well. However, as Δt increases, the linear interpolation may not be appropriate (see Section 5.2.5 for more on discretization error). In most cases, there is agreement between the data and process scales and the linear interpolation performs well.

5.1.5 Heterogeneous Behavior

In practice, there may be many attracting (or repulsion) points. If the attracting point is time-varying and can take on only a finite set of values, $\boldsymbol{\mu}_1^*, \ldots, \boldsymbol{\mu}_j^*, \ldots, \boldsymbol{\mu}_J^*$, that are known in advance, we can modify the model for the position process so that

$$\boldsymbol{\mu}_t = \mathbf{M}\boldsymbol{\mu}_{t-1} + (\mathbf{I} - \mathbf{M})\boldsymbol{\mu}_t^* + \boldsymbol{\varepsilon}_t, \tag{5.7}$$

where $\boldsymbol{\mu}_t^*$ arises from a finite mixture

$$\boldsymbol{\mu}_t^* = \begin{cases} \boldsymbol{\mu}_1^* & \text{with probability } p_1 \\ \vdots & \vdots \\ \boldsymbol{\mu}_J^* & \text{with probability } p_J \end{cases} . \tag{5.8}$$

The probability vector $\mathbf{p} \equiv (p_1, \ldots, p_J)'$ contains the probabilities associated with each attracting point and sums to one. The mixture model in Equation 5.8 could also be implemented with latent binary auxiliary variables so that

$$\boldsymbol{\mu}_t^* = \begin{cases} \boldsymbol{\mu}_1^* & z_{1,t} = 1 \\ \vdots & \vdots \\ \boldsymbol{\mu}_J^* & z_{J,t} = 1 \end{cases} , \tag{5.9}$$

where the vector $\mathbf{z}_t$ contains all zeros and a single one arising from a multinomial distribution: $\mathbf{z}_t \sim \text{MN}(1, \mathbf{p})$. The mixture model in Equation 5.9 can yield computational advantages such as conjugacy in a Bayesian setting (which results in an automatic model fitting algorithm that does not require tuning).

A different approach for incorporating multiple attracting points relies on a temporal change-point model. For example, Figure 5.3 shows two simulated trajectories arising from Equation 5.7 with two attracting points

$$\boldsymbol{\mu}_t^* = \begin{cases} \boldsymbol{\mu}_1^* & t < t^* \\ \boldsymbol{\mu}_2^* & t \geq t^* \end{cases} , \tag{5.10}$$

and t^* is the change point to be estimated. Notice how the trajectory (Figure 5.3d–f) based on $\rho = 0.95$ is almost so smooth that it obscures the fact that a change in attracting point occurred. Longer time series will eventually reveal the change, but the amount of data needed to estimate a change depends on the smoothness.

One approach for adding temporal heterogeneity to the simple random walk models we have discussed thus far is to allow the dynamics to change over time. That is, generically, we could let the propagator matrix from Equation 5.4 vary with time (i.e., $\mathbf{M}_t$). In fact, if $\mathbf{M}_t \equiv \rho_t \mathbf{I}$ and we expect $\rho_t > 0$, we could use $\text{logit}(\rho_t) = \mathbf{x}_t'\boldsymbol{\beta}$ to link the temporal correlation coefficients to a set of time-specific covariates. In essence, this regression formulation for temporal correlation accomplishes two things: (1) it allows for differing degrees of smoothness in the movement at different times and (2) it allows for inference concerning the potential drivers of movement dynamics. For example, if we used a temporal covariate, such as temperature, for x_t in the model $\text{logit}(\rho_t) = \beta_0 + \beta_1 x_t$, then a negative value for β_1 would indicate that the position process $\boldsymbol{\mu}_t$ becomes more steady (i.e., smoother) as temperatures decrease.

We could also use a random effect approach to allow for heterogeneous dynamics. The random effect model could be specified as $\text{logit}(\rho_t) \sim \text{N}(\mu_\rho, \sigma_\rho^2)$. In this case, the time-specific correlations are shrunk back to a general mean μ_ρ (in the logit space)

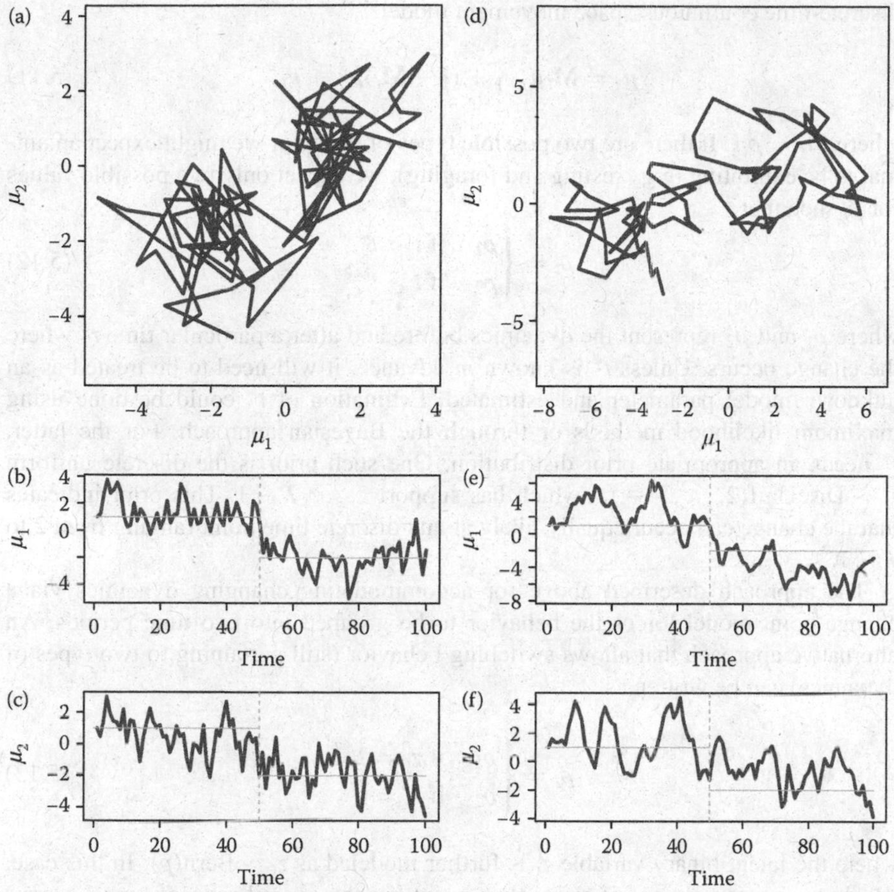

FIGURE 5.3 Joint (a, d) and marginal plots (b, c, e, f) of VAR(1) time series simulated from Equation 5.7 based on attracting points $\boldsymbol{\mu}_1^* = (1, 1)'$ and $\boldsymbol{\mu}_2^* = (-2, -2)'$, with $\sigma^2 = 1$ in both cases. Panels (a–c) show $\boldsymbol{\mu}_t$, $\mu_{1,t}$, and $\mu_{2,t}$ based on $\rho = 0.5$ and panels (d–f) show $\boldsymbol{\mu}_t$, $\mu_{1,t}$, and $\mu_{2,t}$ for $\rho = 0.95$. Horizontal gray lines represent $\boldsymbol{\mu}_1^*$ and $\boldsymbol{\mu}_2^*$ and vertical dashed gray lines represent t^*.

according to their consistency through time as controlled by the variance parameter σ_ρ^2. Both μ_ρ and σ_ρ^2 could be treated as unknown and estimated, or, alternatively, σ_ρ^2 could be tuned to provide the best predictive model.*

Another approach to allow for heterogeneity in the movement dynamics is based on change-point modeling similar to what we used to account for multiple attracting points. Assuming, for simplicity, only one attracting point $\boldsymbol{\mu}^*$, we specify the

* Tuning model parameters to attenuate other model components is a model selection technique called "regularization" (e.g., Hooten and Hobbs 2015).

discrete-time continuous-space movement model

$$\boldsymbol{\mu}_t = \mathbf{M}_t \boldsymbol{\mu}_{t-1} + (\mathbf{I} - \mathbf{M}_t)\boldsymbol{\mu}^* + \boldsymbol{\varepsilon}_t, \tag{5.11}$$

where $\mathbf{M}_t = \rho_t \mathbf{I}$. If there are two possible types of behavior we might expect an animal to be exhibiting (e.g., resting and foraging), we expect only two possible values for ρ_t such that

$$\rho_t = \begin{cases} \rho_1 & \text{if } t < t^* \\ \rho_2 & \text{if } t \geq t^* \end{cases}, \tag{5.12}$$

where ρ_1 and ρ_2 represent the dynamics before and after a particular time t^* where the change occurs. Unless t^* is known in advance, it will need to be treated as an unknown model parameter and estimated. Estimation of t^* could be done using maximum likelihood methods or through the Bayesian approach. For the latter, t^* needs an appropriate prior distribution. One such prior is the discrete uniform $t^* \sim \text{DiscUnif}(2, \ldots, T-1)$, which has support $2, \ldots, T-1$. This prior indicates that the change can occur equally likely at any discrete time point ranging from 2 to $T-1$.*

The approach described above for accommodating changing dynamics via a change-point model forces the behavior to be grouped into two time periods. An alternative approach that allows switching behavior (still pertaining to two types of dynamics) can be written as

$$\rho_t = \begin{cases} \rho_1 & \text{if } z_t = 1 \\ \rho_2 & \text{if } z_t = 0 \end{cases}, \tag{5.13}$$

where the latent binary variable z_t is further modeled as $z_t \sim \text{Bern}(p)$. In this case, the dynamics can change at every time point but the overall ratio of type 1 versus type 2 dynamics is controlled by p. It may be unrealistic to assume that an individual animal could switch back and forth on every time step, so additional smoothing on the switching process could be induced in several ways. A simple approach for smoothing the switching dynamics could be achieved with an HMM such that

$$z_t \sim \begin{cases} \text{Bern}(p_1) & \text{if } z_{t-1} = 1 \\ \text{Bern}(p_0) & \text{if } z_{t-1} = 0 \end{cases}. \tag{5.14}$$

When p_1 is large (i.e., close to one) and p_0 is small (i.e., close to zero), then ρ_t will have a tendency to stay in its current state longer. By contrast, when both p_1 and p_0 are close to 0.5, the model reverts to the simpler case with $p = 0.5$. A stronger assumption would let $p_0 = 1 - p_1$, which is capable of providing appropriate dynamic behavior for some situations. For example, Figure 5.4 shows a simulated trajectory based on the change-point model (5.11) and Equations 5.13 and 5.14 with $p_1 = 1 - p_0 = 0.99$.

* Notice that we do not include times 1 and T in the support for t^*. This is because we would not have enough data to estimate a change point on the boundary of our time series.

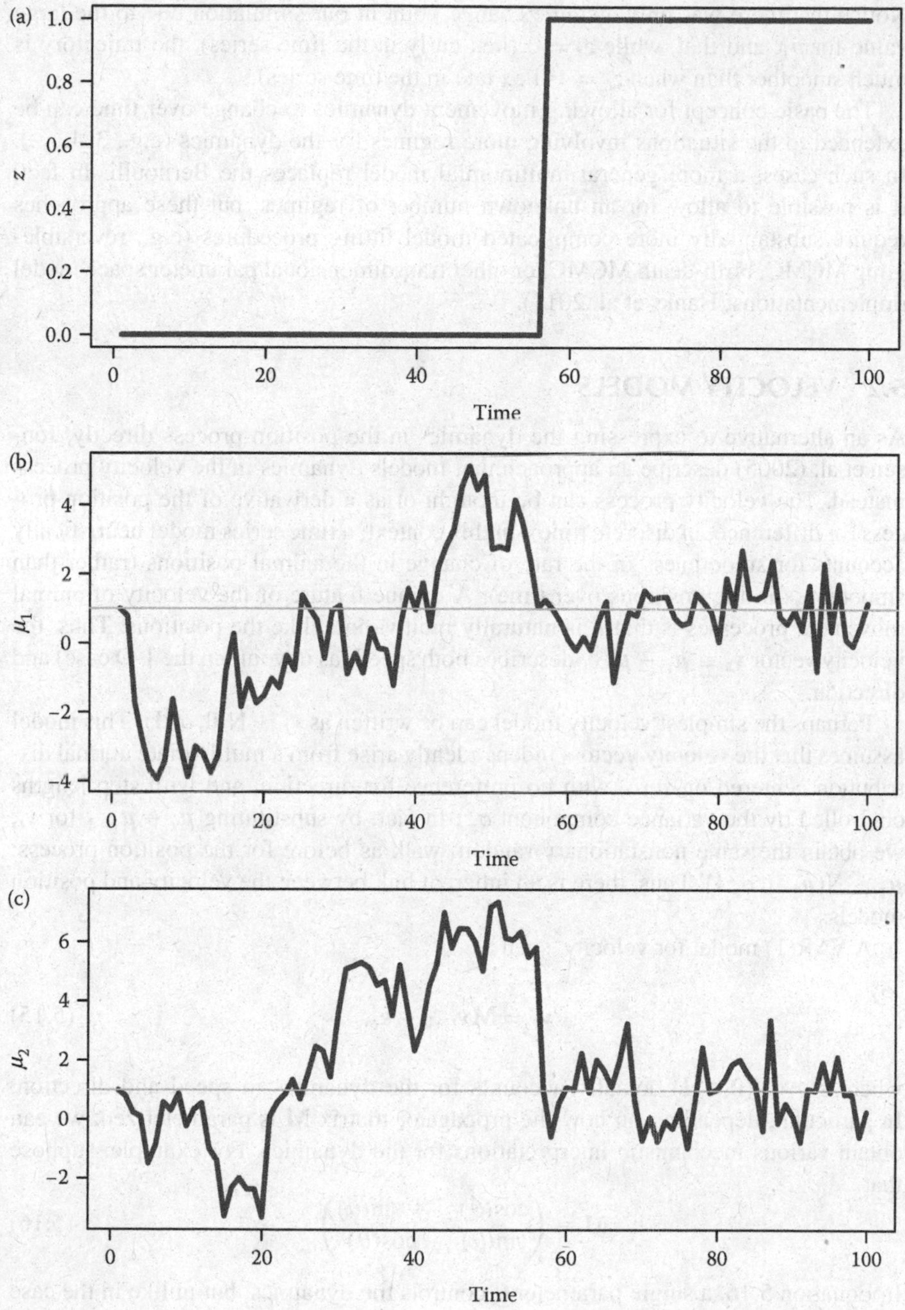

FIGURE 5.4 Hidden Markov process z_t (a) and marginal plots (b, c) of a VAR(1) time series simulated from Equations 5.11, 5.13 and 5.14 with $p = 0.99$, $\boldsymbol{\mu}^* = (1, 1)'$, and $\sigma^2 = 1$ in both cases. Panels (b–c) show $\mu_{1,t}$ and $\mu_{2,t}$ based on $\rho_1 = 0.1$ when $z_t = 1$ and $\rho = 0.99$ when $z_t = 0$.

Notice that there was only a single change point in our simulation due to the large value for p_1, and that, while $z_t = 0$ (i.e., early in the time series), the trajectory is much smoother than when $z_t = 1$ (i.e., late in the time series).

The basic concept for allowing movement dynamics to change over time can be extended to the situations involving more regimes for the dynamics (e.g., $3, 4, \ldots$). In such cases, a more general multinomial model replaces the Bernoulli. In fact, it is possible to allow for an unknown number of regimes, but these approaches require substantially more complicated model fitting procedures (e.g., reversible-jump MCMC, birth-death MCMC, or other transdimensional parameter space model implementations; Hanks et al. 2011).

5.2 VELOCITY MODELS

As an alternative to expressing the dynamics in the position process directly, Jonsen et al. (2005) describe an approach that models dynamics in the velocity process instead. The velocity process can be thought of as a derivative of the position process (or difference, in discrete time). In this context, a time series model heuristically accounts for smoothness in the rate of change in the animal positions (rather than smoothness in the positions over time). A unique feature of the velocity of animal movement processes is that it is naturally multivariate, like the positions. Thus, the velocity vector $\mathbf{v}_t \equiv \boldsymbol{\mu}_t - \boldsymbol{\mu}_{t-1}$ describes both speed (as it would in the 1-D case) and direction.

Perhaps the simplest velocity model can be written as $\mathbf{v}_t \sim \text{N}(\mathbf{0}, \sigma_\varepsilon^2\mathbf{I})$. This model assumes that the velocity vectors independently arise from a multivariate normal distribution centered on zero, with no preference for direction, and with step lengths controlled by the variance component σ_ε^2. In fact, by substituting $\boldsymbol{\mu}_t - \boldsymbol{\mu}_{t-1}$ for $\mathbf{v}_t$, we obtain the same nonstationary random walk as before for the position process: $\boldsymbol{\mu}_t \sim \text{N}(\boldsymbol{\mu}_{t-1}, \sigma_\varepsilon^2\mathbf{I})$. Thus, there is an inherent link between the velocity and position models.

A VAR(1) model for velocity, such as

$$\mathbf{v}_t = \mathbf{M}\mathbf{v}_{t-1} + \boldsymbol{\varepsilon}_t, \tag{5.15}$$

where $\boldsymbol{\varepsilon}_t \sim \text{N}(\mathbf{0}, \sigma_\varepsilon^2\mathbf{I})$ actually accounts for the dynamics in speed and direction. In particular, depending on how the propagator matrix $\mathbf{M}$ is parameterized, we can obtain various mechanistic interpretations for the dynamics. For example, suppose that

$$\mathbf{M} \equiv \begin{pmatrix} \cos(\theta) & -\sin(\theta) \\ \sin(\theta) & \cos(\theta) \end{pmatrix}. \tag{5.16}$$

In Equation 5.16, a single parameter θ controls the dynamics, but unlike in the case where $\mathbf{M} \equiv \rho\mathbf{I}$, the trigonometric specification (5.16) allows θ to control the turning angle from one time to the next and imposes additional correlation between step length and turning angle (McClintock et al. 2014). The turning angle parameter is bounded between $-\pi$ and π; thus, when θ is close to zero, the individual animal will move directly ahead. Conversely, when θ is closer to π or $-\pi$, the animal will turn around 180°. Similarly, $\theta = \pi/2$ and $\theta = -\pi/2$ will turn the animal left and right, respectively. The step length is controlled by the process error variance σ_ε^2, with larger values of σ_ε^2 corresponding to larger step lengths on average.

Given the interpretation of model parameters as controlling turning angles and step lengths in Equation 5.16, the random walk model associated with the velocity process has a decidedly mechanistic feel to it. The random walk model for velocity (5.15) also has a direct relationship with a discrete-time continuous-space model for the position process. To derive this relationship, substitute $\boldsymbol{\mu}_t - \boldsymbol{\mu}_{t-1}$ for $\mathbf{v}_t$ in Equation 5.15 to obtain

$$\boldsymbol{\mu}_t - \boldsymbol{\mu}_{t-1} = \mathbf{M}(\boldsymbol{\mu}_{t-1} - \boldsymbol{\mu}_{t-2}) + \boldsymbol{\varepsilon}_t. \tag{5.17}$$

Then add $\boldsymbol{\mu}_{t-1}$ to both sides and simplify the equation. As we saw in Chapter 3, the result is a VAR(2) model for the position process:

$$\boldsymbol{\mu}_t = (\mathbf{I} + \mathbf{M})\boldsymbol{\mu}_{t-1} - \mathbf{M}\boldsymbol{\mu}_{t-2} + \boldsymbol{\varepsilon}_t, \tag{5.18}$$

where the propagator matrices are $(\mathbf{I} + \mathbf{M})$ for the first-order difference and $-\mathbf{M}$ for the second-order difference. We discussed this result generically in Chapter 3, but now we see how the same basic concept can be helpful in modeling animal movement explicitly. In essence, higher-order dependence in the position process (i.e., longer memory) allows for a useful mechanistic interpretation of the model components.

The parameterization of the propagator matrix $\mathbf{M}$ in Equation 5.16 yields a very restrictive model. A simple extension is

$$\mathbf{M} \equiv \gamma \begin{pmatrix} \cos(\theta) & -\sin(\theta) \\ \sin(\theta) & \cos(\theta) \end{pmatrix}, \tag{5.19}$$

where the parameter γ (for $0 < \gamma < 1$) dampens the contribution of the dynamics in velocity as necessary when γ becomes small. In this new formulation (5.19), the propagator matrix is a function of two unknown variables (i.e., γ and θ) that must be estimated.

Figure 5.5 shows simulated trajectories (i.e., $\boldsymbol{\mu}_t = \sum_{\tau \le t} \mathbf{v}_\tau$) using the velocity VAR(1) model (5.15) for six different parameter scenarios. The trajectories take on very distinct geometric patterns in Figure 5.5d–f; when $\gamma = 1$, the trajectories exhibit all left turns with consistent turning angles. Whereas, when $\gamma = 0.1$ (Figure 5.5a–c), the trajectories exhibit more variability in their turns and step lengths. Realistic animal movement trajectories occur when $-\pi/2 < \theta < \pi/2$ and $\gamma < 1$ for typical temporal resolutions (Δt) associated with most telemetry data.

To fit a Bayesian version of the discrete-time velocity model in Equations 5.15 and 5.19 to data, we specified the priors $\sigma^2 \sim \text{IG}(0.001, 0.001)$, $\theta \sim \text{Unif}(-\pi, \pi)$, and $\gamma \sim \text{Unif}(0, 1)$. To simulate a data set, we used $T = 100$ time steps and let $\theta = \pi/8$, $\gamma = 0.9$, and $\sigma^2 = 1$ (Figure 5.6). Using MCMC to fit the model with 10,000 iterations, the marginal posterior distributions for model parameters are shown in Figure 5.7. Based on the simulated data in Figure 5.6, the model is able to recover the parameters quite well.

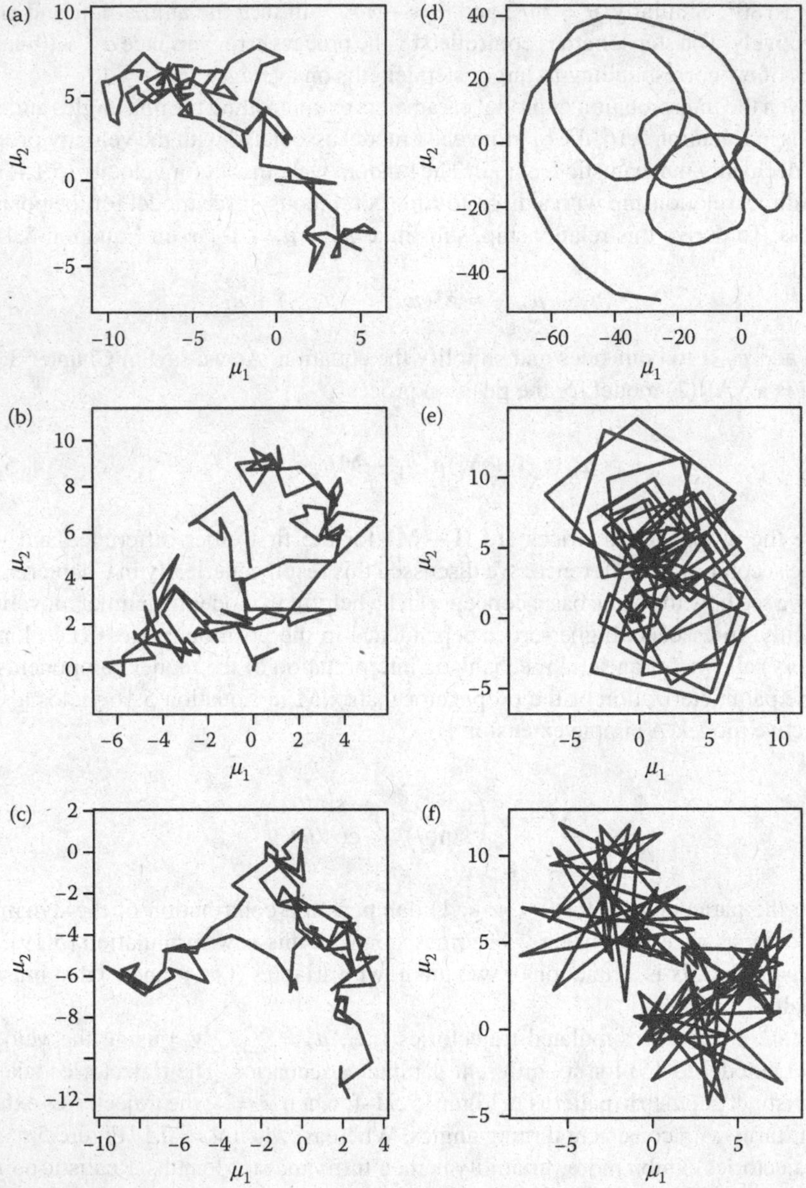

FIGURE 5.5 Simulated position processes (i.e., $\boldsymbol{\mu}_t = \sum_{\tau \leq t} \mathbf{v}_\tau$) using Equation 5.15 for six different parameter scenarios and $T = 100$ time steps: (a) $\theta = 0.1 \cdot \pi$, $\gamma = 0.1$, (b) $\theta = \pi/2$, $\gamma = 0.1$, (c) $\theta = 0.9 \cdot \pi$, $\gamma = 0.1$, (d) $\theta = 0.1 \cdot \pi$, $\gamma = 1$, (e) $\theta = \pi/2$, $\gamma = 1$, and (f) $\theta = 0.9 \cdot \pi$, $\gamma = 1$.

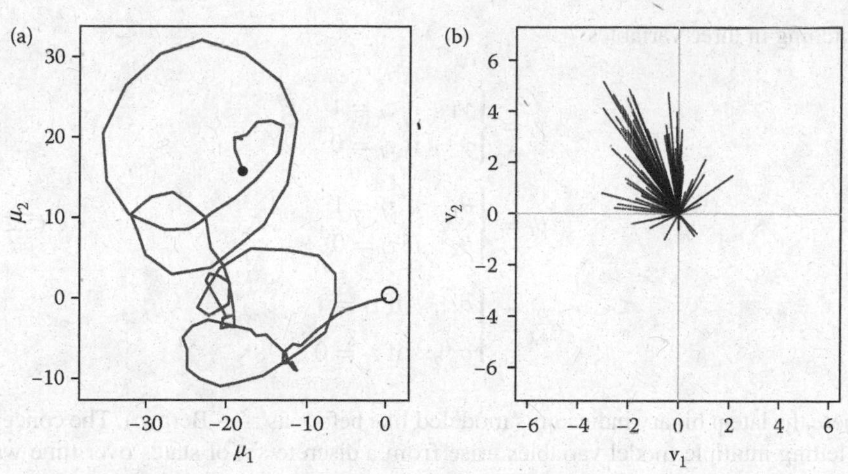

FIGURE 5.6 Simulated position processes (i.e., $\boldsymbol{\mu}_t = \sum_{\tau \leq t} \mathbf{v}_\tau$) using Equation 5.15 for $T = 100$ equally spaced time steps and $\theta = \pi/8$, $\gamma = 0.9$, and $\sigma^2 = 1$. Panel (a) shows the trajectory (i.e., $\boldsymbol{\mu}_t$ or position process) with open and closed circles denoting the starting and ending positions, respectively. Panel (b) shows the velocity vectors ($\mathbf{v}_t$).

As with the first-order dynamic models for the position process, Jonsen et al. (2005) allow this velocity model to contain time-varying dynamics with a switching model similar to Equation 5.13. In this case, several variables could be indexed in time and allowed to arise from a discrete set of possible movement states. For example, in the situation involving two movement states, we could allow for

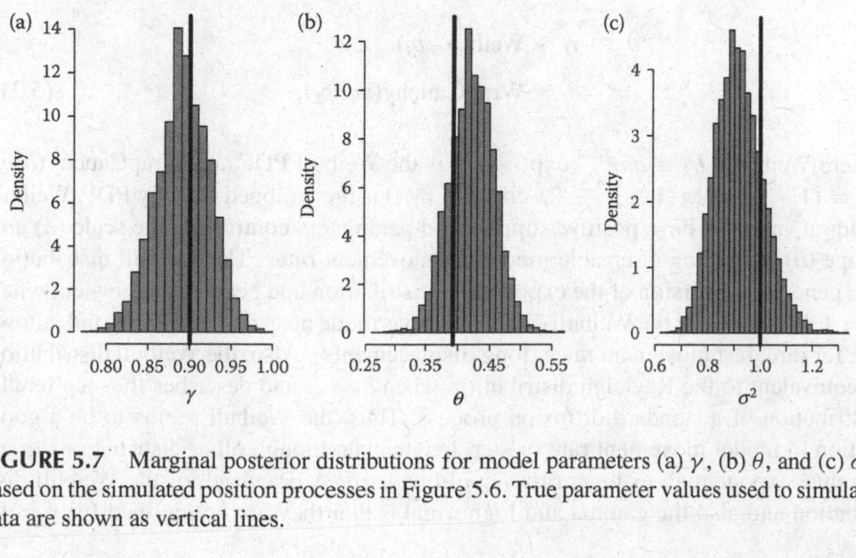

FIGURE 5.7 Marginal posterior distributions for model parameters (a) γ, (b) θ, and (c) σ^2 based on the simulated position processes in Figure 5.6. True parameter values used to simulate data are shown as vertical lines.

switching in three variables:

$$\gamma_t = \begin{cases} \gamma_1 & \text{if } z_t = 1 \\ \gamma_2 & \text{if } z_t = 0 \end{cases},$$
$$\theta_t = \begin{cases} \theta_1 & \text{if } z_t = 1 \\ \theta_2 & \text{if } z_t = 0 \end{cases}, \tag{5.20}$$
$$\sigma^2_{\varepsilon,t} = \begin{cases} \sigma^2_{\varepsilon,1} & \text{if } z_t = 1 \\ \sigma^2_{\varepsilon,2} & \text{if } z_t = 0 \end{cases},$$

where the latent binary indicator is modeled like before as $z_t \sim \text{Bern}(p)$. The concept of letting multiple model variables arise from a discrete set of states over time was introduced by Morales et al. (2004) in the animal movement context. They suggested that animals may alternate among different behaviors, resulting in different movement patterns, and thus, proposed model formulations that allow for state switching behavior.

5.2.1 Modeling Movement Parameters

Morales et al. (2004) provided a similar, but heuristically different, approach to modeling animal movement than what we have described so far. Following Turchin (1998), they considered statistical models for components of discrete-time random walks. For example, let r_t represent the observed daily average movement rate (or displacement distance in a day) and θ_t represent the associated observed turning angle. Then Morales et al. (2004) assumed that their GPS measurement error was negligible for the scale of daily movements of the translocated elk (*Cervus canadensis*) they were studying and specified the data model as

$$r_t \sim \text{Weib}(a_t, b_t),$$
$$\theta_t \sim \text{WrapCauchy}(m_t, \rho_t), \tag{5.21}$$

where $\text{Weib}(r|a, b) \equiv abr^{b-1}\exp(-ar^b)$ is the Weibull PDF and $\text{WrapCauchy}(\theta|m, \rho) \equiv (1-\rho^2)/(2\pi(1+\rho^2-2\rho\cos(\theta-m)))$ is the wrapped Cauchy PDF. Weibull random variables have positive support and parameters controlling the scale (a) and shape (b), providing a sensible model for movement rates. The Weibull distribution is a generalized version of the exponential distribution and becomes equivalent when $b = 1$. When $b < 1$, the Weibull distribution has mode near zero and a long tail, allowing for rare, fast movement rates (long displacements). Also, the Weibull distribution is equivalent to the Rayleigh distribution when $b = 2$, and describes the step length distribution of a standard diffusion process. Thus, the Weibull seems to be a good option to model movement rate or step length even though other distributions (e.g., gamma, exponential, or lognormal) could be used. A drawback of the Weibull distribution and also the gamma and lognormal is that they are not defined for $r_t = 0$.

Thus, zeros in the data have to be ignored or replaced by small numbers. Also, the shape parameter b may not always be identifiable using telemetry data alone. The wrapped Cauchy is a circular distribution with support $-\pi \leq \theta \leq \pi$ and parameters controlling the scale (ρ) and location (m) of probability density on a circle.* The wrapped Cauchy also has the special property that, as $\rho \to 0$, it becomes a uniform distribution on the circle providing equally likely turning angles in any direction.†

Allowing the parameters (e.g., a_t, b_t, m_t, and ρ_t) to vary in time completely would lead to an overfit model with very little learning potential. However, fixing them all in time would not allow for realistic movement behavior. Thus, the strength of the approach proposed by Morales et al. (2004) is in the underlying process that governs the variation in these parameters. Morales et al. (2004) proposed seven different model specifications that provide varying amounts of heterogeneity:

1. "Single": Temporally homogeneous parameters such that $a_t = a$, $b_t = b$, $m_t = m$, and $\rho_t = \rho$ for all t to serve as a baseline to compare to more complex models.
2. "Double": An independent mixture of two movement states such that

$$a_t = \begin{cases} a_1 & \text{if } z_t = 1 \\ a_2 & \text{if } z_t = 0 \end{cases}, \tag{5.22}$$

$$b_t = \begin{cases} b_1 & \text{if } z_t = 1 \\ b_2 & \text{if } z_t = 0 \end{cases}, \tag{5.23}$$

$$m_t = \begin{cases} m_1 & \text{if } z_t = 1 \\ m_2 & \text{if } z_t = 0 \end{cases}, \tag{5.24}$$

$$\rho_t = \begin{cases} \rho_1 & \text{if } z_t = 1 \\ \rho_2 & \text{if } z_t = 0 \end{cases}, \tag{5.25}$$

   and $z_t \sim \text{Bern}(p)$ as in Equation 5.20.
3. "Double with covariates": As an extension of the "double" model, we can let the mixture probability change over time associated with some auxiliary source of temporally varying covariates ($\mathbf{x}_t$). In this case, we let $z_t \sim \text{Bern}(p_t)$ with $\text{logit}(p_t) = \mathbf{x}_t'\boldsymbol{\beta}$. The covariates can describe the habitat type where the animal is located at time t, or time of day, or day of the season, for example. These covariates account for animals that are more likely to move in a certain way when they are located in a particular habitat type or at some particular times of the day or season.

---

* There are numerous other parameterizations of the Weibull and wrapped Cauchy distributions, but these are most similar to those used by Morales et al. (2004).

† Compared to other circular distributions such as the wrapped normal or von Mises, the wrapped Cauchy is more peaked and has heavier tails and thus implies different long-term consequences (Codling et al. 2008).

4. "Double-switch": Like the "double" model, but generalizing the probability of switching back and forth between two states. Using our latent indicator variable approach, this can be written as (5.14)

$$z_t \sim \begin{cases} \text{Bern}(p_1) & \text{if } z_{t-1} = 1 \\ \text{Bern}(p_2) & \text{if } z_{t-1} = 0 \end{cases}. \quad (5.26)$$

   Recall that this is referred to as an HMM in recent animal movement literature.
5. "Switch with covariates": Combining the ideas in the "double-switch" and "double with covariates," we allow the switching probabilities to change over time according to some covariates

$$z_t \sim \begin{cases} \text{Bern}(p_{1,t}) & \text{if } z_{t-1} = 1 \\ \text{Bern}(p_{2,t}) & \text{if } z_{t-1} = 0 \end{cases}, \quad (5.27)$$

   where $\text{logit}(p_{1,t}) = \mathbf{x}_t'\boldsymbol{\beta}_1$ and $\text{logit}(p_{2,t}) = \mathbf{x}_t'\boldsymbol{\beta}_2$. They considered distance to habitat types as covariates modulating the switch from an "exploratory" movement to an "encamped" one.
6. "Switch-constrained": Same as the "Switch with covariates" model but with informative priors on at least a subset of the movement parameters (i.e., a, b, m, or ρ). For example, we could specify a prior that constrains $b_2 > 1$ so that the mode of the distribution for movement rate or step length is away from zero.
7. "Triple-switch": An extension of the "double-switch" model containing three potential movement states rather than two.

Each of these models implies a different form of heterogeneity for the process. The assumptions of each model can be checked formally and, if appropriate, models can be compared to select which among them has the best predictive ability (Morales et al. 2004; Hooten and Hobbs 2015). Of course, each of the scenarios presented by Morales et al. (2004) could be generalized further if the situation dictates (e.g., including additional movement states).

In a Bayesian framework, each of the unknown parameters in the models above needs a distribution and one could proceed as usual in completing the model statement with explicit priors. For example, Morales et al. (2004) used gamma priors for a and b parameters, uniform priors for m, ρ, and p (or p_1 and p_2), and then normal priors for regression coefficients $\boldsymbol{\beta}$ (or $\boldsymbol{\beta}_1$ and $\boldsymbol{\beta}_2$). A potential problem with clustering models such as these is "label switching" (i.e., states may be labeled differently in different model fits). Thus, it is common to define a subset of the parameters for one of the categories or states as a function of others. For example, Morales et al. (2004) set $a_1 = a_2 + \varepsilon$, where ε is the difference between scale parameters and was assigned a truncated normal prior. Thus, state 1 will always have a larger scale parameter, which can help avoid label switching. Alternatively, for the mean step length or

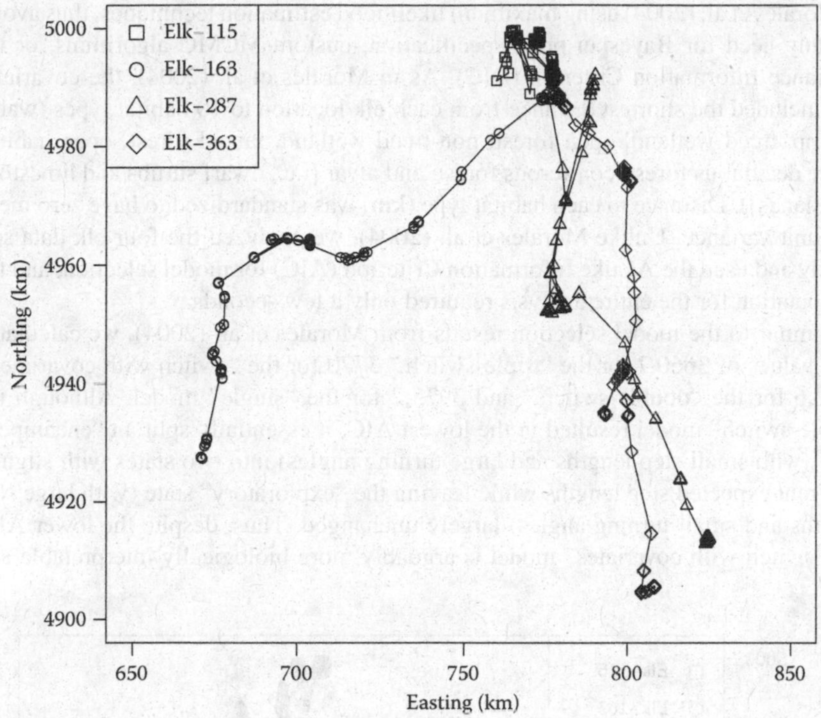

FIGURE 5.8 GPS telemetry data for four individual elk analyzed by Morales et al. (2004).

movement rate of one of the states, Morales et al. (2004) set $m_2 = m_1 + \varepsilon$, yielding the corresponding scale for the Weibull as $a_1 = (m_1/\Gamma(1 + 1/b_1))^{b_1}$.

Morales et al. (2004) demonstrated their discrete-time random walk models using four cow elk (*Cervus canadensis*) GPS telemetry data sets collected in east-central Ontario, Canada (Figure 5.8). Using Bayesian methods, they fit the models using WinBUGS (Lunn et al. 2000) and performed model selection to identify the best predicting model. They also used posterior predictive checks for the temporal autocorrelation of movement rates to justify their use of informative priors for the shape parameter of the Weibull distribution. That is, they simulated trajectories with parameters sampled from the joint posterior and compared the temporal autocorrelation in step length with those from the data.

As previously stated, Bayesian methods are often employed for complicated models that are challenging to fit using non-Bayesian approaches, such as discrete-time movement models that explicitly account for location measurement error or temporally irregular observations (e.g., Jonsen et al. 2005; McClintock et al. 2012). However, because elk are terrestrial and the GPS fixes were obtained at regular time intervals, an analysis similar to that of Morales et al. (2004) can be performed using maximum likelihood methods (e.g., Langrock et al. 2012).

We used the R package "moveHMM" (R Core Team 2013; Michelot et al. 2015) to fit the "single," "double-switch," "switch with covariates," and "triple-switch" HMMs of Morales et al. (2004) using maximum likelihood estimation techniques, thus avoiding any need for Bayesian prior specification, custom MCMC algorithms, or the Deviance Information Criterion (DIC). As in Morales et al. (2004), the covariates (**X**) included the shortest distance from each elk location to 10 habitat types (water, swamp, treed wetland, open forest, non-treed wetland, mixed forest, open habitat, dense deciduous forest, coniferous forest, and alvar [i.e., dwarf shrubs and limestone grasslands]). Distance to each habitat type (km) was standardized to have zero mean and unit variance. Unlike Morales et al. (2004), we analyzed the four elk data sets jointly and used the Akaike Information Criterion (AIC) for model selection, and the computation for the entire analysis required only a few seconds.

Similar to the model selection results from Morales et al. (2004), we calculated AIC values of 3660.7 for the "triple-switch," 3770 for the "switch with covariates," 3790.6 for the "double-switch," and 3975.2 for the "single" model. Although the "triple-switch" model resulted in the lowest AIC, it essentially split an "encamped" state (with small step lengths and large turning angles) into two states with slightly different expected step lengths while leaving the "exploratory" state (with large step lengths and small turning angles) largely unchanged. Thus, despite the lower AIC, the "switch with covariates" model is arguably more biologically interpretable and

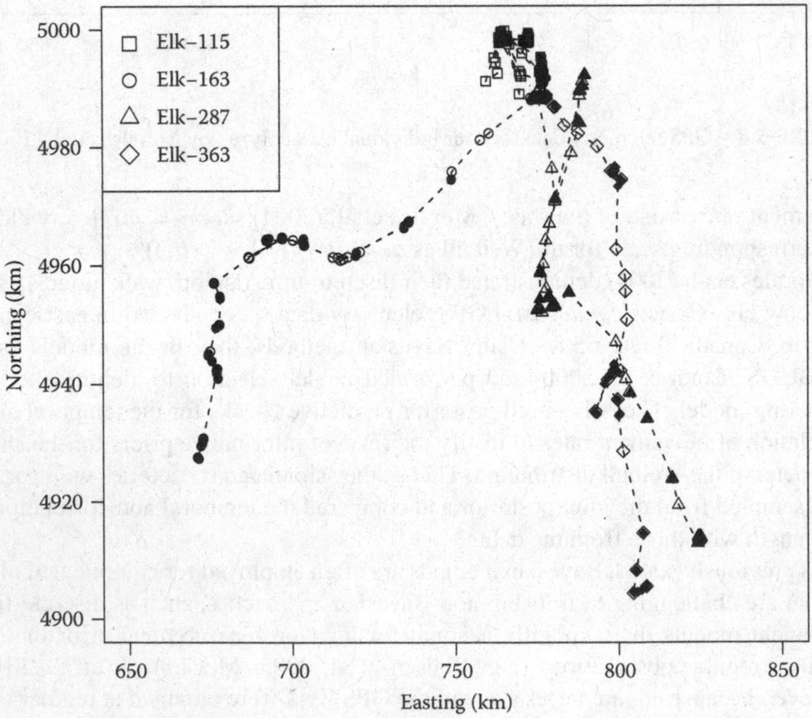

FIGURE 5.9 Estimated elk trajectories and estimated movement states: "encamped" (solid symbols connected by solid lines) and "exploratory" (hollow symbols connected by dashed lines).

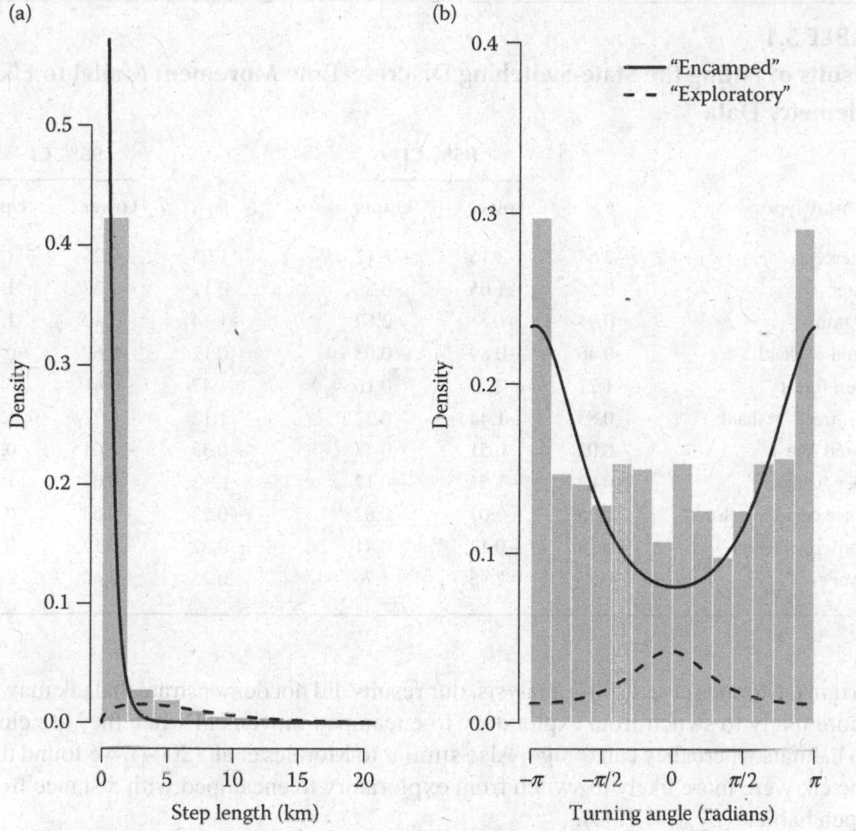

FIGURE 5.10 Estimated distributions for the "encamped" and "exploratory" movement states for (a) step length and (b) turning angle.

meaningful (Morales et al. 2004) with its two distinct "encamped" and "exploratory" states (Figures 5.9 and 5.10).

Using the notation of Equation 5.27, both the "moveHMM" package and Morales et al. (2004) parameterize the "switch with covariates" model in terms of the switching probabilities

$$1 - p_{1,t} = \text{logit}^{-1}\left(\mathbf{x}_t'\boldsymbol{\beta}_1\right); \tag{5.28}$$

that is, the probability of switching from the "encamped" state at time $t-1$ to the "exploratory" state at time t. The second probability,

$$p_{2,t} = \text{logit}^{-1}\left(\mathbf{x}_t'\boldsymbol{\beta}_2\right), \tag{5.29}$$

is the probability of switching from the "exploratory" state at time $t-1$ to the "encamped" state at time t. We found several significant relationships between distance to habitat type and these state switching probabilities in our analysis. As in the

TABLE 5.1
Results of Fitting the State-Switching Discrete-Time Movement Model to Elk Telemetry Data

		95% CI			95% CI	
Habitat Type	$\hat{\beta}_1$	**Lower**	**Upper**	$\hat{\beta}_2$	**Lower**	**Upper**
(Intercept)	−2.64	−3.16	−2.12	1.05	0.25	1.86
Water	−0.25	−1.05	0.56	0.11	−1.27	1.49
Swamp	0.94	−0.8	2.69	−1.14	−3.42	1.14
Treed wetland	−0.46	−0.89	−0.03	0.12	−0.69	0.93
Open forest	−1.27	−2.7	0.16	1.47	−0.61	3.55
Non-treed wetland	−0.83	−1.44	−0.22	1.12	0.32	1.91
Mixed forest	−0.02	−0.51	0.48	−0.33	−1.01	0.36
Open habitat	−0.82	−1.51	−0.12	1.00	0.02	1.99
Dense deciduous forest	0.45	0.07	0.82	−0.23	−0.69	0.24
Coniferous forest	0.14	−0.14	0.41	0.32	−0.12	0.76
Alvar	−0.35	−2.46	1.76	0.25	−2.42	2.92

original Morales et al. (2004) analysis, our results did not demonstrate that elk may be more likely to switch from exploratory to encamped movement when they are close to habitats where they can forage. Also similar to Morales et al. (2004), we found that the elk were more likely to switch from exploratory to encamped with distance from open habitat ($\hat{\beta}_{2,oh} = 1.00$).

Unlike Morales et al. (2004), our joint analysis found that these four elk were more likely to stay in the encamped state when close to dense deciduous forest ($\hat{\beta}_{1,ddf} = 0.45$), more likely to switch from encamped to exploratory when close to non-treed wetland ($\hat{\beta}_{1,ntw} = -0.83$) or treed wetland ($\hat{\beta}_{1,tw} = -0.46$), more likely to switch from encamped to exploratory when close to open habitat ($\hat{\beta}_{1,oh} = -0.82$), and more likely to remain in the exploratory state when close to non-treed wetland ($\hat{\beta}_{2,ntw} = 1.12$) (Table 5.1).

5.2.2 Generalized State-Switching Models

Whereas Morales et al. (2004) set up the essential framework for modeling movement parameters directly, McClintock et al. (2012) generalized this framework to accommodate measurement error, irregular data, and multiple latent states the individual could switch among. McClintock et al. (2012) used the same basic model formulation as Morales et al. (2004) for the "data,"* but modeled step lengths and directions (instead of turning angles). Thus, we retain the notation used in Morales et al. (2004)

* Recall that the data in these models are usually functions of the time series of position data, $\boldsymbol{\mu}_i$.

for r_t, but let ϕ_t represent bearing.* The basic data model structure is

$$r_t \sim \text{Weib}(a_t, b_t), \tag{5.30}$$

$$\phi_t \sim \text{WrapCauchy}(m_t, \rho_t), \tag{5.31}$$

with latent state vector $\mathbf{z}_t$ comprising all zeros and a single one in the element that corresponds to the state for time t. This is a generalization of the model framework presented by Morales et al. (2004) who discussed two or three latent states only. The model proposed by McClintock et al. (2012) allows for any number of states via the dimension of vector $\mathbf{z}_t$. Following Blackwell (1997, 2003), McClintock et al. (2012) allow the latent state to arise from a categorical distribution, which is equivalent to modeling $\mathbf{z}_t$ as a multinomial random vector

$$\mathbf{z}_t \sim \text{MN}(1, \mathbf{p}_t). \tag{5.32}$$

The simplest model for the state probabilities assumes they are static over time and sum to one; that is, $\mathbf{p}_t = \mathbf{p}$, where $\mathbf{p}'\mathbf{1} = 1$. In this case, the state transitions are conditionally independent with certain states being more prevalent than others. The first generalization might be to allow for heterogeneity in the state probabilities with a regression framework. The "mlogit" transformation is one possible way to model multivariate probability vectors and can be written element-wise in terms of the log odds as $\log(p_{t,j}/p_{t,1}) = \mathbf{x}_t'\boldsymbol{\beta}_j$, for $j = 2, \ldots, J$ states. The mlogit transformation properly constrains each $\mathbf{p}_t$ to sum to one and there are $J - 1$ coefficient vectors $\boldsymbol{\beta}_j$ to be estimated.

A more general model that allows for dynamics in the state switching is

$$\mathbf{z}_t \sim \text{MN}(1, \mathbf{P}\mathbf{z}_{t-1}). \tag{5.33}$$

In this case, $\mathbf{P}$ is a transition matrix with columns that sum to one. The elements of $\mathbf{P}$, $p_{j,k}$, control the probability of switching from state k to state j. As McClintock et al. (2012) point out, dynamic multinomial models have become popular in the population modeling literature (e.g., Hobbs et al. 2015) for accommodating demographic changes in populations. The larger the diagonal elements of $\mathbf{P}$ relative to the off-diagonal elements, the more stable the state-switching process $\mathbf{z}_t$ will be.

To allow for centers of attraction and repulsion, McClintock et al. (2012) let the parameters of the model for direction depend on a distance (d_t) between the current location ($\boldsymbol{\mu}_t$) and a point in the domain of interest ($\boldsymbol{\mu}^*$). To achieve this attraction or repulsion, they used a hyperbolic tangent function to link the parameter ρ_t to d_t (i.e., $\rho_t = \tanh(\alpha d_t)$, for scaling parameter α) and let the mean direction m_t be equal to the direction from $\boldsymbol{\mu}_t$ to $\boldsymbol{\mu}^*$. McClintock et al. (2012) utilize the hyperbolic tangent function because it maps the real numbers to those bounded by -1 and 1. Values of $\rho_t < 0$ capture repulsion; thus, if the interest is in attraction only, an alternative link function could be the logit such that $\text{logit}(\rho_t) = \alpha_0 + \alpha_1 d_t$.

* Turning angle and bearing (or direction) are different. Turning angle is the angle between two successive displacement vectors (i.e., moves), whereas bearing is the direction relative to true north.

Furthermore, for mixtures of dynamics and attraction/repulsion on the directions themselves, McClintock et al. (2012) let the mean direction in the wrapped Cauchy distribution be a function of the previous direction and the direction to the center of attraction such that $m_t = u_t\phi_{t-1} + (1 - u_t)m_t^*$.* If the individual is in state j at time t, then m_t^* is the direction from $\boldsymbol{\mu}_t$ to $\boldsymbol{\mu}_t^*$, $\boldsymbol{\mu}_t^* = \boldsymbol{\mu}_j^*$, and $u_t = u_j$, for $0 \leq u_j \leq 1$. McClintock et al. (2012) refer to these models as biased correlated random walks, because they represent a trade-off between systematic, nonrandom movement toward (or away from) a particular location (i.e., biased or directed movement) and short-term directional persistence (i.e., correlated movement). Under this model, movement becomes a biased random walk as $u_t \to 0$. As $u_t \to 1$, movement becomes a correlated random walk. As $\rho_t \to 0$, the model reverts to a simple (i.e., unbiased and uncorrelated) random walk. McClintock et al. (2012) also generalized the basic concept to accommodate multiple potential centers of attraction.†

To allow for exploratory states with directional persistence, McClintock et al. (2012) suggested a mixture framework for the location and scale parameters in the directionality model such that

$$\rho_t = \begin{cases} \psi & \text{if } \mathbf{z}_t \text{ indicates an exploratory state} \\ \tanh(\alpha d_t) & \text{otherwise} \end{cases}, \tag{5.34}$$

$$m_t = \begin{cases} \phi_{t-1} & \text{if } \mathbf{z}_t \text{ indicates an exploratory state} \\ m_t^* & \text{otherwise} \end{cases}. \tag{5.35}$$

If the individual is in an exploratory state, ψ controls directional persistence (with $\psi \to 1$ implying more persistence). Conversely, when the individual is in a state associated with attraction, it will default back to the center of attraction model described previously.

A final generalization suggested by McClintock et al. (2012) is to model the parameters in the step length distribution (a_t and b_t) using a regression framework. They specified a log linear regression for both parameters such that

$$\log(a_t) = \mathbf{x}_t'\boldsymbol{\beta}_a, \tag{5.36}$$

$$\log(b_t) = \mathbf{x}_t'\boldsymbol{\beta}_b, \tag{5.37}$$

where the covariates $\mathbf{x}_t$ could vary among models as well. To force the step length model to correspond to the latent movement state ($\mathbf{z}_t$), we let the coefficients vary in time such that $\boldsymbol{\beta}_{a,t}$ and $\boldsymbol{\beta}_{b,t}$ will be represented by $\boldsymbol{\beta}_{a,j}$ and $\boldsymbol{\beta}_{b,j}$ if $\mathbf{z}_t$ indicates the individual is in state j at time t. The model presented by McClintock et al. (2012)

* The support of m_t is circular; thus, care must be taken when $|\phi_{t-1} - m_t^*| > \pi$. One way to handle this is to compute the weighted average for the Cartesian coordinates and then back-transform: $m_t = \arg(u_t \exp(i\phi_{t-1}) + (1 - u_t)\exp(im_t^*))$.

† Also see Duchesne et al. (2015) and Rivest et al. (2015) for a more general framework that can incorporate additional sources of directional bias in m_t.

could be extended further such that the parameters in the step length distribution are also linked to the distance between the individual and the center of attraction.

To account for measurement error and alignment of the temporally irregular data with an underlying position process at regular time intervals, McClintock et al. (2012) used the same approach described by Jonsen et al. (2005). They used a hierarchical framework (5.5) with a linear weighting of neighboring data time points $\mathbf{s}_i$ like we described in Section 5.1.4:

$$\mathbf{s}_i \sim \text{N}((1 - w_i)\boldsymbol{\mu}_{t-1} + w_i\boldsymbol{\mu}_t, \sigma_s^2\mathbf{I}). \tag{5.38}$$

As before, σ_s^2 represents the measurement error variance, and could be allowed to vary by direction. The weights are a function of the interval between process time points (Δt)

$$w_i = \frac{t - t_i}{\Delta t}. \tag{5.39}$$

Then the movement parameters r_t and ϕ_t are functions of the underlying true positions $\boldsymbol{\mu}_t$ and $\boldsymbol{\mu}_{t-1}$.

McClintock et al. (2012) demonstrated their discrete-time, multistate, biased correlated random walk models using Fastloc-GPS telemetry data collected from a male grey seal (*Halichoerus grypus*) between 9 April and 13 August 2008. The temporally irregular observed locations (i.e., $\mathbf{s}_i$) showed this individual seal generally traveled clockwise among a foraging area (Dogger Bank) in the North Sea and two haul-out sites (Abertay and the Farne Islands) on the eastern coast of Great Britain (Figure 5.11).

While simultaneously accounting for temporal irregularity and measurement error using Equation 5.38, McClintock et al. (2012) fit a movement process model to the grey seal data with five movement behavior states. The behavioral states included three "center of attraction" states (with movement biased toward one of three unknown positions) and two "exploratory" states (with unbiased but potentially directionally persistent movement). Specifically,

$$r_t \sim \text{Weib}\left(a_t, b_t\right), \tag{5.40}$$

$$\phi_t \sim \text{WrapCauchy}\left(m_t, \rho_t\right), \tag{5.41}$$

$$\mathbf{z}_t \sim \text{MN}\left(1, \mathbf{P}\mathbf{z}_{t-1}\right), \tag{5.42}$$

where

$$\log\left(a_t\right) = \begin{cases} \beta^a_{0,z_t} + I_{[0,d_{z_t})}\left(\delta_t\right)\beta^a_{1,z_t} & \text{if } z_t \in \{1,2,3\} \\ \beta^a_{0,z_t} & \text{otherwise} \end{cases}, \tag{5.43}$$

$$\log\left(b_t\right) = \begin{cases} \beta^b_{0,z_t} + I_{[0,d_{z_t})}\left(\delta_t\right)\beta^b_{1,z_t} & \text{if } z_t \in \{1,2,3\} \\ \beta^b_{0,z_t} & \text{otherwise} \end{cases}, \tag{5.44}$$

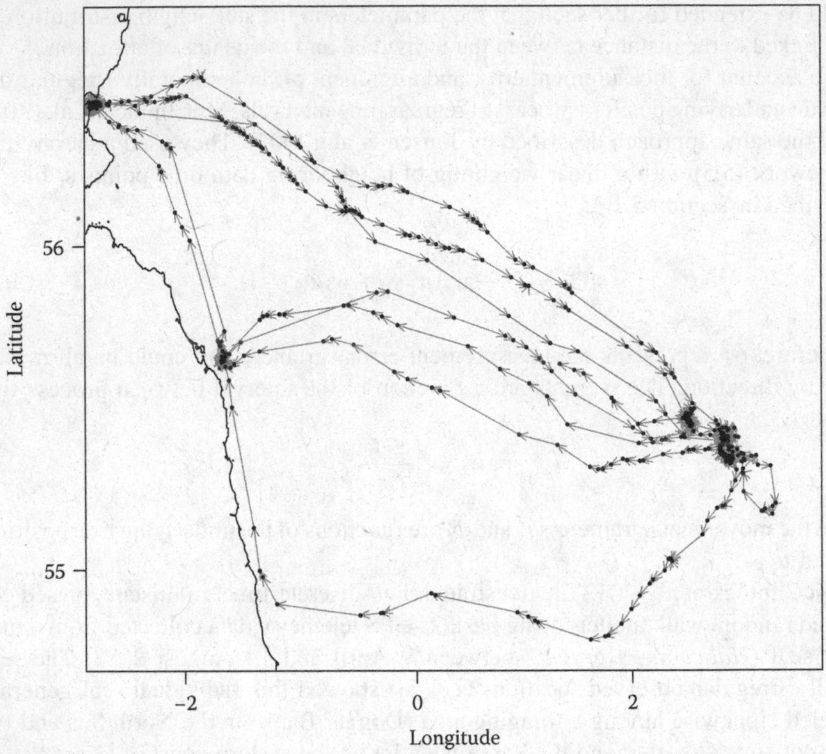

FIGURE 5.11 Grey seal Fastloc-GPS telemetry data ($\mathbf{s}_i$). Arrows indicate direction between successive locations.

$$m_t = \begin{cases} u_t\phi_{t-1} + (1-u_t)\,m_t^* & \text{if } z_t \in \{1,2,3\} \\ \phi_{t-1} & \text{otherwise} \end{cases}, \tag{5.45}$$

$$\text{logit}\,(\rho_t) = \begin{cases} \beta^{\rho}_{0,z_t} + \beta^{\rho}_{1,z_t}\delta_t + \beta^{\rho}_{2,z_t}\delta_t^2 & \text{if } z_t \in \{1,2,3\} \\ \beta^{\rho}_{0,z_t} & \text{otherwise} \end{cases}, \tag{5.46}$$

where $z_t \in \{1, 2, \ldots, Z\}$ indicates which element of $\mathbf{z}_t$ is nonzero (e.g., $z_t = 4$ indicates $\mathbf{z}_t = (0, 0, 0, 1, 0)'$), δ_t is the Euclidean distance between the seal's position at time t $(\boldsymbol{\mu}_t)$ and the current center of attraction $(\boldsymbol{\mu}_t^*)$, d_{z_t} is the threshold distance for a change point in each of the center of attraction states, $I_{[0,d_{z_t})}(\delta_t)$ is an indicator function for $\delta_t \in [0, d_{z_t})$, and all other parameters are defined as in Equations 5.33 through 5.37. The logit link was used to constrain $0 \leq \rho_t < 1$ because the biased movements are relative to centers of attraction. Thus, McClintock et al. (2012) allowed for biased and correlated movements toward three centers of attraction, but also allowed the shape and scale parameters of the Weibull distribution for step length (r_t) to change as a function of δ_t. The strength of bias toward m_t (ρ_t) was allowed to have a quadratic relationship with δ_t in Equation 5.46. In addition to the model parameters, the terms $\boldsymbol{\mu}_t^*$ and d_{z_t} were treated as unknown quantities to be estimated.

McClintock et al. (2012) used a Bayesian model implemented with a reversible-jump MCMC algorithm to fit the model and select among different parameterizations. Parameterizations included a linear or quadratic model for ρ_t (i.e., $\beta^{\rho}_{2,z_t} = 0$ for $z_t \in \{1, 2, 3\}$) and models with no short-term directional persistence (i.e., $u_t = 0$ for $z_t \in \{1, 2, 3\}$ or $\rho_t = 0$ for $z_t \in \{4, 5\}$). McClintock et al. (2012) found strong evidence of biased movement toward the three centers attraction, with estimated locations ($\boldsymbol{\mu}^*$) corresponding to the Farne Islands haul-out site, the Abertay haul-out site, and the Dogger Bank foraging site. They also found strong evidence of shorter step lengths within 5 km of these three centers of attraction, suggesting restricted movement in the vicinity of the haul-out sites and restricted area search while foraging at Dogger Bank. Little evidence was found for short-term directional persistence (i.e., $\rho_t = 0$) for the two exploratory states, but one was characterized by longer expected step lengths (i.e., higher speed) than the other.

Figure 5.12 shows the estimated movement states (z_t) for the interpolated locations ($\boldsymbol{\mu}_t$) corresponding to the Farne Islands haul-out site ("×" symbol), Abertay haul-out

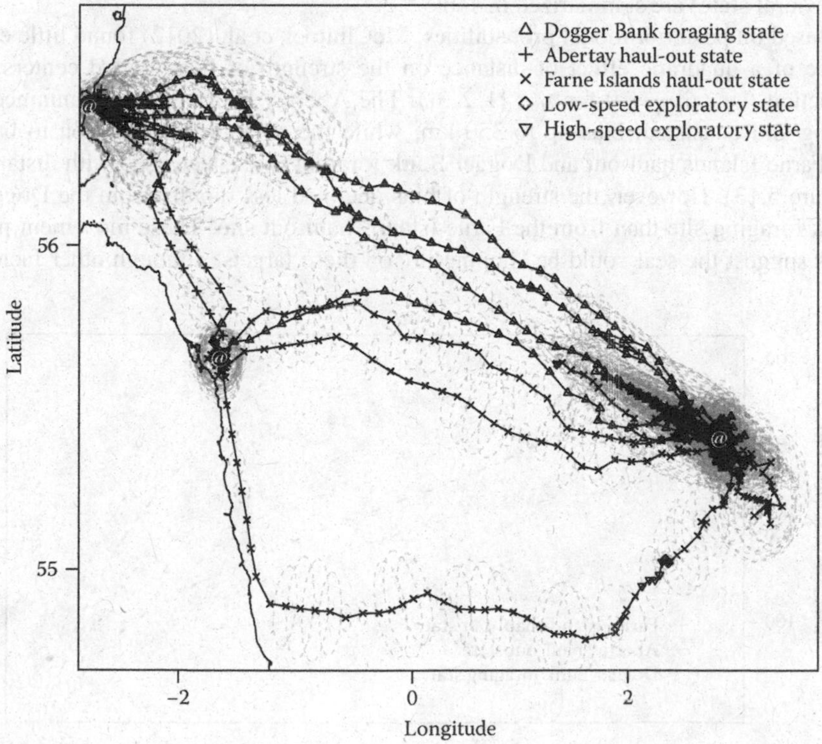

FIGURE 5.12 Estimated grey seal behavioral states. See http://www.esapubs.org/archive/mono/M082/012/appendix-C.htm for an animation of this figure. The dashed ellipses are 95% credible intervals for positions.

TABLE 5.2
Estimated Activity Budgets for the Grey Seal Data

Movement Behavior State	Activity Budget	95% CI Lower	95% CI Upper
Dogger Bank foraging state	0.39	0.37	0.41
Abertay haul-out state	0.27	0.26	0.29
Farne Islands haul-out state	0.17	0.16	0.19
Low-speed exploratory state	0.12	0.1	0.13
High-speed exploratory state	0.05	0.03	0.07

site ("+" symbol), Dogger Bank foraging site ("△" symbol), or spatially unassociated high-speed ("▽") and low-speed ("◇" symbol) exploratory states. The "@" symbols indicate the estimated coordinates of the three centers of attraction ($\boldsymbol{\mu}^*$). Uncertainty in $\boldsymbol{\mu}_t$ is indicated by 95% normal error ellipses (translucent gray dashed lines). The estimated "activity budgets" (i.e., the proportion of time steps allocated to each behavioral state) are summarized in Table 5.2.

Based on posterior model probabilities, McClintock et al. (2012) found little evidence of a quadratic effect of distance on the strength of bias toward centers of attraction (i.e., $\beta^{\rho}_{2,z_t} = 0$ for $z_t \in \{1, 2, 3\}$). The Abertay haul-out site maintained a strong and consistent bias up to 350 km, while the strength of attraction to both the Farne Islands haul-out and Dogger Bank foraging sites decreased with distance (Figure 5.13). However, the strength of bias declined less rapidly from the Dogger Bank foraging site than from the Farne Islands haul-out site. These movement patterns suggest the seal could be "honing in" on these targets, although other factors

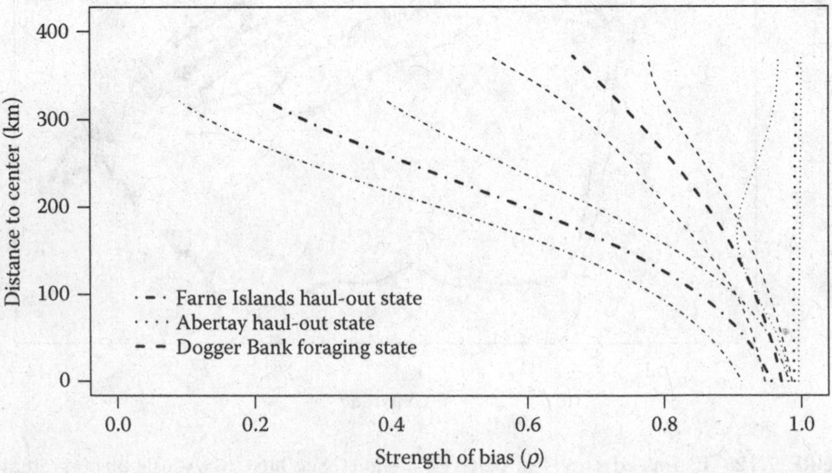

FIGURE 5.13 Estimated grey seal strength of bias curve.

(e.g., ocean currents) are also likely influencing the timing and direction of these movements (Gaspar et al. 2006).

Previous analyses of individual seal movement have been largely limited to simple and correlated random walk models of foraging trips (Jonsen et al. 2005; Johnson et al. 2008a; Breed et al. 2009; Patterson et al. 2010). However, based on posterior estimates and model probabilities, McClintock et al. (2012) found strong evidence that the incorporation of bias toward centers of attraction better explained seal movement than simple or correlated random walks.

Overall, Beyer et al. (2013) demonstrated the effectiveness of relatively simple switching models for estimating behavioral states, but these types of models are rapidly becoming more complicated (e.g., McClintock et al. 2012, 2013; Isojunno and Miller 2015). There is some evidence that these models may not perform well when movement behavioral states are not sufficiently different (Beyer et al. 2013; Gurarie et al. 2016). In general, fitting multistate movement models can be challenging. Thus, care should be taken when implementing complicated multistate models, including appropriate exploratory data analysis (e.g., Gurarie et al. 2016) and model checking (e.g., Morales et al. 2004). As demonstrated for the models of Morales et al. (2004), in the absence of location measurement error, most of the movement process models described by McClintock et al. (2012) can be fit to data in a maximum likelihood framework using HMM fitting machinery. Whether using Bayesian or non-Bayesian methods, the number of latent states is typically fixed *a priori*. Thus, extending generalized state-switching models to an unknown number of latent states remains a promising avenue for future research.

5.2.3 Response to Spatial Features

The models described thus far involve continuous-space discrete-time processes based on either the positions themselves ($\boldsymbol{\mu}_t$) as the response variable (or a noisy version of them, i.e., $\mathbf{s}_i$) or transformations of them, such as velocity ($\mathbf{v}_t$), or movement rates/step lengths (r_t) and turning angles/directions (θ_t, ϕ_t). We discussed the concept of attraction briefly in the context of these models earlier in the chapter and then elaborated on it in the previous section on generalizations of state-switching models. We also presented similar ideas in the different context of spatial point process models in Chapter 4. However, we should note that the concept of modeling the avoidance of obstacles by individual moving particles has a long history in simulation modeling (e.g., engineering and physics). It was brought up specifically in the context of animal movement using statistical models and empirical data by Tracey et al. (2005). They focused on an application involving telemetry data for snakes and their avoidance of obstacles. They noted that many preceding movement modeling approaches assumed "featureless" landscapes, but that an individual's response to landscape features was also important.

Tracey et al. (2005) developed a statistical modeling approach that was quite similar to those presented by Morales et al. (2004) and McClintock et al. (2012) and involved modeling the turning angles (θ_t) and distances (r_t) as well as distances (d_t^*) and angles (m_t^*) to the feature of interest $\boldsymbol{\mu}^*$. They envisioned a basic movement model for the positions as

$$\boldsymbol{\mu}_t = \boldsymbol{\mu}_{t-1} + r_t \mathbf{a}_t, \tag{5.47}$$

where, in their notation, r_t and θ_t are the movement distance and angle between successive positions $\boldsymbol{\mu}_{t-1}$ and $\boldsymbol{\mu}_t$ and $\mathbf{a}_t \equiv (\cos(\theta_t), \sin(\theta_t))'$.

The key to incorporating response to a spatial feature in the model then is to let the movement parameters r_t and θ_t depend on the distance (d_t^*) and angle (m_t^*) to the spatial feature ($\boldsymbol{\mu}^*$). Rather than use a wrapped Cauchy distribution as Morales et al. (2004) and McClintock et al. (2012) did, Tracey et al. (2005) used the von Mises distribution for θ_t (another circular distribution, like the wrapped Cauchy). For distance, they used the gamma distribution* instead of the Weibull, resulting in the model

$$r_t \sim \text{Gamma}(a_t, b_t), \tag{5.48}$$

$$\theta_t \sim \text{vonMises}(m_t, \rho_t). \tag{5.49}$$

In this model (5.49), a_t and b_t are the shape and scale parameters while m_t and ρ_t are the location and concentration parameters. Because the response of an individual to a given feature is of interest, Tracey et al. (2005) model the response angle as $\theta_t - m_t^* \sim \text{vonMises}(m_t, \rho_t)$, rather than the turning angle directly. This modification allows them to consider $m_t = m$ to be fixed and only ρ_t to vary. The basic concept is to let concentration be a function of distance to feature; therefore, $\log(\rho_t) = \alpha_0 + \alpha_1 d_t^*$ will allow for a reduced precision in response angle as d_t^* increases if $\alpha_1 < 0$. Similarly, in the model for distances, Tracey et al. (2005) use moment matching to model the mean angle as $\log(a_t/b_t) = \beta_0 + \beta_1 d_t^*$.† They let the mean of the gamma distribution vary and assume a constant variance, which is estimated separately. This model will let the mean distance or step length decrease as a function of decreasing distance to the feature if $\beta_1 < 0$.

5.2.4 Direct Dynamics in Movement Parameters

Morales et al. (2004) and McClintock et al. (2012) modeled the movement rate (i.e., inverse step length) and turning angle (or bearing) in a form of clustering or mixture model. In that case, the movement process itself was allowed to be dynamic explicitly through its state-switching behavior. Forester et al. (2007) also found it useful to model a movement parameter, but instead specified the model such that the dynamics involved a latent version of the measured parameter itself. Focusing on log step length (y_t), Forester et al. (2007) set up the hierarchical model

$$y_t \sim \text{N}(\tilde{y}_t + \mathbf{w}_t'\boldsymbol{\alpha}, \sigma_y^2), \tag{5.50}$$

$$\tilde{y}_t \sim \text{N}(\rho\tilde{y}_{t-1} + \mathbf{x}_t'\boldsymbol{\beta}, \sigma_{\tilde{y}}^2), \tag{5.51}$$

---

* The gamma distribution has the mathematically elegant property of infinite divisibility, although any practical advantages for discrete-time movement modeling (particularly with respect to choice of Δt) are not well documented.

† Note that Tracey et al. (2005) also explore other models, but we present only the exponential forms here for simplicity.

where $\tilde{y}_t$ is the latent movement parameter (underlying log step length process), $\mathbf{w}_t$ is a vector of covariates involved with the observed log step lengths and $\boldsymbol{\alpha}$ are the associated regression coefficients, ρ controls the smoothness of the latent dynamic process, $\mathbf{x}_t$ are covariates and $\boldsymbol{\beta}$ are regression coefficients associated with the dynamic process, and the variance components σ_y^2 and $\sigma_{\tilde{y}}^2$ control the stochasticity in the model at the observed and latent levels. We also assume that the time intervals are constant so that we are in the typical time series context.

The hierarchical model presented by Forester et al. (2007) has a similar state-space construction as the models for dynamics in velocity (Section 5.2). The difference is that the model proposed by Forester et al. (2007) operates on a univariate functional of velocity (log of inverse speed, or step length). It also contains covariate influences at both the data and process levels.

In fact, Forester et al. (2007) combine the process and observation models and use iterative substitution to show that the mean of y_i can be written as a function of interpretable terms:

$$y_t \sim \mathrm{N}\left(\sum_{c=1}^{C} \rho^c \mathbf{x}_t'\boldsymbol{\beta} + \sum_{c=0}^{C-1} \rho^c(\tilde{y}_t - \rho\tilde{y}_{t-1} - \mathbf{x}_t'\boldsymbol{\beta}) + \rho^C\tilde{y}_{t-C} + \mathbf{w}_t'\boldsymbol{\alpha}, \sigma_y^2\right), \quad (5.52)$$

for C time steps. Forester et al. (2007) explained that the first term in the mean (i.e., $\sum_{c=1}^{C} \rho^c \mathbf{x}_t'\boldsymbol{\beta}$) contains the preceding C "environments" experienced by the individual with strength of past experience attenuated by ρ. For example, when ρ decreases toward zero, the memory of past experiences decreases. Therefore, larger values of ρ indicate longer "memory."* Visually, when viewing the time series, a smoother process will have a larger ρ and a noisier process will have a smaller ρ (approaching zero). Forester et al. (2007) describe the second term involving a sum in Equation 5.52 as similarly attenuated process uncertainty. Essentially, the smaller the process error $(\tilde{y}_t - \rho\tilde{y}_{t-1} - \mathbf{x}_t'\boldsymbol{\beta})$, the more the covariates $(\mathbf{x}_t)$ can influence the movement process.

At first glance, the covariates in both the measurement and process models (5.51) might appear to be redundant. However, as Forester et al. (2007) explain, we can think of this as a multiscale model in that the covariate effects at the data level $(\mathbf{w}_t'\boldsymbol{\alpha})$ have an immediate effect on y_t, whereas the process covariate effects $(\mathbf{x}_t'\boldsymbol{\beta})$ have a longer-term effect on y_t because they accumulate at a rate controlled by ρ. Thus, discrepancies among $\boldsymbol{\alpha}$ and $\boldsymbol{\beta}$ can indicate multiscale dynamics in the process.

The model described by Forester et al. (2007) is completely linear (except for the initial log transformation of the step lengths), and thus, can be fit using maximum likelihood and Kalman filtering methods to estimate the latent process $(\tilde{y}_t)$. However, a Bayesian implementation of the model is straightforward. In the Bayesian situation, we just need to specify priors for the unknown parameters $\boldsymbol{\alpha}$, $\boldsymbol{\beta}$, ρ, σ_y^2, and $\sigma_{\tilde{y}}^2$. If Gaussian priors are used for $\boldsymbol{\alpha}$, $\boldsymbol{\beta}$, and ρ, while inverse gamma priors are used for

* In time series, memory has a different definition than this, but, for consistency, we maintain Forester's use of the term here to help with visualization.

the variance components (σ_y^2 and $\sigma_{\tilde{y}}^2$), the full-conditional distributions will all be conjugate and an MCMC algorithm can be easily constructed with all Gibbs updates.

As previously mentioned, the hierarchical step length model (5.51) is closely related to the vector autoregressive models for velocity. In fact, we can specify a multivariate model for velocity using the same approach:

$$\mathbf{v}_t \sim \text{N}(\tilde{\mathbf{v}}_t + \mathbf{W}_t\boldsymbol{\alpha}, \sigma_y^2\mathbf{I}), \tag{5.53}$$

$$\tilde{\mathbf{v}}_t \sim \text{N}(\mathbf{M}\tilde{\mathbf{v}}_{t-1} + \mathbf{X}_t\boldsymbol{\beta}, \sigma_{\tilde{y}}^2\mathbf{I}). \tag{5.54}$$

Following Forester et al. (2007), we combine these conditional models (5.54) for a heuristic about memory. Using iterative substitution, we arrive at

$$\mathbf{v}_t \sim \text{N}\left(\sum_{c=1}^{C}\mathbf{M}^c\mathbf{X}_t\boldsymbol{\beta} + \sum_{c=0}^{C-1}\mathbf{M}^c(\tilde{\mathbf{v}}_t - \mathbf{M}\tilde{\mathbf{v}}_{t-1} - \mathbf{X}_t\boldsymbol{\beta}) + \mathbf{M}^C\tilde{\mathbf{v}}_{t-C} + \mathbf{W}_t\boldsymbol{\alpha}, \sigma_v^2\mathbf{I}\right), \tag{5.55}$$

where $\mathbf{M}$ raised to the power c represents an iterative matrix product of order c. In this hierarchical vector autoregression, the term $\sum_{c=1}^{C}\mathbf{M}^c\mathbf{X}_t\boldsymbol{\beta}$ represents the memory process that Forester et al. (2007) described. However, in this multivariate process, the "memory" can assume more complicated forms. For example, depending on the structure of $\mathbf{M}$, the memory of past experiences can decay as a damped spiral. This flexibility certainly allows for additional realism in the process, but it remains to be seen whether empirical evidence supports the need for extra generality in the model.

5.2.5 Patch Transitions

The approach proposed by Forester et al. (2007) considers a form of memory or inertia in step length, but this type of memory will not result in stable home range patterns or in animals revisiting certain areas of the landscape. To account for such large-scale properties of movement trajectories, one can model memory in the location of movement targets and movement bias (e.g., Merkle et al. 2014; Avgar et al. 2013, 2015).

For example, Morales et al. (2016) assumed that the landscape comprises a network of patches and movement decisions involve the choice of the next patch to visit. We use the term "decision" in a broad sense, because the movement from one patch to another may actually represent a decision by the animal (e.g., returning to a previously visited place on purpose), but it could also be less deliberate (e.g., when an animal finds a patch during exploration).

If we represent movement as a sequence of patch identities visited by the animal, a possible model for the next patch to visit is a multinomial

$$\mathbf{z}_{i+1} \sim \text{MN}(1, \mathbf{p}_i), \tag{5.56}$$

where $\mathbf{z}_{i+1}$ represents the identity for the next patch to visit and $\mathbf{p}_i$ is a vector of probabilities that each patch will be chosen as the next place to visit. In a model without memory, we can assume that the probability of moving from one patch to another is affected by between-patch distances

$$d_{j|k} = \exp\left(-\left(\frac{r_{k,j}}{\alpha}\right)^{\beta}\right), \tag{5.57}$$

$$p_{ji} = \frac{d_{j|k}}{\sum_{l \neq k} d_{l|k}}, \tag{5.58}$$

where $d_{j|k}$ is the propensity of choosing patch j given that the animal is now at patch k, which is located at distance $r_{k,j}$. The case of $j = k$ is excluded by definition because a move is defined as the displacement from one patch to another. This propensity changes with distance as a function of a scale parameter α and a shape parameter β, both of which need to be estimated from the observed sequence of patch to patch movements. This model can be expanded to include patch-level covariates, such as the area (A) of the patches, yielding

$$d_{j|k} = \exp\left(-\left(\frac{r_{k,j}}{\alpha}\right)^{\beta}\right), \tag{5.59}$$

$$\log(c_j) = \beta_0 + \beta_1 A_j, \tag{5.60}$$

$$p_{ji} = \frac{d_{j|k} c_j}{\sum_{l \neq k} d_{l|k} c_j}. \tag{5.61}$$

For a simple movement model with memory, Morales et al. (2016) considered the case in which the probability of visiting a particular foraging patch increased with the number of previous visits to that patch. Also, the probability of visiting a new patch (i.e., a patch where the total number of previous visits after i moves is equal to zero) is a decreasing function of the total number of unique patches visited so far. To represent the memory effects, we can write

$$m_j = \begin{cases} \exp(-\gamma u_i) & \text{if } v_{j,i} = 0 \\ 1 - \exp\left(-\left(\frac{v_{j,i}}{a}\right)^{b}\right) & \text{if } v_{j,i} > 0 \end{cases}, \tag{5.62}$$

where u_i is the number of unique patches visited so far, and $v_{j,i}$ holds the number of previous visits to patch j. The parameter γ controls how quickly the individual avoids choosing new patches as u_i increases. We combine these values with the effect of distance from current patch location to other patches and standardize to obtain

$$p_{ji} = \frac{d_{j|k} m_j}{\sum_{l \neq k} d_{l|k} m_j}. \tag{5.63}$$

Morales et al. (2016) analyzed data from elk newly translocated to the Rocky Mountain foothills near Alberta, Canada between December through February, during 2000–2002, from three neighboring areas: Banff National Park, Cross Conservation Area, and Elk Island National Park. The capture, handling, release, and fates of these animals was described by Frair et al. (2007). A total of 20 elk individuals were selected for this study and were fitted with GPS collars that recorded one location every hour for up to 11 months. Foraging patches were delimited combining dry/mesic and wet meadows, shrubland, clear cuts, and reclaimed herbaceous classes. The GPS telemetry data were transformed into patch-to-patch movement sequences. Figure 5.14 shows an example of the spatial distribution and size of foraging patches and a simplified elk trajectory for one of the tracked elk.

After specifying priors for unknown parameters, we fit the above models to the elk data and computed DIC values of 353.53 for a model considering distances and patch areas and 386.12 for the model considering distance and number of visits. The DIC scores suggest that the model with distance and area of patches has a better predictive ability. However, if we simulate trajectories with the fitted models, we see that the model without memory implies that animals keep visiting new patches as they

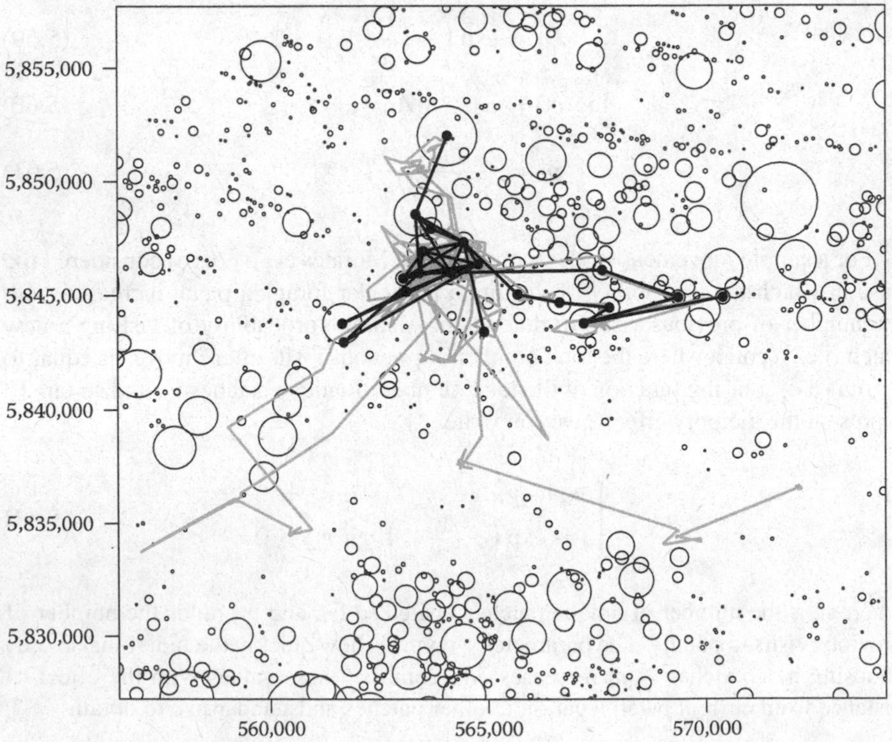

FIGURE 5.14 Example of elk trajectory (in gray) simplified to a sequence of patch-to-patch movements (black). Foraging patches are represented as circles with diameter proportional to patch area.

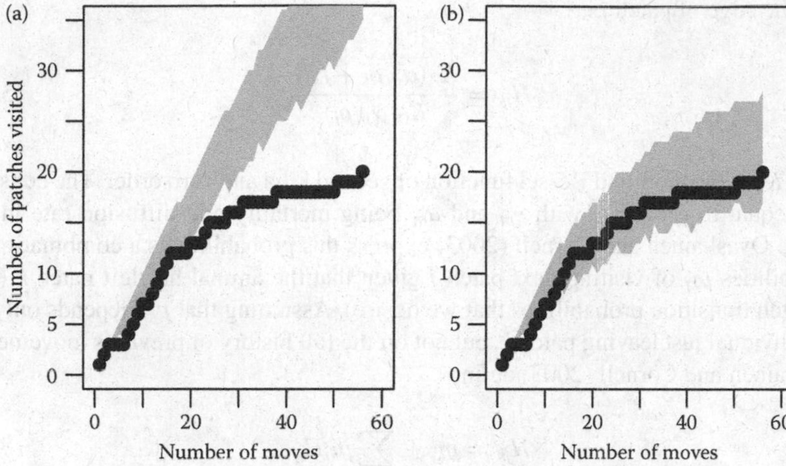

FIGURE 5.15 Posterior predictive check on the number of unique patches visited by an elk released into an unfamiliar landscape. Black dots show the observed increase of patches used by the animal as they move. Gray shades show the 90% credible intervals from data simulated using parameters sampled from the posterior distribution of a model that included distance from current location to all patches and their area (panel a), and a model that considered the effect of distance and the number of previous visits to all patches (using Equations 5.62 and 5.63) (panel b).

move on the landscape (Figure 5.15a). In contrast, the model that takes into account the history of patch visits (Figure 5.15b) results in similar saturation pattern of the number of unique patches that the animal visits as it moves.

This simple example illustrates the importance of memory in movement patterns but also the importance of checking for emergent properties of movement trajectories when assessing model fit and comparing alternative models. The model considering distance from current location to all available patches and their area performs well in modeling the identity of the next patch visited by the animal but has no way to prevent the animal from wandering through the whole network of patches. In contrast, including the number of previous visits results in a form of reinforcement in the movement path and a more restricted use of space. Even though these patterns are expected from theoretical grounds, their relevance is apparent when we compare simulated trajectories from the fitted models.

A drawback of the patch transition models we just presented is that they do not take into account the potential shadowing effects of nearby patches. That is, even without memory involved, a particular patch can be less visited than expected by distance and area effects because it is near other patches that compete as possible destinations. Modeling movement in highly fragmented landscapes (where distances between patches are large compared to the size of patches), Ovaskainen and Cornell (2003) derived patch-to-patch movement probabilities, taking into account the spatial configuration of the patch network. They showed that, if movement is modeled as a simple diffusion, the probability that an individual leaving patch k will eventually reach (before dying) a patch of radius ρ_j, given that the animal is at a distance r_{kj}

from the edge of patch j is

$$H_{jk} = \frac{K_0(\alpha_m \rho_j + r_{kj})}{K_0(\alpha_m \rho_j)}, \tag{5.64}$$

where K_0 is the modified Bessel function of second kind and zero order. The constant α_m is equal to $\sqrt{c_m/a_m}$, with c_m and a_m being mortality and diffusion rate in the matrix. Ovaskainen and Cornell (2003) express this probability as a combination of probabilities p_{kj} of visiting next patch j given that the animal has left patch k (i.e., the patch transition probabilities that we desire). Assuming that p_{kj} depends only on the individual just leaving patch k, but not on the full history of previous movements, Ovaskainen and Cornell (2003) define

$$H_{jk} = p_{kj} + \sum_{i \neq j} p_{ki} H_{ij}. \tag{5.65}$$

For example, if the network is composed of just three patches, an individual leaving from patch 1 can eventually reach patch 2 by either going there directly (p_{12}) or going to patch 3 first and then, eventually, going from patch 3 to patch 2 ($p_{13}H_{32}$).

We can write $\mathbf{Hp} = \mathbf{h}$, where $\mathbf{H}$ is a matrix containing the values obtained from Equation 5.65 and with the diagonal elements equal to one. The vector $\mathbf{h}$ has the same values (i.e., probabilities of eventually getting to patch j) but, as we condition on actually emigrating from a patch, we set $h_{kj} = 0$ for all $k = j$. Ovaskainen and Cornell (2003) used a linear solver to obtain the patch transition probabilities (p_{kj}), which take into account the spatial configuration of the network. The probability of an animal dying or leaving the patch network, given that it has just left patch k, is equal to $1 - \sum_{k \neq j} p_{kj}$.

Ovaskainen (2004) and Ovaskainen et al. (2008) used this approach to fit heterogeneous movement models to butterfly capture–recapture data. In principle, it is possible to use this approach replacing the diffusion result (5.64) with a generic equation such as Equation 5.57 and consider the effect of previous visits by adding weights to the transition probabilities derived from distance and area effects. However, this approach is probably inaccurate when patches are close to each other or when patch shapes or movement imply that we cannot ignore the location where animals are leaving or entering patches.

5.2.6 Auxiliary Data

Our emphasis thus far has been on inference from telemetry location data, but advances in animal-borne technology now facilitate the collection of vast amounts of other types of biotelemetry data. For example, biologging devices can record environmental data (e.g., temperature or altitude), proximity to conspecifics (e.g., Ji et al. 2005), time-at-depth and other dive profile information for marine animals (e.g., Higgs and Ver Hoef 2012), high-frequency acceleration (e.g., Shepard et al. 2008), and even stomach temperature (e.g., Austin et al. 2006). Until recently, these rich (and often interrelated) biotelemetry data were typically analyzed independently of one another (e.g., LeBoeuf et al. 2000; Austin et al. 2006; Jonsen et al. 2007).

As illustrated previously in this chapter, mechanistic discrete-time multistate movement models often aim to associate different types of movement with distinct behavioral states (e.g., Morales et al. 2004; Jonsen et al. 2005; McClintock et al. 2012), but such inference is typically drawn from location data only. There is mounting evidence that inferring animal behavior based on horizontal displacement alone can be difficult and problematic (Gaspar et al. 2006; Patterson et al. 2008; McClintock et al. 2013). Thus, the vast majority of discrete-time behavior-switching movement models are limited to two dissimilar behavior states, such as "encamped" (short step lengths, low directional persistence) and "exploratory" (long step lengths, high directional persistence), as in Morales et al. (2004) and Jonsen et al. (2005). However, integrated multistate movement models using both animal location and auxiliary data can improve our ability to identify and characterize a broader class of biologically meaningful behavioral states (Patterson et al. 2009; McClintock et al. 2013).

Incorporating auxiliary data into the discrete-time models covered thus far is relatively straightforward. The typical approach is to specify a conditional likelihood for the auxiliary data (ω) and combine it with the conditional likelihood for the location data into a joint conditional likelihood. For example, McClintock et al. (2013) built on the framework of Morales et al. (2004) and McClintock et al. (2012) to specify a multistate movement model with three states for harbor seals (*Phoca vitulina*) by combining the conditional likelihood components for step length (r_t), bearing (ϕ_t), and the proportion of each time step spent diving below 1.5 m ($0 \leq \omega_t \leq 1$) such that

$$r_t \sim \text{Weib}\left(a_t, b_t\right), \tag{5.66}$$

$$\phi_t \sim \text{WrapCauchy}\left(\rho_t, \phi_{t-1}\right), \tag{5.67}$$

$$\omega_t \sim \text{Beta}\left(\eta_t, \delta_t\right), \tag{5.68}$$

where η_t and δ_t are the (state-dependent) shape parameters for the beta distribution, $\mathbf{z}_t \sim \text{MN}(1, \mathbf{P}\mathbf{z}_{t-1})$, and $\mathbf{z}_t \equiv (z_{1,t}, z_{2,t}, z_{3,t})'$.* As before for a_t, b_t, and ρ_t (5.22 through 5.25), we have $\eta_t = \boldsymbol{\eta}'\mathbf{z}_t$ and $\delta_t \equiv \boldsymbol{\delta}'\mathbf{z}_t$, where $\boldsymbol{\eta} \equiv (\eta_1, \eta_2, \eta_3)'$ and $\boldsymbol{\delta} = (\delta_1, \delta_2, \delta_3)'$. For this particular model, $z_{1,t} = 1$ indicates the "resting" state (characterized by short step lengths and smaller values for ω_t), $z_{2,t} = 1$ indicates the "foraging" state (moderate step lengths, low directional persistence, and larger values for ω_t), $z_{3,t} = 1$ indicates the "transit" state (long step lengths, high directional persistence, and larger values for ω_t; Figures 5.16 and 5.17).

Adopting a Bayesian framework, McClintock et al. (2013) used simple prior constraints on a_t, b_t, ρ_t, η_t, and δ_t to reflect the expected relationships for the three

* This model belongs to the general class of multivariate HMMs (e.g., Zucchini et al. (2016), pp. 138–141). In fact, the basic movement process models of Morales et al. (2004), Jonsen et al. (2005), and McClintock et al. (2012) can all be considered multivariate HMMs because they all consist of multiple data sets assumed to arise from a Markov process with a finite number of hidden states (e.g., r_t and θ_t constitute the two data sets in the movement process model proposed by Morales et al. 2004).

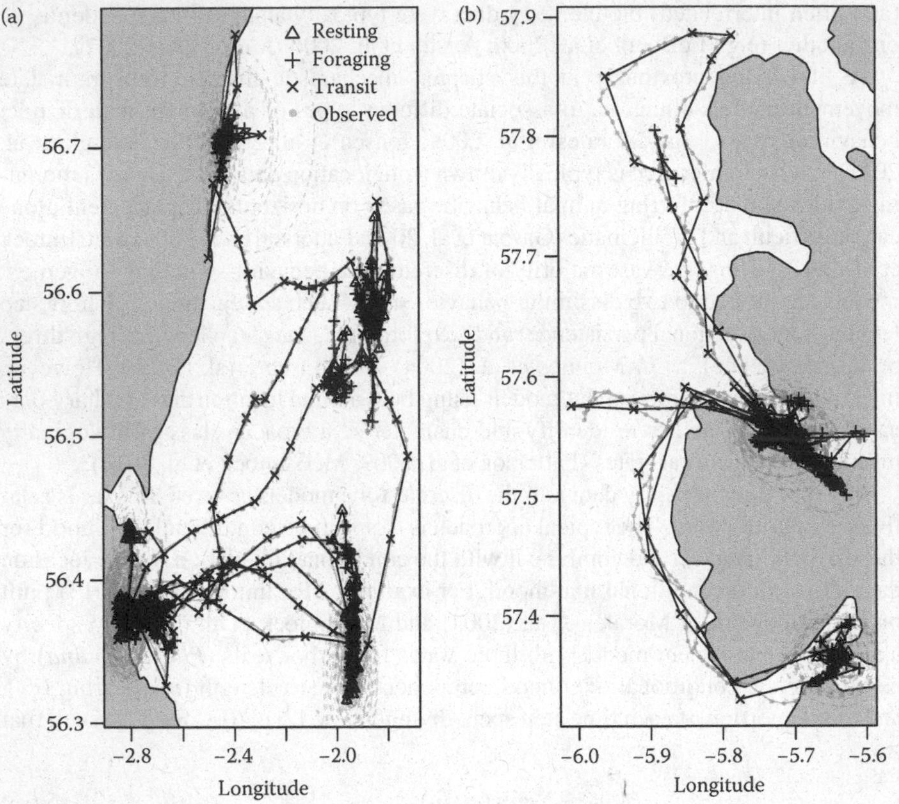

FIGURE 5.16 Predicted locations and movement behavior states for two harbor seals in the United Kingdom: (a) a male in southeastern Scotland and (b) a female in northwestern Scotland. Estimated movement states for the predicted locations correspond to "resting" ("△" symbol), "foraging" ("+" symbol), and "transit" ("×" symbol) movement behavior states. Light gray points indicate observed locations ($\mathbf{s}_i$). Uncertainty in predicted locations are indicated by 95% credible ellipses (dashed translucent gray lines).

behavioral states (e.g., $\rho_1, \rho_2 < \rho_3$). Using this approach, they were able to detect significant differences in the proportion of time harbor seals allocated to each behavioral state (i.e., "activity budgets") in the pre- and postbreeding seasons (Table 5.3).

McClintock et al. (2013) also demonstrated the dangers of attempting to estimate the "resting," "foraging," and "transit" movement behaviors based on horizontal trajectory alone (i.e., r_t and ϕ_t only). They found that 33% of time steps with $\omega_t > 0.5$ were assigned to the "resting" state when inferred from horizontal trajectory alone, but only 1% of these were assigned to "resting" when inferred from both horizontal trajectory and the auxiliary dive data using their integrated model. Similarly, they found that 46% of time steps with $\omega_t < 0.5$ were assigned to "foraging" or "transit" based on trajectory alone, but only 12% of these time steps were assigned to "foraging" or "transit" when using the auxiliary dive data. Owing to the difficulty

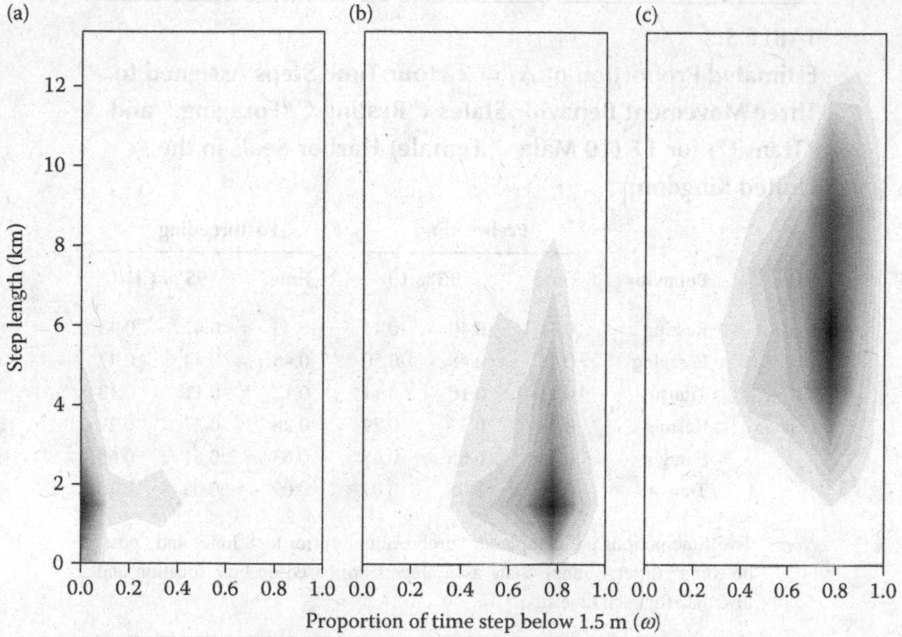

FIGURE 5.17 Estimated bivariate densities of harbor seal step length (r_t) and proportion of time step spent diving below 1.5 m (ω) from McClintock et al. (2013). Densities were estimated for three distinct movement behavior states (i.e., (a) "resting," (b) "foraging," and (c) "transit"), where darker shades indicate higher relative densities. Time steps are $\Delta_t = 2$ h.

of distinguishing more than two behavior states from horizontal trajectory alone, the incorporation of auxiliary data is becoming commonplace when >2 behavioral states are of interest (McClintock et al. 2013, 2014, 2015; Russell et al. 2014, 2015; Isojunno and Miller 2015).

Figure 5.16b demonstrates an important consideration for discrete-time movement models when the observed location data are temporally irregular. Notice that there were many observed locations during some of the "transit" 2 h time steps, but the temporally regular predicted movement path diverges somewhat from the observed locations. Clearly, discretization of the movement path can introduce additional error in the fit of the observed locations ($\mathbf{s}_i$) to the estimated true locations ($\boldsymbol{\mu}_i$) that is not attributable to location measurement error. Discretization error is often reduced by choosing smaller time steps (Δ_t), but generally, with smaller Δ_t comes greater computational burden. Thus, with temporary irregular $\mathbf{s}_i$, we are often posed with a trade-off between choice of Δ_t and discretization error when fitting discrete-time movement models.

McClintock et al. (2013, 2015) and Russell et al. (2014, 2015) constitute "pseudo 3-D" movement models in the sense that they utilize horizontal trajectory and discrete vertical categories to characterize >2 behavioral states for diving animals. However, animal-borne tags equipped with accelerometers are enabling great strides toward

TABLE 5.3
Estimated Proportion of $\Delta_t = 2$ Hour Time Steps Assigned to Three Movement Behavior States ("Resting," "Foraging," and "Transit") for 17 (10 Male, 7 Female) Harbor Seals in the United Kingdom

		Prebreeding			Postbreeding		
Sex	Behavior	Time	95% CI		Time	95% CI	
Male	Resting	0.41	0.40	0.41	0.43	0.42	0.43
	Foraging	0.49	0.48	0.50	0.45	0.44	0.47
	Transit	0.11	0.10	0.11	0.12	0.11	0.13
Female	Resting	0.28	0.27	0.29	0.28	0.27	0.29
	Foraging	0.66	0.65	0.67	0.63	0.61	0.64
	Transit	0.06	0.05	0.07	0.09	0.08	0.10

Note: Two time periods are compared: "prebreeding" (prior to 1 June) and "postbreeding" (after 1 June). State assignments are based on both location and dive data for each time step.

discrete-time 3-D movement models in continuous space (e.g., Lapanche et al. 2015). Even with a single behavioral state, continuous-space 3-D approaches are complicated, data hungry, computationally intensive, and still in their infancy. Although much work remains to be done, thinking in 3-D holds much promise for our shared goal of building more realistic and biologically meaningful movement models.

5.2.7 Population-Level Inference

In Chapters 3 and 4, we briefly introduced the concept of population-level inference with the individual animal as the "sample unit" in studies of animal populations. Analogous hierarchical methods for borrowing strength at some level among individual- or group-level parameters have been developed for discrete-time movement models. Jonsen et al. (2003) introduced the idea in the context of Bayesian discrete-time velocity models, and Jonsen et al. (2006) later extended the approach of Jonsen et al. (2005) to make population-level inference about diel variation in travel rates of migrating leatherback turtles (*Dermochelys coriacea*). Langrock et al. (2012) and McClintock et al. (2013) extended the approaches of Morales et al. (2004) and McClintock et al. (2012) to population-level hierarchical models using maximum likelihood and Bayesian methods, but it can often be difficult to fit hierarchical models with many individual-level random effects using maximum likelihood methods (Altman 2007; Patterson et al. 2009; Langrock et al. 2012; but see Hooten et al. 2016). Langrock et al. (2014) modeled the group dynamic behavior of reindeer (*Rangifer tarandus*), where individual-level movements were found to be weakly influenced by a latent group centroid (i.e., herding behavior).

Although population-level movement models have appeared infrequently in the movement modeling literature, other examples of population-level analyses using hierarchical discrete-time movement models include Eckert et al. (2008), McClintock et al. (2015), and Scharf et al. (In Press). Jonsen (2016) presents evidence supporting several of the compelling reasons for using hierarchical population-level discrete-time movement models, including substantially less bias and uncertainty in parameter estimates (particularly when locations are subject to measurement error). The additional complexity and computational burden of hierarchical population-level models are certainly among the reasons they have received little use thus far, but perhaps the most significant hurdle is the current lack of user-friendly software for implementing them (but see Jonsen et al. 2006; Langrock et al. 2012; McClintock et al. 2013, 2015, and Langrock et al. 2014 for custom code provided by the authors).

5.3 ADDITIONAL READING

We introduced models for discrete-time processes in Chapter 3 and referenced classical texts in these areas (e.g., Shumway and Stoffer 2006; Brockwell and Davis 2013). Despite the fact that movement trajectories and telemetry data are naturally multivariate and that sets them apart from much of the classical time series topics, they also have an inherent mechanistic dynamic structure to them. Therefore, statistical models for trajectories are most closely related to the methods described in spatio-temporal statistics (e.g., Wikle and Hooten 2010; Cressie and Wikle 2011). They differ from most spatio-temporal models in that they are often implemented in a lower-dimensional space (i.e., two spatial components). Many spatio-temporal models are designed for continuous spatial settings, or discrete spatial settings with many components. However, discrete-time animal movement models are often formulated with important mechanisms in mind as opposed to many time series models that are purely phenomenological.

While we touched on some of the computational techniques that have been developed for time series analysis, we only scratched the surface of what is available. As discrete-time movement modeling continues to increase in value, the computational efficiency of animal movement model implementation will also need to improve. Kalman filtering methods are well known and trusted for inference in maximum likelihood settings, but many Bayesian implementations for hierarchical animal movement models are slower because of the inherent sampling-based algorithms that are used (e.g., MCMC, Hamiltonian Monte Carlo [HMC]). When location measurement error and missing data are negligible, Franke et al. (2006), Holzmann et al. (2006), Patterson et al. (2009), and Langrock et al. (2012) have made use of HMM specifications that can improve computational efficiency dramatically, allowing more discrete-time animal movement models to become operational.

At the time of this writing, the fields of spatial and spatio-temporal statistics have been buzzing with reduced-rank approaches for implementing statistical models based on large data sets. Buderman et al. (2016) and Hooten and Johnson (2016) presented approaches for using basis functions (Hefley et al. 2016a) to facilitate both mechanistic and computationally efficient animal movement models, but these are more aimed at continuous-time settings like those presented in Chapter 6. Other computational techniques, such as the use of sparse matrix storage and manipulation (e.g., Rue et al. 2009), will become essential for discrete-time animal movement modeling. In time, more of these types of computational approaches will be borrowed from time series and adapted for use in the analysis of telemetry data.

As we discussed in Chapter 1, a fundamental characteristic of animal movement is that individuals interact with each other, both within and among species. Many approaches have been proposed for modeling interactions among individuals in populations and communities (e.g., Deneubourg et al. 1989; Couzin et al. 2002; Eftimie et al. 2007; Giuggioli et al. 2012), and while most of them are purely mathematical or statistically *ad hoc*, formal statistical models are now being developed regularly. Delgado et al. (2014) presented a linear mixed model approach, modeling "sociability" (the difference between observed and null proximity metrics) as a function of random individual and temporal effects. Delgado et al. (2014) relate their approach to that used in step selection functions (SSFs). Also in the context of SSFs, Potts et al. (2014b) provided a concise review of approaches for studying interactions and suggested that many can be considered in a step selection framework. More recently, Russell et al. (2016a) and Scharf et al. (In Press) have developed formal hierarchical movement models that provide inference for interactions. Russell et al. (2016a) focused on interactions using point process formulations and Scharf et al. (In Press) developed a discrete-time movement model to provide inference for animal social networks.

Finally, similar discrete-time models have been proposed in other branches of ecology, for example, Clark (1998) and Clark et al. (2003) for implementations of integro-difference models based on dispersal kernels. Such models could be modified for the animal movement setting and fit using telemetry data.

6 Continuous-Time Models

The value in considering animal movement processes in discrete-time contexts is undeniable. The discrete-time context is valuable because (1) a wealth of tools can be borrowed from the time series literature, (2) the dynamics are easily conceptualized in discrete time, and finally, (3) we are implementing models digitally on computers, thus we must discretize the procedure at some prespecified resolution regardless. Therefore, discrete-time models are sensible and practical.

One could argue, however, that the true process of movement really occurs in both continuous time and continuous space. That is, the term "continuous time" refers to the fact that the process is defined for any time in the interval $[0, T]$. Thus, there is value in constructing statistical animal movement models from the continuous-time perspective, even though we may end up discretizing the implementation.

6.1 LAGRANGIAN VERSUS EULERIAN PERSPECTIVES

Discrete-time models for animal movement appeal to an inherently simple heuristic. This may be due, at least in part, to the algorithmic nature of the models. Turchin (1998) outlines a path from discrete-time individual-based (i.e., Lagrangian) to continuous-time population-level (i.e., Eulerian) models. He begins with a simple recurrence equation and ends up at an elegant and well-known partial differential equation (PDE). We adapt that line of analysis here to illustrate the connection between the two schools of thought (i.e., discrete vs. continuous). In the statistical literature pertaining to animal movement, these ideas also appear in Hooten and Wikle (2010) and Hooten et al. (2013a). In the mathematical literature pertaining to animal movement, they have also served as a basis for describing efficient approaches to implementation (Garlick et al. 2011, 2014).

The fundamental idea in developing a Langrangian movement model is to start from a simple set of first principles. To most easily communicate the basic ideas, we begin with a 1-D discrete spatial domain (Figure 6.1). In this setting, an individual animal can move left or right one unit of space, or stay where they currently are during time step t. That is, an animal at location μ can move left, right, or remain where it is with probabilities $\phi_L(\mu, t)$, $\phi_R(\mu, t)$, and $\phi_N(\mu, t)$, respectively (where $\phi_L(\mu, t) + \phi_R(\mu, t) + \phi_N(\mu, t) = 1$). Then the probability of the animal occupying location μ at time t is

$$\begin{aligned} p(\mu, t) = {} & \phi_L(\mu + \Delta\mu, t - \Delta t) p(\mu + \Delta\mu, t - \Delta t) \\ & + \phi_R(\mu - \Delta\mu, t - \Delta t) p(\mu - \Delta\mu, t - \Delta t) \\ & + \phi_N(\mu, t - \Delta t) p(\mu, t - \Delta t), \end{aligned} \tag{6.1}$$

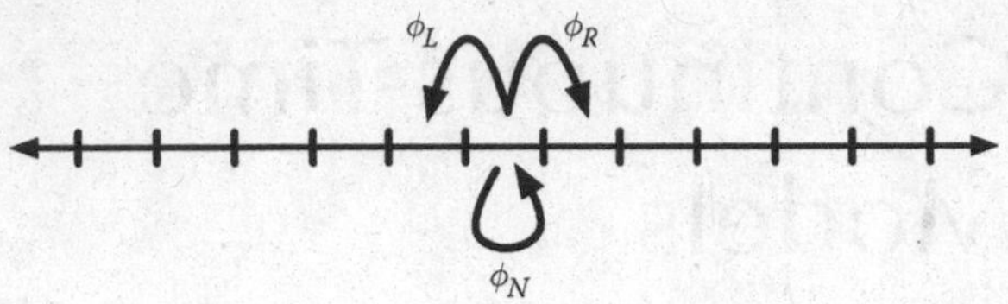

FIGURE 6.1 One-dimensional spatial domain with movement probabilities for a move left (ϕ_L), move right (ϕ_R), and no move (ϕ_N).

where the Δ notation refers to changes in time and space (i.e., $+\Delta\mu$ represents the change in spatial location in the positive direction, for a 1-D spatial domain). If we ultimately seek an Eulerian model on the probability of occupancy, $p(\mu, t)$, we need to replace the Δ notation with differential notation. Turchin (1998) proceeds by expanding each of the probabilities in a Taylor series, truncating to remove higher-order terms, and then substituting the truncated expansions back into Equation 6.1. The Taylor series expansion yields a recurrence equation involving partial derivatives:

$$\begin{aligned} p = (\phi_L + \phi_N + \phi_R)p - \Delta t(\phi_L + \phi_N + \phi_R)\frac{\partial p}{\partial t} - \Delta t p \frac{\partial}{\partial t}(\phi_L + \phi_N + \phi_R) \\ - \Delta\mu(\phi_R - \phi_L)\frac{\partial p}{\partial \mu} - \Delta\mu p \frac{\partial}{\partial \mu}(\phi_R - \phi_L) + \frac{\Delta\mu^2}{2}(\phi_L + \phi_R)\frac{\partial^2 p}{\partial \mu^2} \\ + \Delta\mu^2 \frac{\partial p}{\partial \mu}\frac{\partial}{\partial \mu}(\phi_L + \phi_R) + p\frac{\Delta\mu^2}{2}\frac{\partial^2}{\partial \mu^2}(\phi_L + \phi_R) + \cdots, \end{aligned} \tag{6.2}$$

where we have defined $p \equiv p(\mu, t)$, $\phi_L \equiv \phi_L(\mu, t)$, $\phi_N \equiv \phi_N(\mu, t)$, and $\phi_R \equiv \phi_R(\mu, t)$ to simplify the expressions. Combining like terms and truncating off higher-order terms in Equation 6.2 results in a PDE of the form

$$\frac{\partial p}{\partial t} = -\frac{\partial}{\partial \mu}(\beta p) + \frac{\partial^2}{\partial \mu^2}(\delta p), \tag{6.3}$$

where $\beta = \Delta\mu(\phi_R - \phi_L)/\Delta t$ and $\delta = \Delta\mu^2(\phi_R + \phi_L)/2\Delta t$. The resulting model in Equation 6.3 is Eulerian and known as the Fokker–Planck or Kolmogorov equation (e.g., Risken 1989; Barnett and Moorcroft 2008).* We can scale up to the population level and consider the spatial intensity $u(\mu, t)$ of some number of total animals (N), by letting $u(\mu, t) \equiv Np(\mu, t)$. In this context, assuming for the moment that there is no advection (i.e., drift or bias) component (i.e., $\beta = 0$), we have the

* Some mathematicians may object to the use of δ for a diffusion coefficient, preferring instead, D or μ, but we use δ to stay consistent with the rest of the mathematical notation in this book.

ecological diffusion equation:

$$\frac{\partial u}{\partial t} = \frac{\partial^2}{\partial \mu^2}(\delta u), \tag{6.4}$$

where the process of interest is $u \equiv u(\mu, t)$, and $\delta \equiv \delta(\mu, t)$ represents the diffusion coefficients that could vary over space and time. In the animal movement context, the diffusion parameter (δ) represents animal motility. One could arrive at an alternative reduction of the Fokker–Planck equation by assuming that $\delta = 0$, thus implying that animal movement is driven by advection only. Though, perhaps less intuitive, we might expect such behavior in wind- or water-advected populations (e.g., egg dispersal in a river system) or in cases where there is strong attraction or repulsion to spatial features.

There are other ways to derive the ecological diffusion model in Equation 6.4 (Turchin 1998); however, we feel that this perspective may be directly beneficial to those modeling spatio-temporal population dynamics, as the recent literature suggests (e.g., Wikle and Hooten 2010; Cressie and Wikle 2011; Lindgren et al. 2011). The properties of ecological diffusion (6.4) are different than those of plain or Fickian diffusions. The fundamental difference is that the diffusion coefficient (δ) appears on the inside of the two spatial derivatives rather than between them (Fickian, $\partial u/\partial t = (\partial/\partial \mu)\delta(\partial/\partial \mu)u$) or on the outside (plain, $\partial u/\partial t = \delta(\partial^2/\partial \mu^2)u$). Ecological diffusion describes a much less smooth process $u(\mu, t)$ than Fickian or plain diffusion, and allows for motility-driven congregation to sharply differ among neighboring habitat types. In some areas, animals may move slow, perhaps to forage, whereas in other areas, they move fast, as in exposed terrain. The resulting behavior shows a congregative effect in areas of low motility (i.e., $\delta \downarrow$) and a dispersive effect in areas of high motility (i.e., $\delta \uparrow$). In fact, depending on the boundary conditions, the steady-state solution implies that u is proportional to the inverse of δ.

The Lagrangian–Eulerian connection in ecological diffusion directly relates to the continuous-versus discrete-time formulations in animal movement models. We presented the Lagrangian–Eulerian connection for one particular scenario only, but similar approaches can be used to connect many other specifications for movement models in both Lagrangian and Eulerian contexts. The Taylor series expansion (6.2) suggests that the discrete-time model is more general because we are truncating higher-order terms to arrive at the continuous formulation. However, the continuous model allows for more compact notation and facilitates a continuous mathematical analysis, which can have advantages from an implementation perspective. In fact, Garlick et al. (2011) and Hooten et al. (2013a) show that aspects of the resulting continuous model (6.4) can be exploited to yield approximate solutions that are highly efficient to obtain numerically. Specifically, Garlick et al. (2011) and Hooten et al. (2013a) use a type of perturbation theory called the method of multiple scales, or homogenization, to arrive at an approximate solution to the PDE that is fast enough that it can be used iteratively in a statistical algorithm for large spatial and temporal domains. Such improvements in computational efficiency may not be possible using the discrete-time model (6.1) directly.

6.2 STOCHASTIC DIFFERENTIAL EQUATIONS

In the previous section, we indicated that there are advantages to formulating a deterministic movement probability in continuous time.* To compare and contrast the Lagrangian and Eulerian perspectives, we began with a deterministic discrete-time Lagrangian model for probability and scaled up to an Eulerian model in continuous time. We now show how to convert a stochastic discrete-time model (i.e., like those from the Chapter 5) to continuous time. We begin with the simple discrete-time random walk model we presented in Chapter 5. In this case, we use the notation **b**, instead of $\boldsymbol{\mu}$, to represent position, for reasons that will become apparent later. Thus, the discrete-time random walk model is

$$\mathbf{b}(t_i) = \mathbf{b}(t_{i-1}) + \boldsymbol{\varepsilon}(t_i), \tag{6.5}$$

where we explicitly use the parenthetical function notation (e.g., $\mathbf{b}(t_i)$) that depends on time directly, rather than the subscript notation (e.g., $\mathbf{b}_i$), and the change in time is $\Delta_i = t_i - t_{i-1}$. In this case, to let the individual step lengths correspond to time intervals between $\mathbf{b}(t_i)$ and $\mathbf{b}(t_{i-1})$, we let the displacement vectors $\boldsymbol{\varepsilon}(t_i)$ depend on Δ_i so that $\boldsymbol{\varepsilon}(t_i) \sim \text{N}(\mathbf{0}, \Delta_i \mathbf{I})$. For large gaps in time, the displacement distance (i.e., step length) of the individual during that time period will be larger on average. For simplicity, we consider the case where all the time intervals are equal (i.e., $\Delta_i = \Delta t, \forall i$).

An alternative way to write the model for the current position $\mathbf{b}(t_i)$ is as a sum of individual steps $\mathbf{b}(t_i) - \mathbf{b}(t_{i-1})$ beginning with the initial position at the origin $\mathbf{b}(t_0) = (0, 0)'$ and $t_0 = 0$ such that

$$\mathbf{b}(t_i) = \sum_{j=1}^{i} \mathbf{b}(t_j) - \mathbf{b}(t_{j-1}) \tag{6.6}$$

$$= \sum_{j=1}^{i} \boldsymbol{\varepsilon}(t_j). \tag{6.7}$$

For example, Figure 6.2 shows two simulated realizations of a Brownian motion process based on 1000 time steps to accentuate the necessary computational discretization. Forcing the time intervals between positions to be increasingly small (i.e., $\Delta t \to 0$) puts the model into a continuous-time setting. Then the individual steps become small but the sum is over infinitely many random quantities. Thus, the continuous time model arises as the limit

$$\mathbf{b}(t_i) = \lim_{\Delta t \to 0} \sum_{j=1}^{i} \boldsymbol{\varepsilon}(t_j), \tag{6.8}$$

* The standard PDE setting allows the probability of individual presence to evolve over time dynamically, but it is deterministic itself.

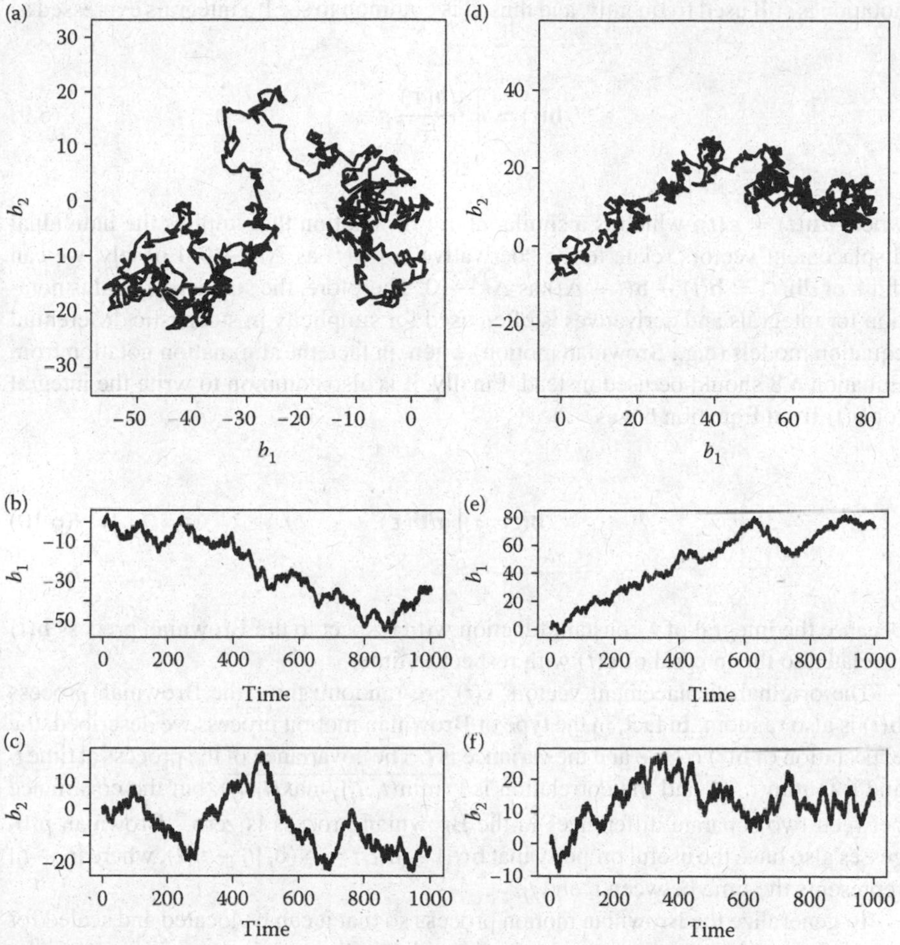

FIGURE 6.2 Joint (a, d) and marginal plots (b, c, e, f) of two simulated Brownian motion processes based on $\Delta t = 1$, and $n = 1000$ in both cases. Panels (a–c) and (d–f) show $\mathbf{b}(t)$, $b_1(t)$, and $b_2(t)$.

which resembles the operator known as the Ito integral from stochastic calculus. The resulting sequence $\mathbf{b}(t)$, for all t, is known as the Weiner process or Brownian motion.* It is more comfortable to write Equation 6.8 using traditional integral notation, but because we are "integrating" over a random quantity, the traditional deterministic integral notation is not technically valid. Nonetheless, the traditional

* Hence, the **b** notation stands for "Brownian." Note that we use a lowercase bold **b** to stay consistent with our vector notation; however, it is common to see an uppercase B in related literature.

notation is still used frequently, and thus, it is common to see Ito integrals expressed as

$$\mathbf{b}(t) = \int_0^t \frac{d\mathbf{b}(\tau)}{d\tau} d\tau, \tag{6.9}$$

where $d\mathbf{b}(t) = \boldsymbol{\varepsilon}(t)$, which is a similar abuse of notation that implies the individual displacement vectors relate to the "derivative" of $\mathbf{b}(t)$ as $\Delta t \to 0$. Loosely, we can think of $d\mathbf{b}(t) = \mathbf{b}(t) - \mathbf{b}(t - \Delta t)$ as $\Delta t \to 0$. Therefore, the standard calculus notation for integrals and derivatives is often used for simplicity in stochastic differential equation models (e.g., Brownian motion) when, in fact, the summation notation from Equation 6.8 should be used instead. Finally, it is also common to write the integral for $\mathbf{b}(t)$ from Equation 6.9 as

$$\mathbf{b}(t) = \int_0^t d\mathbf{b}(\tau), \tag{6.10}$$

because the integral of a constant function with respect to the Brownian process $\mathbf{b}(t)$ is related to the integral of $\boldsymbol{\varepsilon}(t)$ with respect to time.

The original displacement vectors $\boldsymbol{\varepsilon}(t)$ are random; thus, the Brownian process $\mathbf{b}(t)$ is also random. In fact, in the type of Brownian motion process we described, the expectation of $\mathbf{b}(t)$ is zero and the variance is t. The covariance of the process at time t_i and t_j is $\min(t_i, t_j)$ and the correlation is $\sqrt{\min(t_i, t_j)/\max(t_i, t_j)}$, but the covariance between two separate differences in the Brownian process is zero.* Brownian processes also have the useful property that $\mathbf{b}(t_i) - \mathbf{b}(t_j) \sim \text{N}(\mathbf{0}, |t_i - t_j|\mathbf{I})$, where $|t_i - t_j|$ represents the time between t_i and t_j.

To generalize the Brownian motion process so that it can be located and scaled for a specific position process $\boldsymbol{\mu}(t_i)$, we begin with the discrete-time specification again, such that

$$\boldsymbol{\mu}(t_i) = \boldsymbol{\mu}(t_{i-1}) + \boldsymbol{\varepsilon}(t_i). \tag{6.11}$$

Then, to relocate the process, we assume that the initial position is $\boldsymbol{\mu}(0)$ and, to scale the process, we let the displacement vectors $\boldsymbol{\varepsilon}(t_i) \sim \text{N}(\mathbf{0}, \sigma^2 \Delta t \mathbf{I})$, where σ^2 stretches or shrinks the trajectory in space. Using the Brownian notation, this model results in

$$\boldsymbol{\mu}(t) = \boldsymbol{\mu}(0) + \sigma \mathbf{b}(t), \tag{6.12}$$

for any time t. Figure 6.3 shows the simulated Brownian motion processes based on the same $\boldsymbol{\varepsilon}(t_i)$ from Figure 6.3, but relocated using Equation 6.12 and initial position at $\boldsymbol{\mu}(0) = (100, 100)'$.

---

* This is known as the "independent increments" property and arises from the fact that each difference in Brownian processes represents an $\boldsymbol{\varepsilon}$ and they are independent Gaussian random variables.

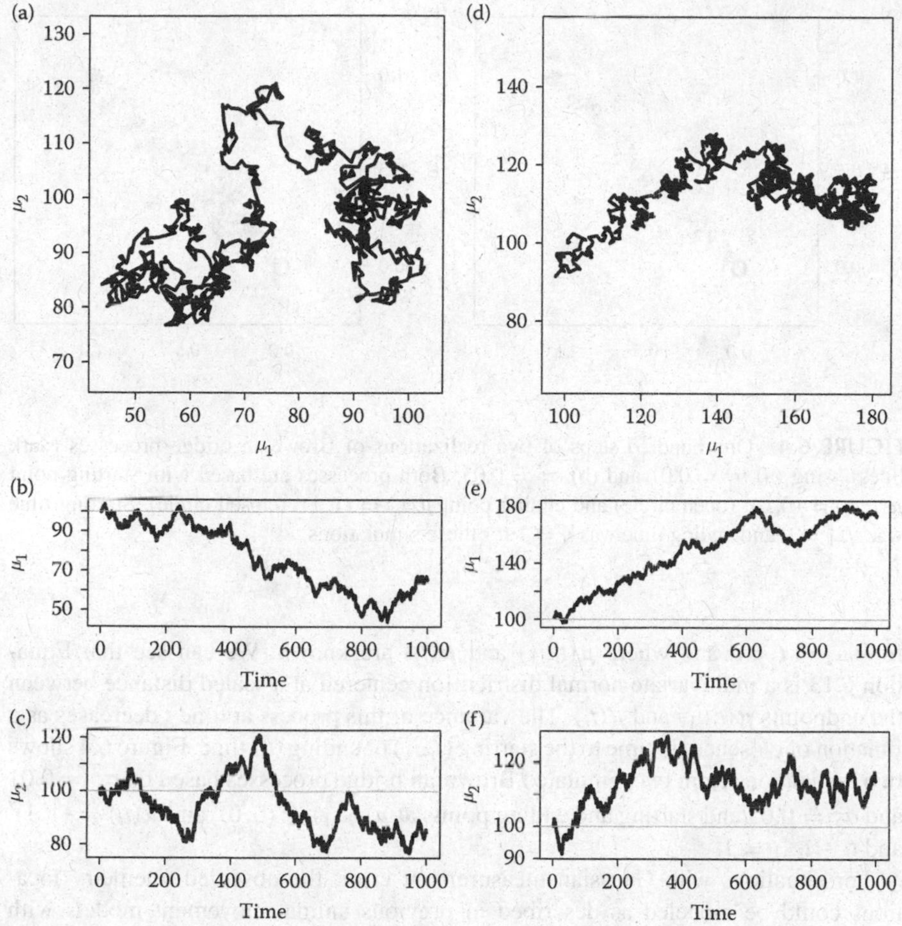

FIGURE 6.3 Joint (a, d) and marginal plots (b, c, e, f) of two simulated Brownian motion processes based on $\boldsymbol{\mu}(0) = (100, 100)'$, $\Delta t = 1$, $\sigma^2 = 1$, and $n = 1000$ in both cases. Panels (a–c) and (d–f) show $\boldsymbol{\mu}(t)$, $\mu_1(t)$, and $\mu_2(t)$. Horizontal gray lines correspond to the initial position $\boldsymbol{\mu}(0)$.

6.3 BROWNIAN BRIDGES

The term "Brownian bridge" has become popular in animal movement modeling, in large part, because of a series of papers and associated software to fit statistical models to telemetry data. A Brownian bridge is a Brownian motion process with known and fixed starting and ending times and locations. Reverting to the position notation $\boldsymbol{\mu}(t)$ we used previously, Horne et al. (2007) describe the Brownian bridge as multivariate normal random process such that

$$\boldsymbol{\mu}(t) \sim \mathrm{N}\left(\boldsymbol{\mu}(t_{i-1}) + \frac{t - t_{i-1}}{t_i - t_{i-1}}(\boldsymbol{\mu}(t_i) - \boldsymbol{\mu}(t_{i-1})), \frac{(t - t_{i-1})(t_i - t)}{t_i - t_{i-1}}\sigma^2\right) \tag{6.13}$$

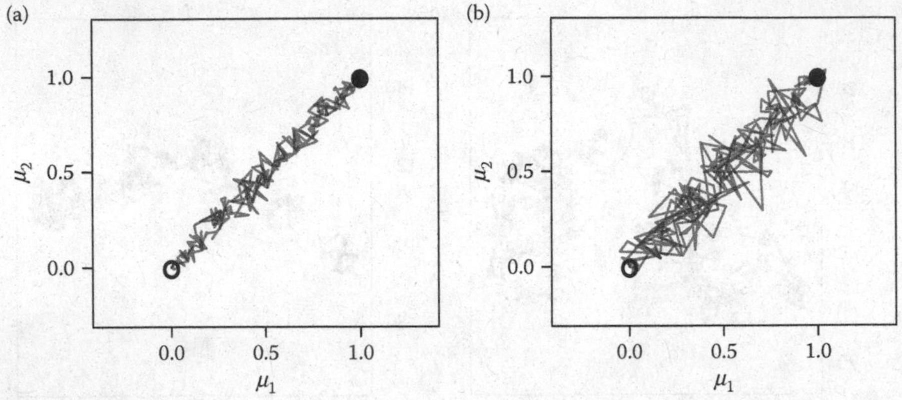

FIGURE 6.4 One hundred steps of two realizations of Brownian bridge processes (dark lines) using (a) $\sigma_1^2 = 0.01$ and (b) $\sigma_2^2 = 0.05$. Both processes are based with starting point $\boldsymbol{\mu}(t_{i-1}) = (0,0)'$ (open circle) and ending point $\boldsymbol{\mu}(t_i) = (1,1)'$ (closed circle). Starting time was $t_{i-1} = 0$ and ending time was $t_i = 1$ for these simulations.

for $t_{i-1} < t < t_i$ and where $\boldsymbol{\mu}(t_{i-1})$ and $\boldsymbol{\mu}(t_i)$ are known. We can see that Equation 6.13 is a multivariate normal distribution centered at a scaled distance between the endpoints $\boldsymbol{\mu}(t_{i-1})$ and $\boldsymbol{\mu}(t_i)$. The variance of this process at time t decreases as a function of closeness in time to the starting (t_{i-1}) or ending (t_i) time. Figure 6.4 shows two realizations from two simulated Brownian bridge processes based on $\sigma_1^2 = 0.01$ and $\sigma_2^2 = 0.05$ and starting and ending points at $\boldsymbol{\mu}(t_{i-1}) = (0,0)'$ and $\boldsymbol{\mu}(t_i) = (1,1)'$ and $t_i - t_{i-1} = 1$.

For situations with Gaussian measurement error, the observed telemetry locations could be modeled as described in previous animal movement models with $\mathbf{s}(t_i) \sim \text{N}(\boldsymbol{\mu}(t_i), \sigma_s^2\mathbf{I})$ for $i = 1, \ldots, n$. This adds a natural hierarchical structure to the model. However, most common methods for implementing these models integrate over the Brownian motion process $(\boldsymbol{\mu}(t))$ to fit the model using likelihood methods.

Horne et al. (2007) propose an approach that conditions on every other observation as an endpoint, using the middle locations as data to fit the Brownian bridge model. Their idea was to exploit the independence property using triplets of the data. After passing through the data once, they cycle back through it again after shifting the triplets, ultimately yielding a "sample size" of approximately $n/2$ observations. Despite the fact that this procedure results in a computationally efficient method for fitting models to telemetry data, Pozdnyakov et al. (2014) suggest several potential problems that could arise with it. First, the method Horne et al. (2007) described for forming the likelihood produces a bias in the estimation of the movement variance (σ^2) that increases as the measurement error variance (σ_s^2) increases. Second, the movement and measurement error variances are not identifiable in the likelihood, especially with equal time intervals between observations. Third, only approximately half of the data are used to fit the model.

Pozdnyakov et al. (2014) demonstrate that the variance of the observed telemetry location is

$$\mathrm{var}(\mathbf{s}(t_i)) = \sigma^2 t_i \mathbf{I} + \sigma_s^2 \mathbf{I} \tag{6.14}$$

and covariance is

$$\mathrm{cov}(\mathbf{s}(t_i), \mathbf{s}(t_j)) = \sigma^2 \min(t_i, t_j). \tag{6.15}$$

Thus the covariance matrix for the joint telemetry data is dense (completely filled with nonzero elements). However, the covariance matrix for the observed velocities (i.e., $\mathbf{s}(t_i) - \mathbf{s}(t_{i-1})$) is tri-diagonal, but not diagonal, meaning not all off-diagonal elements of the matrix are zero. In fact, the measurement error variance occurs on the off-diagonals, which implies that the non-diagonal nature of the covariance matrix for the joint process becomes increasingly important as the measurement error increases. The diagonal elements of the covariance matrix for the joint velocity process are equal to $\sigma^2(t_i - t_{i-1}) + \sigma_s^2$. Pozdnyakov et al. (2014) suggest using the joint distribution of all velocities (which is multivariate normal) as the likelihood to fit the Brownian motion model instead of the Brownian bridge methods proposed by Horne et al. (2007), and claims the approach is just as easy to implement.

Thus, rather than condition on an incremental sequence of endpoints, there is value in modeling the animal movement process as a true dynamic continuous-time process. We return to the covariance modeling perspective of Pozdnyakov et al. (2014) for a broader class of movement models based on continuous-time stochastic processes in the sections that follow.

Other applications of Brownian bridge models for telemetry data include Liu et al. (2014) and Liu et al. (2015), who use Brownian bridges in a hierarchical model to characterize dead-reckoned paths of marine mammals. Liu et al. (2014) developed a computationally efficient Bayesian melding approach for path reconstruction that provides improved inference as compared with linear interpolation procedures.

6.4 ATTRACTION AND DRIFT

Brownian motion ($\mathbf{b}(t)$) results in smoother trajectories than white noise ($\boldsymbol{\varepsilon}(t)$) because it is an integrated quantity. That is the reason why Brownian motion is often chosen as a framework for modeling animal movement in continuous time. However, as presented in Equation 6.8, Brownian motion is not a very flexible model for movement because it lacks drift and attraction components.

It is straightforward to create more flexible models for animal movement using the general procedure from the previous sections. Recall that we can convert a discrete-time process into a continuous-time process using the following steps:

1. Specify the stochastic recursion

$$\boldsymbol{\mu}(t_i) = \boldsymbol{\mu}(0) + \sum_{j=1}^{i} (\boldsymbol{\mu}(t_j) - \boldsymbol{\mu}(t_{j-1})). \tag{6.16}$$

2. Specify the standard parametric conditional discrete-time model for $\boldsymbol{\mu}(t_j)$.

3. Substitute the model for $\boldsymbol{\mu}(t_j)$ into the right-hand side of Equation 6.16.
4. Take the limit of $\boldsymbol{\mu}(t_i)$ as $\Delta t \to 0$ to obtain the Ito integral representation of the continuous-time process.
5. If desired, rewrite the model in terms of the Ito derivative of $\boldsymbol{\mu}(t)$.

To demonstrate this procedure, suppose we wish to add point-based attraction to the Brownian motion process. In this case, recall the discrete-time model for attraction from Chapter 5:

$$\boldsymbol{\mu}(t_i) = \mathbf{M}\boldsymbol{\mu}(t_{i-1}) + (\mathbf{I} - \mathbf{M})\boldsymbol{\mu}^* + \boldsymbol{\varepsilon}(t_i), \tag{6.17}$$

where $\mathbf{M}$ is the VAR(1) propagator matrix, $\boldsymbol{\mu}^*$ is the attracting location, and $\boldsymbol{\varepsilon}(t) \sim \text{N}(\mathbf{0}, \sigma^2 \Delta t \mathbf{I})$. Substituting this conditional discrete-time model into Equation 6.16 for $\boldsymbol{\mu}(t_j)$ results in

$$\boldsymbol{\mu}(t_i) = \boldsymbol{\mu}(0) + \sum_{j=1}^{i} (\boldsymbol{\mu}(t_j) - \boldsymbol{\mu}(t_{j-1})) \tag{6.18}$$

$$= \boldsymbol{\mu}(0) + \sum_{j=1}^{i} (\mathbf{M}\boldsymbol{\mu}(t_{j-1}) + (\mathbf{I} - \mathbf{M})\boldsymbol{\mu}^* + \boldsymbol{\varepsilon}(t_j) - \boldsymbol{\mu}(t_{j-1})) \tag{6.19}$$

$$= \boldsymbol{\mu}(0) + \sum_{j=1}^{i} ((\mathbf{M} - \mathbf{I})(\boldsymbol{\mu}(t_{j-1}) - \boldsymbol{\mu}^*) + \boldsymbol{\varepsilon}(t_j)) \tag{6.20}$$

$$= \boldsymbol{\mu}(0) + \sum_{j=1}^{i} (\mathbf{M} - \mathbf{I})(\boldsymbol{\mu}(t_{j-1}) - \boldsymbol{\mu}^*) + \sum_{j=1}^{i} \boldsymbol{\varepsilon}(t_j). \tag{6.21}$$

We recognize the last term, $\sum_{j=1}^{i} \boldsymbol{\varepsilon}(t_j)$, from the previous section, as the building block of Brownian motion. Thus, taking the limit of the right-hand side as $\Delta t \to 0$ results in the Ito integral equation

$$\boldsymbol{\mu}(t) = \boldsymbol{\mu}(0) + \int_0^t (\mathbf{M} - \mathbf{I})(\boldsymbol{\mu}(\tau) - \boldsymbol{\mu}^*)\, d\tau + \int_0^t \sigma\, d\mathbf{b}(\tau) \tag{6.22}$$

$$= \boldsymbol{\mu}(0) + \int_0^t (\mathbf{M} - \mathbf{I})(\boldsymbol{\mu}(\tau) - \boldsymbol{\mu}^*)\, d\tau + \sigma \mathbf{b}(t). \tag{6.23}$$

The integral Equation 6.23 contains three components: the quantity $\boldsymbol{\mu}(0)$, which provides the proper starting position, the attracting process $\int_0^t (\mathbf{M} - \mathbf{I})(\boldsymbol{\mu}(\tau) - \boldsymbol{\mu}^*)\, d\tau$, and the scaled Brownian motion process $\sigma \mathbf{b}(t)$. Finally, by Ito differentiating both sides of Equation 6.23, we arrive at the stochastic differential equation (SDE) for

Brownian motion with attraction

$$d\boldsymbol{\mu}(t) = (\mathbf{M} - \mathbf{I})(\boldsymbol{\mu}(t) - \boldsymbol{\mu}^*)dt + \sigma\, d\mathbf{b}(t) \tag{6.24}$$

$$= (\mathbf{M} - \mathbf{I})(\boldsymbol{\mu}(t) - \boldsymbol{\mu}^*)dt + \boldsymbol{\varepsilon}(t). \tag{6.25}$$

Note that $\boldsymbol{\varepsilon}(t) \sim \text{N}(\mathbf{0}, \sigma^2 dt\mathbf{I})$ and the form in Equation 6.25 is common in the SDE literature, but it can also be written as

$$\frac{d\boldsymbol{\mu}(t)}{dt} = (\mathbf{M} - \mathbf{I})(\boldsymbol{\mu}(t) - \boldsymbol{\mu}^*) + \frac{\boldsymbol{\varepsilon}(t)}{dt}, \tag{6.26}$$

with the usual derivative with respect to time ($d\boldsymbol{\mu}(t)/dt$) on the left-hand side. Now we recognize Equation 6.26 as a differential equation with an additive term corresponding to differentiated Brownian motion. This is what sets SDEs apart from deterministic differential equations with additive error. The "error" term (i.e., $\boldsymbol{\varepsilon}(t)$) itself is wrapped up in the derivative of the position process $\boldsymbol{\mu}(t)$.

We can rewrite the stochastic integral equation (SIE) (6.23) in words as

$$\text{Position} = \text{starting place} + \text{cumulative drift} + \text{cumulative diffusion}. \tag{6.27}$$

The cumulative drift integrates (i.e., adds up) the drift process, which, in the case of Equation 6.23, are the propagated displacements from the attracting point $\boldsymbol{\mu}^*$. The cumulative diffusion integrates the uncorrelated steps or "errors" to arrive at a correlated movement process (described earlier as Brownian motion). Together, these two components combine to provide a realistic continuous-time movement model for animals such as central place foragers. However, the expression in Equation 6.27 also provides a very general way to characterize many different SIE models by modifying the drift and diffusion components directly.*

Figure 6.5 shows two simulated stationary SDE processes arising from Equation 6.25 assuming $\mathbf{M} = \rho\mathbf{I}$. As in the discrete-time models in Chapter 5, the stochastic process in Figure 6.5a ($\rho = 0.75$) is less smooth than that in Figure 6.5d ($\rho = 0.99$), but both processes are attracted to the point $\boldsymbol{\mu}^* = (0, 0)'$.

We began with a simple Brownian motion process with no attraction in the previous section and we added a drift term to it, resulting in a more flexible model for true animal position processes. The resulting SIE (6.23) is not Brownian, but rather contains a Brownian component. In fact, the SDE in Equation 6.26 represents one way to specify an Ornstein–Uhlenbeck (OU) process (Dunn and Gipson 1977; Blackwell 1997).

6.5 ORNSTEIN–UHLENBECK MODELS

In the preceding section, we noted that Brownian motion with attraction (6.26) is referred to as an OU process. In fact, we derived it by differentiating an SIE (6.23) that originated from a sequence of heuristic arguments based on the sum of infinite

* Drift is often referred to as bias or advection in the PDE literature.

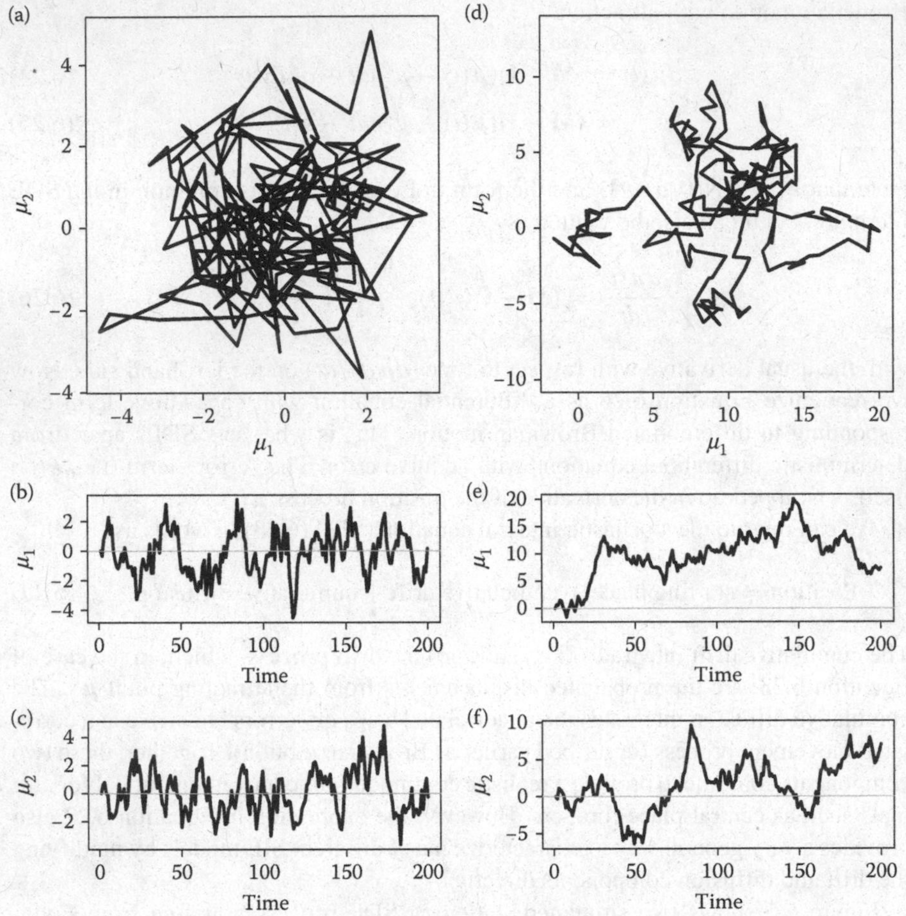

FIGURE 6.5 Two simulated stationary SDE processes (dark lines) using (a–c) $\rho = 0.75$ and (d–f) $\rho = 0.99$. Both processes are based on attracting point $\boldsymbol{\mu}^* = (0, 0)'$ and variance $\sigma^2 = 1$. Panels (a) and (d) show the joint process $\boldsymbol{\mu}(t)$ while panels (b–c) and (e–f) show the marginal processes $\mu_1(t)$ and $\mu_2(t)$.

steps. However, the OU process is often expressed in exponential notation (e.g., Dunn and Gipson 1977; Blackwell 2003; Johnson et al. 2008a).

To arrive at the OU expression involving exponentials, we note that it is more common in mathematical modeling to start with the SDE involving the velocity process and then "solve" it to find the position process $\boldsymbol{\mu}(t)$. To demonstrate how solutions to the SDE are typically derived, we begin with a simplified SDE based on Equation 6.26 in 1-D space and with attractor $\mu^* = 0$, Brownian variance $\sigma^2 = 1$, and autocorrelation parameter θ, such that

$$d\mu(t) = -\theta\mu(t)\,dt + db(t). \tag{6.28}$$

One solution technique involves a variation of parameters method. In this case, multiply both sides of Equation 6.28 by $e^{\theta t}$ and then integrate both sides from 0 to t. The $e^{\theta t}$ term actually simplifies the required integration and allows for an analytical solution. Thus, multiplying both sides of Equation 6.28 by $e^{\theta t}$ results in

$$e^{\theta t}\, d\mu(t) = -\theta e^{\theta t} t\mu(t)\, dt + e^{\theta t}\, db(t). \tag{6.29}$$

Then, integrating both sides of Equation 6.29 from 0 to t yields

$$\int_0^t e^{\theta\tau}\, d\mu(\tau)\, d\tau = -\theta \int_0^t e^{\theta\tau}\mu(\tau)\, d\tau + \int_0^t e^{\theta\tau}\, db(\tau). \tag{6.30}$$

The integral on the left-hand side of Equation 6.30 can be solved using integration by parts:

$$\int_0^t e^{\theta\tau}\, d\mu(\tau)\, d\tau = e^{\theta t}\mu(t) - \mu(0) - \int_0^t \mu(\tau)\theta e^{\theta\tau}\, d\tau. \tag{6.31}$$

Substituting Equation 6.31 back into Equation 6.30 yields

$$e^{\theta t}\mu(t) - \mu(0) - \int_0^t \mu(\tau)\theta e^{\theta\tau}\, d\tau = -\theta \int_0^t e^{\theta\tau}\mu(\tau)\, d\tau + \int_0^t e^{\theta\tau}\, db(\tau), \tag{6.32}$$

which, after some algebra, simplifies to

$$\mu(t) = \mu(0)e^{-\theta t} + \int_0^t e^{-\theta(t-\tau)}\, db(\tau). \tag{6.33}$$

The resulting solution has several interesting properties. First, notice that, as $t \to \infty$, the first term on the right-hand side of Equation 6.33 goes away (i.e., $\mu(0)e^{-\theta t} \to 0$). This result implies that, as the period of time increases, the initial position has less effect on the solution for $\mu(t)$. Second, the integral on the right-hand side is a convolution of $\exp(-\theta(t-\tau))$ with a white noise process (Iranpour et al. 1988). To determine the mean and variance of this random variable, we return to the infinite summation representation of the Ito integral. Thus,

$$\int_0^t e^{-\theta(t-\tau)}\, db(\tau) = \lim_{\Delta t \to 0} \sum_{j=1}^{i} e^{-\theta(t-\tau)}(b(t_j) - b(t_{j-1})), \tag{6.34}$$

where $t_0 = 0$ and $t_i = t$. For any Δt, $\sum_{j=1}^{i} e^{-\theta(t-t_j)}(b(t_j) - b(t_{j-1}))$ is a weighted sum of independent normal random variables with mean zero and variances

$\sigma^2 e^{-2\theta(t-t_j)} \Delta t$; therefore, the variance of Equation 6.34 is

$$\lim_{\Delta t \to 0} \sum_{j=1}^{i} \sigma^2 e^{-2\theta(t-t_j)} \Delta t = \int_0^t \sigma^2 e^{-2\theta(t-\tau)} \, d\tau \tag{6.35}$$

$$= \frac{\sigma^2}{2\theta}(1 - e^{-2\theta t}), \tag{6.36}$$

Another common way to express the OU process is using conditional distribution notation. Dunn and Gipson (1977) use conditional distribution notation in their seminal paper on OU processes as models for animal movement. In the context of our simple 1-D OU process, for $t > \tau$, we can write

$$\mu(t)|\mu(\tau) \sim \mathrm{N}\left(\mu(\tau)e^{-\theta(t-\tau)}, \frac{\sigma^2}{2\theta}\left(1 - e^{-2\theta(t-\tau)}\right)\right). \tag{6.37}$$

Thus, as the time gap increases between $\mu(t)$ and $\mu(\tau)$, the conditional process reverts to zero and the variance converges to σ^2. However, with small $|t - \tau|$, $\mu(t)$ will be closer to $\mu(\tau)$. Understanding stochastic processes in terms of covariance will become important in the following sections.

Figure 6.6 shows two 1-D conditional univariate stochastic processes simulated from Equation 6.37 based on two different values for θ. Figure 6.6a shows the conditional process based on a relatively large $\theta = 1$, while Figure 6.6b shows the conditional process based on a much smaller $\theta = 0.001$. While both processes are conditioned on $\mu(\tau) = 1$, the conditional process in Figure 6.6a shows very little memory of $\mu(\tau) = 0$, the process in Figure 6.6b clearly indicates longer-range dependence on $\mu(\tau) = 1$.

6.6 POTENTIAL FUNCTIONS

In a series of papers on the statistical modeling of trajectories, Brillinger and colleagues described a more flexible drift component for SIE/SDE movement models of the form presented in Equation 6.27. Borrowing a concept from fluid mechanics called the "potential function," Brillinger (2010) describes how it can be used as a drift component in SIE/SDE movement models to account for both static and dynamic attractors and the possible effect of covariates on movement in a continuous-time context.

Generalizing the SDE (6.26) from the previous section, we can write

$$\frac{d\boldsymbol{\mu}(t)}{dt} = \mathbf{g}(\boldsymbol{\mu}(t)) + \frac{\boldsymbol{\varepsilon}(t)}{dt}, \tag{6.38}$$

where the function $\mathbf{g}(\boldsymbol{\mu}(t))$ acts as the drift component of the SDE model and we assume $\boldsymbol{\varepsilon}(t) \sim \mathrm{N}(\mathbf{0}, \sigma^2 dt\mathbf{I})$ in this section. In the previous section, we arrived at the functional form $\mathbf{g}(\boldsymbol{\mu}(t)) = (\mathbf{M} - \mathbf{I})(\boldsymbol{\mu}(t) - \boldsymbol{\mu}^*)$ for the drift component based

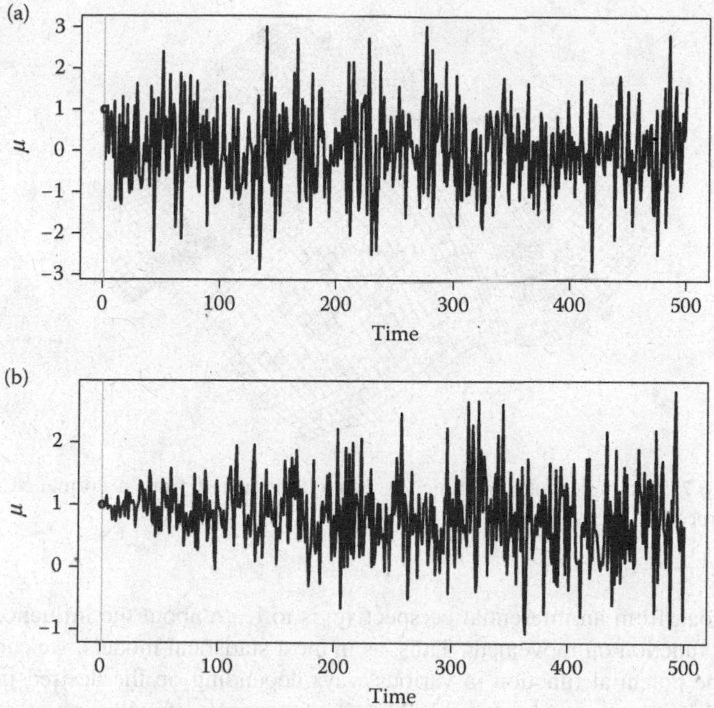

FIGURE 6.6 Two 1-D simulated conditional processes (dark lines) from Equation 6.37 based on $\sigma^2 = 1$, $\tau = 1$ (vertical gray line), $\mu(\tau) = 1$ (open circle). (a) $\theta = 1$ and (b) $\theta = 0.001$.

on a conversion from the discrete- to the continuous-time model. Preisler et al. (2004) explains several ways to generalize the drift to better mimic realistic animal movement. For example, a potential function could be used for drift such that $\mathbf{g}(\boldsymbol{\mu}(t)) = -\nabla p(\boldsymbol{\mu}(t))$ (Brillinger et al. 2001).* The potential function $p(\boldsymbol{\mu}(t))$ can be a function in space alone or in both time and space, depending on model assumptions. It may also be a function of other information (e.g., covariates, known points of attraction) and parameters. The potential function is often referred to as a "force field" that acts on the animal, controlling its movement. The potential function can be visualized as a hilly surface in the geographical space of the study area (like a topographic map; Figure 6.7) upon which a ball could be placed that represents the individual animal of interest. The ball will naturally roll downhill on the surface and the speed at which it rolls relates to the steepness of the surface. Thus, a derivative of the surface in the direction the ball rolls is negatively correlated with the speed of the animal, providing a heuristic for the general SDE in Equation 6.38.

* Several notational issues arise here. First, ∇ refers to the gradient operator; thus, $\nabla p(\boldsymbol{\mu}(t)) = (dp/d\mu_1, dp/d\mu_2)'$. Second, we use p to represent the potential function because the first letter of potential is p. In many of the papers by Billinger and Preisler, H is used for the potential function, $\mathbf{r}$ is used for position, and $\boldsymbol{\mu}$ is used for drift. Yes, this can be confusing at first, but to remain consistent with other literature and our expressions thus far, a notational change is necessary.

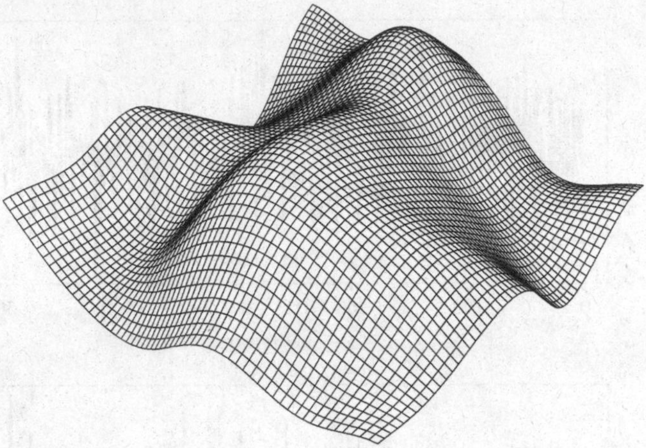

FIGURE 6.7 Example potential function $p(\boldsymbol{\mu}(t))$ simulated from a correlated Gaussian random process.

Our goal, from an inferential perspective, is to learn about the influences of the potential function on movement. Thus, as in most statistical models, we can parameterize the potential function in various ways depending on the desired inference. If the goal is to learn about the influence of a single attracting point on movement, we can retain the SDE model from the previous section, or we could use the potential concept directly, letting $p(\boldsymbol{\mu}(t), \boldsymbol{\mu}^*) \equiv \frac{1}{2}(\boldsymbol{\mu}(t) - \boldsymbol{\mu}^*)'(\boldsymbol{\mu}(t) - \boldsymbol{\mu}^*)$, the L_2 norm associated with distance between $\boldsymbol{\mu}(t)$ and $\boldsymbol{\mu}^*$. Using this potential function, we arrive at $\mathbf{g}(\boldsymbol{\mu}(t)) = -(\mu_1 - \mu_1^*, \mu_2 - \mu_2^*)' = -(\boldsymbol{\mu}(t) - \boldsymbol{\mu}^*)$ for a gradient field. The resulting gradient field implies that the mean structure for the velocity $d\boldsymbol{\mu}(t)/dt$ will be zero when $\boldsymbol{\mu}(t)$ is close to the attracting point $\boldsymbol{\mu}^*$, imposing no particular directional bias on movement when the animal is near the central place. As the animal ventures far from the attracting point $\boldsymbol{\mu}^*$, the mean structure implied by the gradient will bias movement back toward the central place (Figure 6.8a).

We can attenuate the attractive force by incorporating a multiplicative term that decreases the velocity as needed. For example, if we use $\mathbf{g}(\boldsymbol{\mu}(t)) \equiv -(1 - \rho)(\boldsymbol{\mu}(t) - \boldsymbol{\mu}^*)$ such that $0 < \rho < 1$, we arrive at the same SDE model as Equation 6.26. In that case, the propagator matrix is $\mathbf{M} \equiv \rho\mathbf{I}$ and a unity autocorrelation parameter (i.e., $\rho = 1$) will remove the attractive effect completely, allowing the individual to wander aimlessly. As in the time series context, values of ρ less than one will ensure the individual's path is stationary over time, forcing the animal to move toward the central place $\boldsymbol{\mu}^*$ eventually. For example, Figure 6.8b shows the potential function obtained by integrating $\mathbf{g}(\boldsymbol{\mu}(t))$ based on $\rho = 0.5$ and $\boldsymbol{\mu}^* = (0.5, 0.5)'$.

The potential function in Figure 6.8b is flatter than that in Figure 6.8a because the autocorrelation is stronger ($\rho = 0.5$ vs. $\rho = 0$ in Figure 6.8a). As $\rho \to 1$, the potential function becomes perfectly flat, allowing the individual to move without an attracting force.

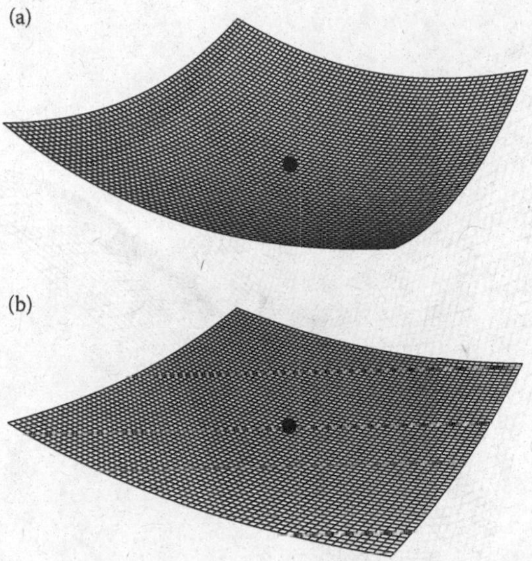

FIGURE 6.8 (a) Potential surface $p(\boldsymbol{\mu}(t), \boldsymbol{\mu}^*) = \frac{1}{2}(\boldsymbol{\mu}(t) - \boldsymbol{\mu}^*)'(\boldsymbol{\mu}(t) - \boldsymbol{\mu}^*)$ based on a single attracting point $\boldsymbol{\mu}^*$ (black circle). (b) Potential surface $p(\boldsymbol{\mu}(t), \boldsymbol{\mu}^*) = (1-\rho)/2 \cdot (\boldsymbol{\mu}(t) - \boldsymbol{\mu}^*)'(\boldsymbol{\mu}(t) - \boldsymbol{\mu}^*)$ based on a single attracting point $\boldsymbol{\mu}^*$ (black circle) and $\rho = 0.5$.

The number of attracting points can be increased easily by letting the potential function be a sum or product of several individual functions. For example, in the case of two additive attractors, $\boldsymbol{\mu}_1^*$ and $\boldsymbol{\mu}_2^*$, we have

$$p(\boldsymbol{\mu}(t), \boldsymbol{\mu}_1^*, \boldsymbol{\mu}_2^*) = p_1(\boldsymbol{\mu}(t), \boldsymbol{\mu}_1^*) + p_2(\boldsymbol{\mu}(t), \boldsymbol{\mu}_2^*)$$
$$= -\frac{1}{2}[\boldsymbol{\mu}(t)|\boldsymbol{\mu}_1^*, \sigma_1^2] - \frac{1}{2}[\boldsymbol{\mu}(t)|\boldsymbol{\mu}_2^*, \sigma_2^2]. \quad (6.39)$$

where $[\boldsymbol{\mu}(t)|\boldsymbol{\mu}_1^*, \sigma_1^2]$ and $[\boldsymbol{\mu}(t)|\boldsymbol{\mu}_2^*, \sigma_2^2]$ are bivariate Gaussian density functions with means $\boldsymbol{\mu}_1^*$ and $\boldsymbol{\mu}_2^*$ and variances σ_1^2 and σ_2^2. The potential function in Equation 6.39 results in a complicated gradient function ($\mathbf{g}(\boldsymbol{\mu}(t))$) with a saddle point between the two attracting points (Figure 6.9).

Another way to specify the potential function is to let it be a polynomial and interaction function of the elements of position (e.g., Kendall 1974; Brillinger 2010). For example, the potential function

$$p(\boldsymbol{\mu}(t), \boldsymbol{\beta}) = \beta_1\mu_1(t) + \beta_2\mu_2(t) + \beta_3\mu_1^2(t) + \beta_4\mu_2^2(t) + \beta_5\mu_1(t)\mu_2(t), \quad (6.40)$$

will allow for learning about the best-fitting elliptical home range by estimating the coefficients $\boldsymbol{\beta}$.

One approach to account for boundaries to movement is to let the potential function be time-varying and depend on a region $\mathcal{R}$. For example, we can let $p(\boldsymbol{\mu}(t), \gamma, \mathcal{R}) \equiv \gamma/d_{\min}(t)$, where $d_{\min}(t) = \min_{\boldsymbol{\mu}^*}(\boldsymbol{\mu}(t) - \boldsymbol{\mu}^*)'(\boldsymbol{\mu}(t) - \boldsymbol{\mu}^*)$ is the

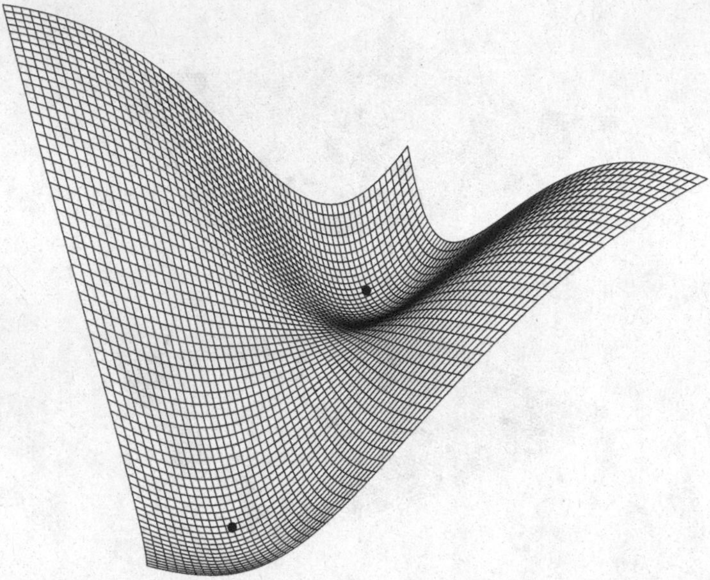

FIGURE 6.9 Potential surface $p(\boldsymbol{\mu}(t), \boldsymbol{\mu}^*) = -[\boldsymbol{\mu}(t)|\boldsymbol{\mu}_1^*, \sigma_1^2]/2 - [\boldsymbol{\mu}(t)|\boldsymbol{\mu}_2^*, \sigma_2^2]/2$, where the overall potential function is an average of potential functions that are negative bivariate Gaussian density functions with means $\boldsymbol{\mu}_1^*$ and $\boldsymbol{\mu}_2^*$ (black circles) and equal variances ($\sigma_1^2 = \sigma_2^2$).

squared distance to the closest point in $\mathcal{R}$ from the current position $\boldsymbol{\mu}(t)$. In this specification, if $\gamma > 0$, the drift term will push the animal out of the region $\mathcal{R}$, which is particularly effective for marine species.* An alternative approach to account for boundaries is to specify the potential function such that it has higher potential outside of a boundary. For example, suppose there are two activity centers within a circular bounded region $\mathcal{R}^c$ (e.g., a pond or crater with two divots; Figure 6.10). A corresponding potential function can be specified as

$$p(\boldsymbol{\mu}(t)) = \begin{cases} -\theta_1\left(\frac{1}{2}[\boldsymbol{\mu}(t)|\boldsymbol{\mu}_1^*, \sigma_1^2] + \frac{1}{2}[\boldsymbol{\mu}(t)|\boldsymbol{\mu}_2^*, \sigma_2^2]\right) & \text{if } \boldsymbol{\mu}(t) \in \mathcal{R}^c \\ \theta_2\sqrt{(\boldsymbol{\mu}(t) - \boldsymbol{\mu}_3^*)'(\boldsymbol{\mu}(t) - \boldsymbol{\mu}_3^*)} & \text{if } \boldsymbol{\mu}(t) \in \mathcal{R} \end{cases}, \quad (6.41)$$

where $\boldsymbol{\mu}_3^*$ is the overall space use center and the multipliers θ_1 and θ_2 control the strength of boundary and attraction. Figure 6.11 shows a simulated trajectory based on the potential function in Equation 6.41. The simulated individual trajectory generally is attracted to $\boldsymbol{\mu}_1^*$ and $\boldsymbol{\mu}_2^*$ and, if it wanders outside of $\mathcal{R}^c$, it slides back in due to the steepness of potential at the boundary.

---

* Recall that there are other ways to account for boundaries to movement in the point process modeling framework (e.g., Brost et al. 2015).

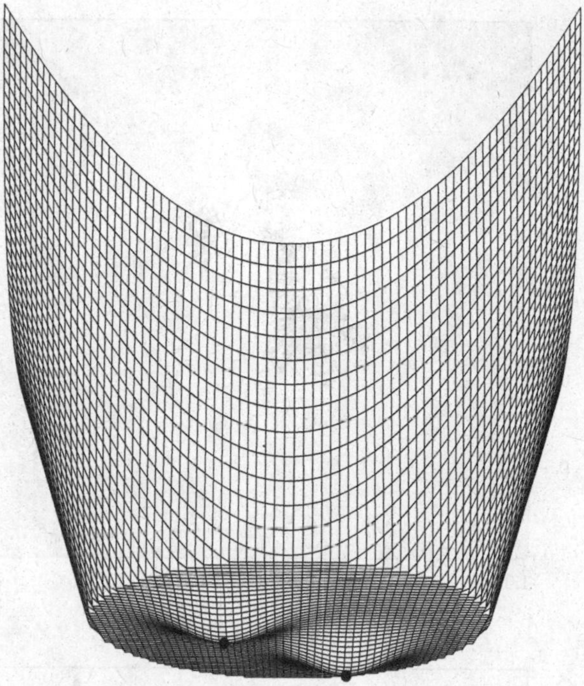

FIGURE 6.10 Potential function $p(\boldsymbol{\mu}(t))$ based on two attracting points $\boldsymbol{\mu}_1^*$ and $\boldsymbol{\mu}_2^*$ (black circles) and a steeply rising boundary condition delineating a circular region of space use.

There is no reason why the form of the potential function is limited to a function of points in geographical space. In fact, it could be a function of covariates $\mathbf{x}(\boldsymbol{\mu}(t))$. For example, the potential function $p(\boldsymbol{\mu}(t), \boldsymbol{\beta}) \equiv \mathbf{x}(\boldsymbol{\mu}(t))'\boldsymbol{\beta}$ takes on a multiple regression form and implies that certain linear combinations of spatially explicit covariates should influence the velocity of an individual's movement. These covariates could also vary in time and include things such as soil moisture, ambient temperature, or other dynamic environmental factors. Regression specifications for potential functions have been used in many different models and applications, including discrete-space animal movement (Hooten et al. 2010b; Hanks et al. 2011, 2015a), disease transmission (Hooten and Wikle 2010; Hooten et al. 2010a), invasive species spread (Broms et al. 2016), and landscape genetics and connectivity models (Hanks and Hooten 2013).

To implement SDE models based on potential functions, Brillinger (2010) suggests a statistical model specification similar to

$$\boldsymbol{\mu}(t_i) - \boldsymbol{\mu}(t_{i-1}) = (t_i - t_{i-1})\mathbf{g}(\boldsymbol{\mu}(t_{i-1})) + \sqrt{t_i - t_{i-1}}\boldsymbol{\varepsilon}(t_i), \tag{6.42}$$

where $\boldsymbol{\varepsilon}(t_i) \sim \text{N}(0, \sigma^2\mathbf{I})$ and the left-hand side of Equation 6.42 is the velocity vector from $\boldsymbol{\mu}(t_i)$ to $\boldsymbol{\mu}(t_{i-1})$. This specification can be useful when the data are collected at a fine temporal resolution and there is little or no measurement error. For example,

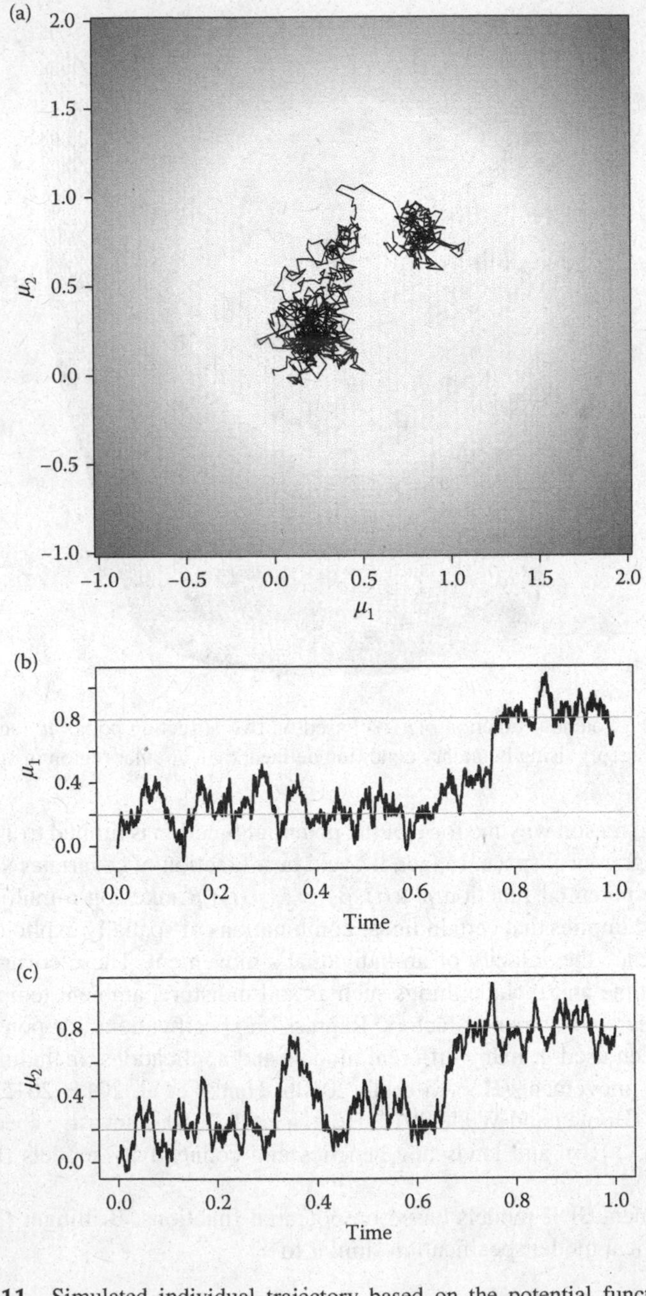

FIGURE 6.11 Simulated individual trajectory based on the potential function in Equation 6.41, which is composed of two attracting points and a steeply rising boundary condition delineating a circular region of space use. Panel (a) shows the potential function (background image) with joint trajectory simulation (black line). Panels (b) and (c) show the marginal positions over time; attracting points are show as horizontal gray lines in the marginal plots.

we fit a Bayesian SDE model in Equation 6.42, based on the quadratic–interaction potential function in Equation 6.40 to the mountain lion telemetry data described in Chapter 4 (e.g., Figure 4.1). We used a multivariate Gaussian prior for the coefficients in the potential function $\boldsymbol{\beta} \sim \text{N}(\mathbf{0}, 100 \cdot \mathbf{I})$ and a uniform distribution for standard deviation associated with the Brownian motion component of the model ($\sigma \sim \text{Unif}(0, 100)$). The posterior distribution for this SDE model is

$$[\boldsymbol{\beta}, \sigma | \{\mathbf{v}(t_i), \forall i\}] \propto \left(\prod_{i=2}^{n} [\mathbf{v}(t_i) | \mathbf{g}(\boldsymbol{\mu}(t_{i-1}), \boldsymbol{\beta}), \sigma^2] \right) [\boldsymbol{\beta}][\sigma], \tag{6.43}$$

where $\mathbf{v}(t_i) \equiv \boldsymbol{\mu}(t_i) - \boldsymbol{\mu}(t_{i-1})$. Table 6.1 shows the posterior summary statistics for all parameters resulting from fitting the quadratic–interaction potential function SDE to the mountain lion telemetry data. While the 95% credible intervals for β_1, β_2, and β_4 do not overlap zero, those for β_3 and β_4 do. Thus, the posterior potential function is an upward concave shape with elliptical isopleths stretched in the north–south orientation. Figure 6.12 shows the posterior mean and standard deviation of the potential function, the shape of which concurs with our interpretation of the parameter estimates (Table 6.1). The Bayesian implementation of the SDE model facilitates inference for the potential function, regardless of how complicated it is. For the mountain lion SDE model, we see that the uncertainty in the potential function increases away from the center of data. Inference for the potential function can be useful for understanding spatial regions where the data provide insight about environmental factors influencing animal movement.

The stochastic process model based on a potential function in Equation 6.42 can be embedded into a hierarchical statistical framework in the same way that many other physical processes have been modeled (e.g., Wikle and Hooten 2010). For example, if we use the SIE representation of the model

$$\boldsymbol{\mu}(t_i) = \boldsymbol{\mu}(0) + \int_0^{t_i} \mathbf{g}(\boldsymbol{\mu}(\tau))\, d\tau + \int_0^{t_i} \sigma\, d\mathbf{b}(\tau), \tag{6.44}$$

TABLE 6.1
Posterior Summary Statistics for the Parameters in the Mountain Lion Potential Function SDE

Parameter	Mean	SD	95% CI
β_1	1.933	9.839	(−18.305, 20.886)
β_2	1.975	9.606	(−16.615, 20.796)
β_3	41.50	4.946	(31.797, 51.279)
β_4	9.167	2.902	(3.434, 15.060)
β_5	−2.678	4.923	(−12.320, 6.848)
σ	9.845	0.519	(8.878, 10.894)

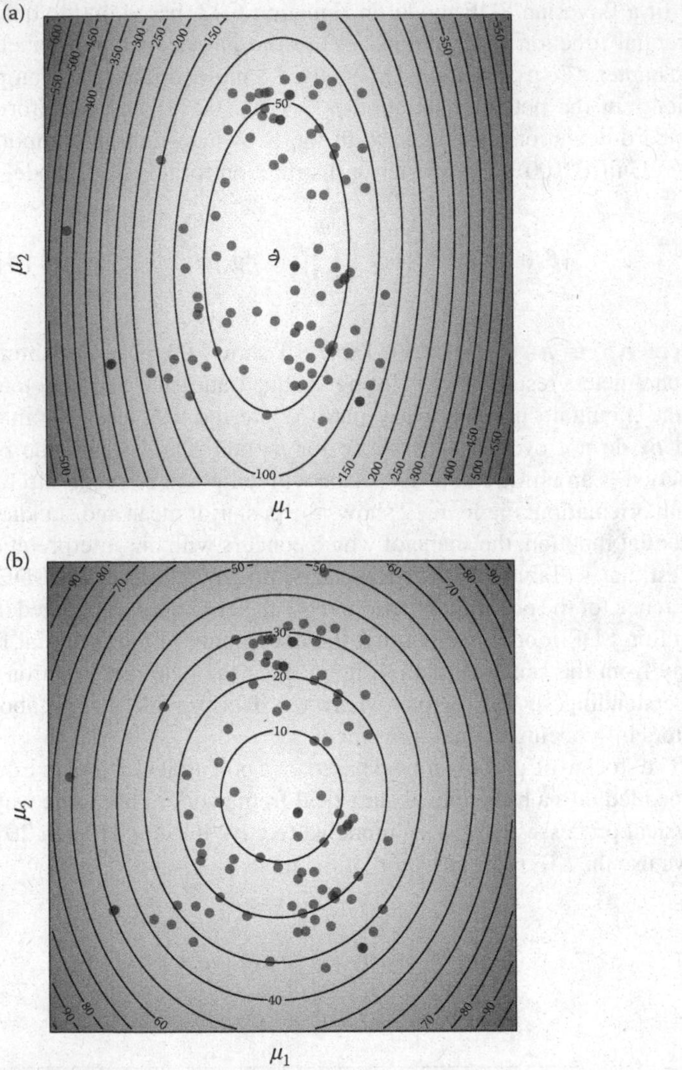

FIGURE 6.12 Posterior (a) mean and (b) standard deviation of the potential function based on fitting the Bayesian SDE model (using the potential function specification in Equation 6.40) to the mountain lion telemetry data (dark points). Isopleth contours are shown as dark lines.

it provides a natural "solution" for $\boldsymbol{\mu}(t)$ and facilitates straightforward use as a process model in a larger hierarchical framework. Assuming Gaussian measurement error and observed telemetry locations $\mathbf{s}(t_i)$, we can use the previously discussed data model

$$\mathbf{s}(t_i) \sim \text{N}(\boldsymbol{\mu}(t_i), \sigma_s^2\mathbf{I}). \tag{6.45}$$

The combination of Equations 6.44 and 6.45 forms a state-space model and can be implemented from a likelihood (if the process model could be integrated out) or Bayesian perspective. A few complications can arise when implementing the hierarchical model:

1. Approximation of $\mathbf{g}(\boldsymbol{\mu}(t))$ for analytically intractable model forms.
2. Evaluation of the process model for a particular set of data and parameter values.

The SIE in Equation 6.44 can be discretized for computational purposes to simulate a stochastic process based on potential functions. We showed how to derive continuous-time stochastic trajectory models earlier in this chapter. In contrast, to discretize Equation 6.44, we can use the temporal difference equation

$$\boldsymbol{\mu}(t_i) = \boldsymbol{\mu}(0) + \sum_{j=2}^{i-1}(t_j - t_{j-1})\mathbf{g}(\boldsymbol{\mu}(t_{j-1})) + \sum_{j=1}^{i}\sqrt{t_j - t_{j-1}}\boldsymbol{\varepsilon}(t_j), \tag{6.46}$$

where the potential function can be approximated using a spatial difference equation

$$\mathbf{g}(\boldsymbol{\mu}(t)) \approx -\frac{1}{2\Delta\mu}\begin{pmatrix} p((\mu_1(t)+\Delta\mu, \mu_2(t))') - p((\mu_1(t)-\Delta\mu, \mu_2(t))') \\ p((\mu_1(t), \mu_2(t)+\Delta\mu)') - p((\mu_1(t), \mu_2(t)-\Delta\mu)') \end{pmatrix}. \tag{6.47}$$

6.7 SMOOTH BROWNIAN MOVEMENT MODELS

While Brownian motion plays an important role in forming a mechanistic foundation for basic movement processes in continuous time, it is not very smooth and its utility for modeling animal movement directly has been questioned. Thus, in this section, we generalize the standard SDE/SIE approaches to modeling animal movement based on Brownian motion. Our generalization explicitly allows the process itself to be smoother than standard Brownian motion, but is still grounded in the same principles.

A natural way to smooth a noisy process is to integrate it. Integrals or sums are inherently smoother than their derivatives. Thus, we provide a general approach, based on integrating the Brownian motion process itself, to yield the necessary smoothness in a continuous-time correlated random walk (CTCRW) model. Johnson et al. (2008b) developed a specific movement model that falls into a broader class of smooth stochastic processes. They model the velocity directly as an OU process and then integrate to yield the position process. Hooten and Johnson (2016) generalized the approach of Johnson et al. (2008b) to express the CTCRW model as a convolution. The convolution approach allows us to frame the model as a Gaussian process similar to what is often used in spatial statistics and functional data analysis for time series; hence we refer to this class of movement models as "functional movement models" (FMMs) (Buderman et al. 2016; Hooten and Johnson 2016). FMMs can yield many benefits in terms of flexibility and computational efficiency.

6.7.1 Velocity-Based Stochastic Process Models

The OU processes described in the previous sections have been used to model movement directly in the position space (e.g., Dunn and Gipson 1977; Blackwell 2003). However, as we have seen, the standard OU process (and Brownian motion) often results in a quite noisy simulated animal movement path. In such cases, it might be desirable to use a smoother form of process model that still relies on the solid foundation of Brownian mechanics. As alluded to earlier, we can integrate the Brownian motion process to smooth it. Thus, Johnson et al. (2008a) presented an OU model for the velocity associated with animal movement, which was then integrated over time to yield the position process.

To show how the velocity model of Johnson et al. (2008a) fits into a larger class of stochastic animal movement models, we begin with the simple Brownian motion process and then proceed on to more complicated and useful models for movement. Recall from Equation 6.9 that Brownian motion $\mathbf{b}(t)$ can be expressed as an integral of white noise (or random uncorrelated random variables). Thus, if we integrate Brownian motion itself, with respect to time, we have

$$\eta(t) = \int_0^t \sigma \mathbf{b}(\tau)\, d\tau, \tag{6.48}$$

where $\eta(t)$ is a slightly simpler version of the integrated stochastic process that was proposed by Johnson et al. (2008a). For example, Figure 6.13 shows a Brownian motion process ($\mathbf{b}(t)$) and the associated integrated Brownian motion process ($\eta(t)$).

The integrated Brownian motion model in Equation 6.48 can be likened to that of Johnson et al. (2008a) by relating $\eta(t)$ to the position process by $\boldsymbol{\mu}(t) = \boldsymbol{\mu}(0) + \eta(t)$. If an integral of a process yields the position, then the process being integrated is related to velocity. Thus, the idea of Johnson et al. (2008a) was to model the velocity process as an SDE and integrate it to yield a more appropriate (i.e., smoother) model for animal movement. This basic concept already had a precedent, as Jonsen et al. (2005) proposed the same idea, but in the discrete-time framework we described in the previous chapter.*

The velocity modeling approach proposed by Johnson et al. (2008a) requires a strict relationship between $\mathbf{b}(t)$ and $\boldsymbol{\mu}(t)$, but also suggests a more general framework for modeling movement. To show this, we define the function

$$h(t,\tau) = \begin{cases} 1 & \text{if } 0 < \tau \le t \\ 0 & \text{if } t < \tau \le T \end{cases}. \tag{6.49}$$

* Also, integrated temporal models are common in time series and known as ARIMA models, as described in Chapter 3.

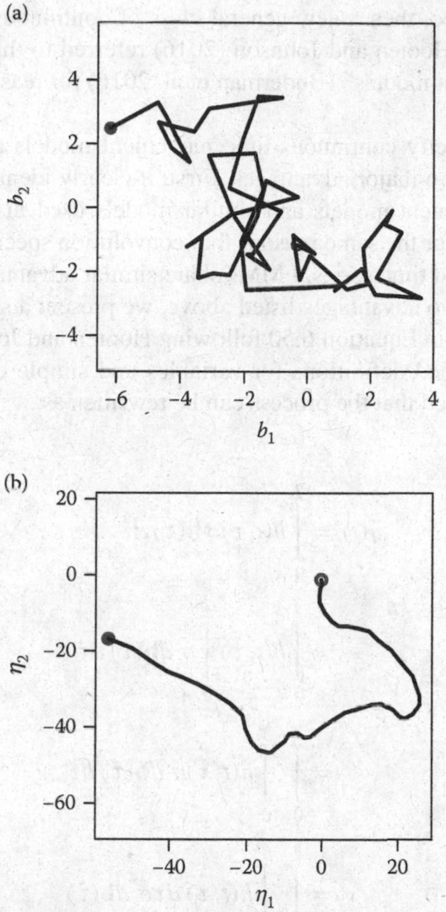

FIGURE 6.13 Simulated (a) Brownian motion process ($\mathbf{b}(t)$) and (b) integrated Brownian motion process ($\eta(t)$). Only 50 time steps are shown to illustrate the difference in smoothness. Starting and ending positions are denoted by open and closed circles.

If we substitute Equation 6.49 into Equation 6.48, the velocity-based Brownian motion model appears as the convolution

$$\eta(t) = \int_0^T h(t, \tau)\sigma \mathbf{b}(\tau)\, d\tau. \tag{6.50}$$

The convolution in Equation 6.50 is the key to recognizing a more general class of stochastic process models for animal movement.* For example, if $h(t, \tau)$ is a continuous function such that $0 \leq t \leq T$, $0 \leq \tau \leq T$ and with finite positive integral

* Recall that a convolution is an integral function of the form: $\int g(x, y)f(y)\, dy$.

$0 < \int_0^T h(t,\tau)\,d\tau < \infty$, then a new general class of continuous-time animal movement models arises. Hooten and Johnson (2016) referred to this class of models as "functional movement models" (Buderman et al. 2016) for reasons that will become clear.

The ability to specify continuous-time movement models as convolutions (i.e., Equation 6.50) has two major advantages. First, it clearly identifies the connections among animal movement models and similar models used in spatial statistics and time series. Second, for the same reasons that convolution specifications are popular in spatial statistics and time series, FMMs share similar advantageous properties.

To illustrate the two advantages listed above, we present a simple analysis of the new FMM presented in Equation 6.50 following Hooten and Johnson (2016). Using the previously specified definitions for variables and simple calculus, Hooten and Johnson (2016) showed that the process can be rewritten as

$$\boldsymbol{\eta}(t) = \int_0^T h(t,\tau)\sigma\mathbf{b}(\tau)\,d\tau \tag{6.51}$$

$$= \int_0^T h(t,\tau)\int_0^\tau \sigma\,d\mathbf{b}(\tilde{\tau})\,d\tau \tag{6.52}$$

$$= \int_0^T \int_0^\tau h(t,\tau)\sigma\,d\mathbf{b}(\tilde{\tau})\,d\tau \tag{6.53}$$

$$= \int_0^T \int_{\tilde{\tau}}^T h(t,\tau)\,d\tau\sigma\,d\mathbf{b}(\tilde{\tau}) \tag{6.54}$$

$$= \int_0^T \tilde{h}(t,\tilde{\tau})\sigma\,d\mathbf{b}(\tilde{\tau}) \tag{6.55}$$

where a step-by-step description for the above is as follows:

1. Equation 6.51: Begin with the convolution model from Equation 6.50.
2. Equation 6.52: Write the Brownian term, $\mathbf{b}(\tau)$, in its integral form.
3. Equation 6.53: Move the function $h(t,\tau)$ inside both integrals. Note that $0 < \tilde{\tau} < \tau$ and $0 < \tau < T$.
4. Equation 6.54: Switch the order of integration, paying careful attention to the limits of integration. That is, $\tilde{\tau} < \tau < T$ and $0 < \tilde{\tau} < T$.
5. Equation 6.55: Define $\tilde{h}(t,\tilde{\tau}) = \int_{\tilde{\tau}}^T h(t,\tau)\,d\tau$ resulting in a convolution of white noise.

Returning to the advantages of this FMM approach, the expression in Equation 6.55 has the same form described in spatial statistics as a "process convolution" (or kernel convolution; e.g., Barry and Ver Hoef 1996; Higdon 1998; Lee et al. 2005; Calder 2007). The process convolution has been instrumental in many fields, but especially in statistics for allowing for both complicated and efficient representations of covariance structure. Covariance structure in time series and spatial statistics is a critical tool for modeling dependence in processes. Thus, it seems reasonable that the same idea can be helpful in the context of modeling animal movement.

There are three main computational advantages to using the convolution perspective in continuous-time movement models. First, it is clear from Equation 6.55 that we never have to simulate Brownian motion; rather, we can operate on it implicitly by transforming the function $h(t, \tau)$ to $\tilde{h}(t, \tau)$ via integration and convolving $\tilde{h}(t, \tau)$ with white noise directly. This is exactly the same way that covariance models for spatial processes have been developed.

As an example, we let $h(t, \tau)$ be the Gaussian kernel. The Gaussian kernel is probably the most commonly used function in kernel convolution methods. If we first normalize the kernel so that it integrates to one for $0 < \tau < T$, we have a truncated normal PDF for the function such that $h(t, T) \equiv \text{TN}(\tau, t, \phi^2)_0^T$. We can then convert it to the required function $\tilde{h}(t, \tilde{\tau})$ with

$$\tilde{h}(t, \tilde{\tau}) = \int_{\tilde{\tau}}^{T} h(t, \tau)\, d\tau \tag{6.56}$$

$$= 1 - \int_{0}^{\tilde{\tau}} h(t, \tau)\, d\tau. \tag{6.57}$$

When the kernel function is the truncated normal PDF, the calculation in Equation 6.57 results in a numerical solution for the new kernel function $\tilde{h}(t, \tau)$ by subtracting the truncated normal CDF from one, a trivial calculation in any statistical software. With respect to the time domain, this kernel looks different than most kernels used in time series or spatial statistics (Figure 6.14j). Rather than being unimodal and symmetric, it has a sigmoidal shape equal to one at $t = 0$ and nonlinearly decreasing to zero at $t = T$. In effect, the new kernel in Equation 6.57 is accumulating the white noise up to near time t and then including a discounted amount of white noise ahead of time t.

The options for kernel functions are limitless. Each kernel results in a different stochastic process model for animal movement. In fact, we have already seen that this class of movement models is general enough to include that proposed by Johnson et al. (2008a), but it also includes the original unsmoothed Brownian motion process if we let $h(t, \tau)$ be a point mass function at $\tau = t$ and zero elsewhere (Figure 6.14a). The point mass kernel function can also be achieved by taking the limit as $\phi \to 0$ of

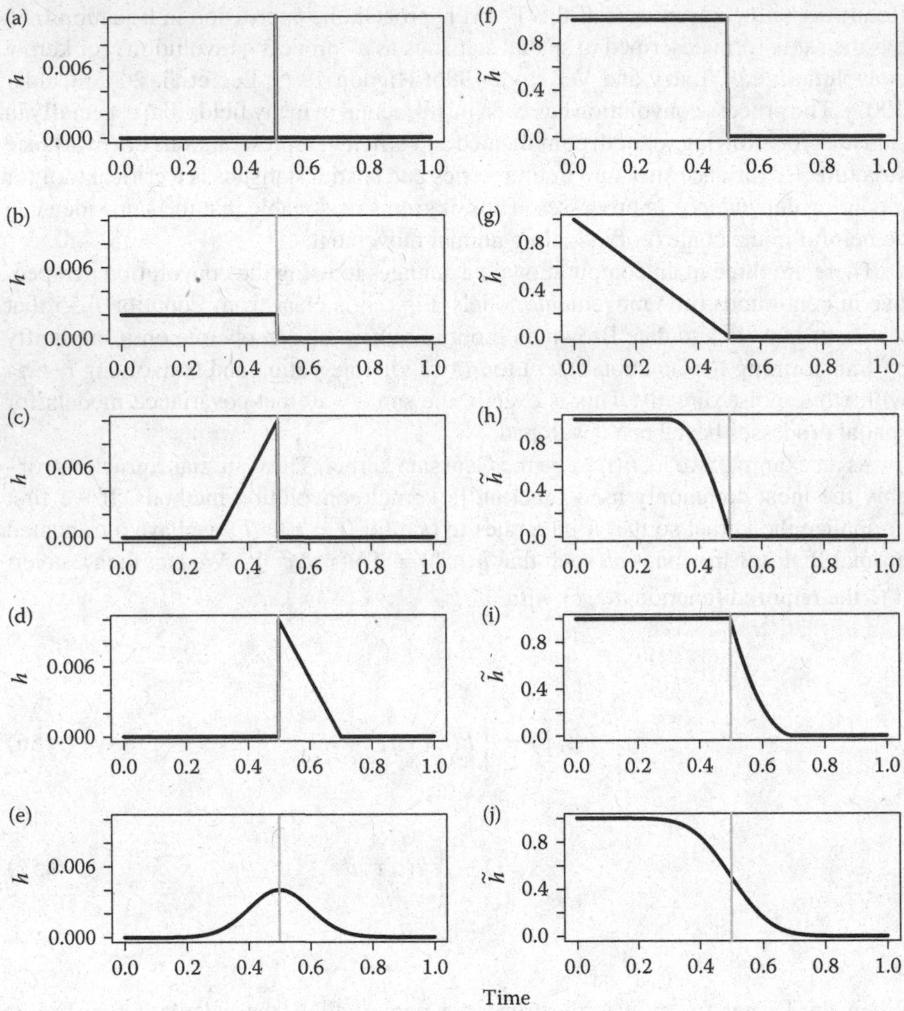

FIGURE 6.14 Example kernels $h(t, \tau)$ (a–e) and resulting integrated kernels $\tilde{h}(t, \tau)$ (f–j). The first row (a,f) results in the regular Brownian motion, row two (b,g) is equivalent to that used by Johnson et al. (2008a), rows three (c,h) through five (e,j) are more common in time series and spatial statistics. The vertical gray line indicates time t for the particular kernel shown; in this case, $t = 0.5$.

our Gaussian kernel, resulting in the integrated kernel function

$$\tilde{h}(t, \tau) = \begin{cases} 1 & \text{if } 0 < \tau \leq t \\ 0 & \text{if } t < \tau \leq T \end{cases}. \tag{6.58}$$

We can imagine the integrated kernel in Equation 6.58 summing up all of the past velocities to obtain the current position (Figure 6.14f). This precisely follows the procedure we described for specifying SIEs (6.8) in the previous sections. In Figure 6.14f, the steep drop at $\tau = t$ is what provides the original Brownian motion process its roughness. Whereas, when we use a non-point mass function for $h(t, \tau)$, we arrive at a smoother stochastic process model for movement.

We describe a few different kernel functions in more detail to examine their implications for resulting animal movement behavior. In doing so, it is simplest to interpret the $h(t, \tau)$ and $\tilde{h}(t, \tau)$ functions directly. For example, using the direct integration of velocity as proposed by Johnson et al. (2008a) results in the $\tilde{h}(t, \tau)$ in Figure 6.14g. The individual's position is an accumulation of its past steps. The steps are noisy themselves in direction and length but have some general momentum. In the case of the "tail up" kernel shown in Figure 6.14c, the influence of past steps on the current position decays linearly with time (Figure 6.14h).* In this case, the individual's position is more strongly a function of recent steps than steps in the distant past. The opposite is true with the "tail down" model shown in Figure 6.14d, where future steps also influence the position (Figure 6.14i). Heuristically, we might interpret the resulting movement behavior as perception driven. That is, the individual may have an awareness of a distant destination that drives their movement. Finally, the Gaussian kernel discussed earlier in Figure 6.14e indicates a symmetric mixture of previous and future velocities, suggesting the perception of former and future events by the individual.

6.7.2 Functional Movement Models and Covariance

FMMs provide a rich class of continuous-time models to characterize animal movement. Using the basic model (6.50) presented in the previous section, we can choose from a large set of possible smoothing kernels ($h(t, \tau)$) for Brownian motion and arrive at the appropriate form for the kernel ($\tilde{h}(t, \tau)$) that is convolved with white noise. At that point, we can use $\tilde{h}(t, \tau)$ directly to construct the proper covariance function for the joint process as a whole. In fact, for a 1-D movement process $\eta(t)$, the covariance function can be calculated (e.g., Paciorek and Schervish 2006) as the convolution of the kernels

$$\text{cov}(\eta(t_1), \eta(t_2)) = \sigma \int_0^T \tilde{h}(t_1, \tau)\tilde{h}(t_2, \tau)\, d\tau, \tag{6.59}$$

at any two times, t_1 and t_2.

One benefit of the covariance function (6.59) is that, for a finite subset of n times $\{t_1, \ldots, t_n\}$ and process $\boldsymbol{\eta} \equiv (\eta_1, \ldots, \eta_n)'$, the joint probability model can be expressed as

$$\boldsymbol{\eta} \sim \text{N}(\mathbf{0}, \sigma^2 \Delta t \tilde{\mathbf{H}}\tilde{\mathbf{H}}'), \tag{6.60}$$

where $\mathbf{0}$ is an $n \times 1$ vector of zeros and $\tilde{\mathbf{H}}$ is referred to as a matrix of basis functions with the ith row equal to $\tilde{h}(t_i, \tau)$ for all τ. This method for constructing the covariance matrix and defining a correlated process is similar to that recommended

* The tail up and tail down terminology arises from the statistical literature concerning spatial covariance models for stream networks (Ver Hoef and Peterson 2010).

in spatial statistics (e.g., Paciorek and Schervish 2006; Ver Hoef and Peterson 2010). The resulting model for the position process in this FMM for smooth Brownian motion is

$$\boldsymbol{\mu} \sim \mathrm{N}(\mu(0)\mathbf{1}, \sigma^2 \Delta t \tilde{\mathbf{H}}\tilde{\mathbf{H}}'). \tag{6.61}$$

It may not be immediately apparent how this expression (6.61) is helpful. In fact, it is often more intuitive to model the process from the first moment (i.e., mean dynamical structure) rather than the second moment (Wikle and Hooten 2010). However, the joint form with dependence imposed through the matrix $\tilde{\mathbf{H}}\tilde{\mathbf{H}}'$ can be useful for computational reasons. When the integral in Equation 6.59 cannot be used to analytically compute the necessary covariance matrix, we can still use the outer product of the matrices explicitly (i.e., $\tilde{\mathbf{H}}\tilde{\mathbf{H}}'$). However, the true covariance requires the number of columns of $\tilde{\mathbf{H}}$ to approach infinity, which, under approximation, can lead to computational difficulties. Higdon (2002) suggested a finite process convolution as an approximation. In the finite approximation, $\tilde{\mathbf{H}}$ could be reduced to m columns. The reduction of columns in $\tilde{\mathbf{H}}$ implies that there are m "knots" (spaced Δt apart) in the temporal domain that anchor the basis functions (i.e., kernels), and thus, only m white noise terms are required so that $\boldsymbol{\eta} \approx \tilde{\mathbf{H}}\boldsymbol{\varepsilon}$, where $\tilde{\mathbf{H}}$ is an $n \times m$ matrix and $\boldsymbol{\varepsilon} \equiv (\varepsilon(t_1), \ldots, \varepsilon(t_j), \ldots, \varepsilon(t_m))'$ is an $m \times 1$ vector. As we discussed in Chapter 2, a finite approximation of the convolution is also sometimes referred to as a reduced-rank method (Wikle 2010a). Reduced-rank methods for representing dependence in statistical models can improve computational efficiency substantially, and have become popular in spatial and spatio-temporal statistics for large data sets (Cressie and Wikle 2011).

To illustrate how the kernel functions relate to covariance, we simplified the movement process so that it is 1-D in space. This approach can also be generalized to higher dimensions. In the more typical 2-D case, we form the vector $\boldsymbol{\eta} \equiv (\eta(1,t_1), \ldots, \eta(1,t_n), \eta(2,t_1), \ldots, \eta(2,t_n))'$ by stacking the temporal vectors for each coordinate. Then the joint model can be written as

$$\boldsymbol{\eta} \sim \mathrm{N}(\mathbf{0}, \sigma^2 \Delta t(\mathbf{I} \otimes \tilde{\mathbf{H}}\tilde{\mathbf{H}}')), \tag{6.62}$$

where $\mathbf{I}$ is a 2×2 identity matrix.* As in the previous section, this new joint specification assumes that $\tilde{\mathbf{H}}$ contains the appropriate set of basis vectors for both coordinates (longitude and latitude). This assumption can also be generalized to include different types or scales of kernels for each direction. In fact, the original SIE for $\boldsymbol{\eta}(t)$ from Equation 6.50 can be rewritten as

$$\boldsymbol{\eta}(t) = \int_0^T \mathbf{H}(t,\tau)\sigma \mathbf{b}(\tau)\, d\tau. \tag{6.63}$$

where $\mathbf{H}(t,\tau)$ is a 2×2 matrix function. If both diagonal elements are equal to the previous $h(t,\tau)$ with zeros on the off-diagonals, we have an equivalent expression to

* Recall that an identity matrix has ones on the diagonal and zeros elsewhere. Also, the $\otimes$ symbol denotes the Kronecker product, which multiplies every element of $\mathbf{I}$ by $\tilde{\mathbf{H}}\tilde{\mathbf{H}}'$ to form a new matrix.

Equation 6.50. However, if the diagonal elements of $\mathbf{H}(t,\tau)$ are different, $h_1(t,\tau)$ and $h_2(t,\tau)$, we allow for the possibility of different types of movement in each direction. This might be appropriate when the individual animal behavior relates strongly to a linearly oriented habitat (e.g., movement corridor) or seasonal behavior (e.g., migration). Allowing the off-diagonals of $\mathbf{H}(t,\tau)$ to be nonzero functions can introduce additional flexibility, such as off-axis home range shapes.

6.7.3 Implementing Functional Movement Models

FMMs can be implemented from the first- or second-order perspective. As an example of the second-order implementation, consider the hierarchical model

$$\mathbf{s} \sim \text{N}(\mathbf{K}\boldsymbol{\mu}, \mathbf{I} \otimes \boldsymbol{\Sigma}_s), \tag{6.64}$$

where the joint latent movement process is an FMM such that

$$\boldsymbol{\mu} \sim \text{N}(\boldsymbol{\mu}(0) \otimes \mathbf{1}, \sigma^2(\mathbf{I} \otimes \tilde{\mathbf{H}})(\mathbf{I} \otimes \tilde{\mathbf{H}})'). \tag{6.65}$$

Note that we omit the Δt notation in this section because the variance term σ^2 can account for the grain of temporal discretization. The observed telemetry data ($\mathbf{s}$, which is $2n \times 1$-dimensional) and position process ($\boldsymbol{\mu}$, which is $2m \times 1$-dimensional) are stacked vectors of coordinates in longitude and latitude. The mapping matrix $\mathbf{K}$ is composed of zeros and ones that isolate the positions in $\boldsymbol{\mu}$ at times when data are available so that $\mathbf{K}\boldsymbol{\mu}$ is temporally matched with $\mathbf{s}$. The correlation matrix in Equation 6.65, $(\mathbf{I} \otimes \tilde{\mathbf{H}})(\mathbf{I} \otimes \tilde{\mathbf{H}})'$, involves Kronecker products to account for the fact that we are modeling the bivariate position process jointly. We use a simple measurement error variance specification such that $\boldsymbol{\Sigma}_s \equiv \sigma_s^2\mathbf{I}$ and Gaussian basis functions in $\tilde{\mathbf{H}}$ that are parameterized with a single range parameter ϕ.

If we condition on the initial state $\boldsymbol{\mu}(0)$, the full hierarchical model is composed of the unknown quantities: $\boldsymbol{\mu}$, σ_s^2, σ^2, and ϕ. In a Bayesian implementation of the hierarchical model, each of the unknown quantities would need to be sampled in an MCMC algorithm. However, we can use Rao-Blackwellization and integrate out the latent process $\boldsymbol{\mu}$, resulting in a much more stable algorithm. The resulting integrated likelihood is multivariate normal such that

$$\mathbf{s} \sim \text{N}(\mathbf{K}(\boldsymbol{\mu}(0) \otimes \mathbf{1}), \mathbf{I} \otimes \boldsymbol{\Sigma}_s + \sigma^2\mathbf{K}(\mathbf{I} \otimes \tilde{\mathbf{H}})(\mathbf{I} \otimes \tilde{\mathbf{H}})'\mathbf{K}'). \tag{6.66}$$

We relied on the reparameterization of the covariance matrix in Equation 6.66 suggested by Diggle and Ribeiro (2002) such that

$$\mathbf{I} \otimes \boldsymbol{\Sigma}_s + \sigma^2\mathbf{K}(\mathbf{I} \otimes \tilde{\mathbf{H}})(\mathbf{I} \otimes \tilde{\mathbf{H}})'\mathbf{K}') = \sigma_s^2\left(\mathbf{I} + \sigma_{\mu/s}^2\mathbf{K}(\mathbf{I} \otimes \tilde{\mathbf{H}})(\mathbf{I} \otimes \tilde{\mathbf{H}})'\mathbf{K}'\right). \tag{6.67}$$

Hooten and Johnson (2016) used an inverse gamma prior for σ_s^2, a uniform prior for $\sigma_{\mu/s}$, and a discrete uniform prior for the range parameter ϕ. The discrete uniform prior allows us to precalculate the matrix $\mathbf{K}(\mathbf{I} \otimes \tilde{\mathbf{H}})(\mathbf{I} \otimes \tilde{\mathbf{H}})'\mathbf{K}'$ and perform the necessary operations (e.g., inverses) off-line so that the MCMC algorithm only has to access the results without having to recompute them. To illustrate the inference obtained using an FMM, we simulated data from a stochastic process based on the FMM in Equations 6.64 and 6.65. We simulated $n = 300$ observations on the time domain (0, 1) using the "true" parameter values: $\phi = 0.005$, $\sigma_s^2 = 0.001$, and $\sigma^2 = 0.01$. Fitting the model in Equation 6.66 to the simulated data results in the marginal posterior histograms in Figure 6.15, which indicate that the Bayesian FMM is able to recover the model parameters quite well in our simulation example. Note that the Brownian motion variance parameter σ^2 is a derived quantity in our model because of the reparameterization in Equation 6.67.

The FMM based on the integrated likelihood specification in Equation 6.66 does not explicitly provide direct inference for the latent movement process $\boldsymbol{\mu}$. However, we can obtain MCMC samples for $\boldsymbol{\mu}$ using a secondary algorithm. The full-conditional distribution for $\boldsymbol{\mu}$ given all other parameters and data is

$$[\boldsymbol{\mu}|\cdot] = \text{N}(\boldsymbol{\Sigma}_{\mu|\cdot}(\mathbf{K}'\boldsymbol{\Sigma}_s^{-1}\mathbf{s} + \boldsymbol{\Sigma}_\mu^{-1}(\boldsymbol{\mu} \otimes \mathbf{1})), \boldsymbol{\Sigma}_{\mu|\cdot}), \tag{6.68}$$

where $\boldsymbol{\Sigma}_\mu \equiv \sigma^2(\mathbf{I} \otimes \tilde{\mathbf{H}})(\mathbf{I} \otimes \tilde{\mathbf{H}})'$ and $\boldsymbol{\Sigma}_{\mu|\cdot} \equiv (\mathbf{K}'\boldsymbol{\Sigma}_s^{-1}\mathbf{K} + \boldsymbol{\Sigma}_\mu^{-1})^{-1}$. Figure 6.16 shows the simulated data as well as posterior position process in comparison to the true, unobserved position process $\boldsymbol{\mu}$. Notice that the uncertainty in the position process ($\boldsymbol{\mu}$) increases in the larger gaps between observed data.

6.7.4 Phenomenological Functional Movement Models

The FMM model specification in Equations 6.64 and 6.65 provide a general framework for modeling continuous-time trajectories. In the preceding sections, we specified FMMs that are grounded in mechanistic first principles (e.g., Brownian motion), but the same framework can be used to specify and implement phenomenological models based solely on smoothing the data optimally. Buderman et al. (2016) presented a phenomenological framework based on regression spline specifications of

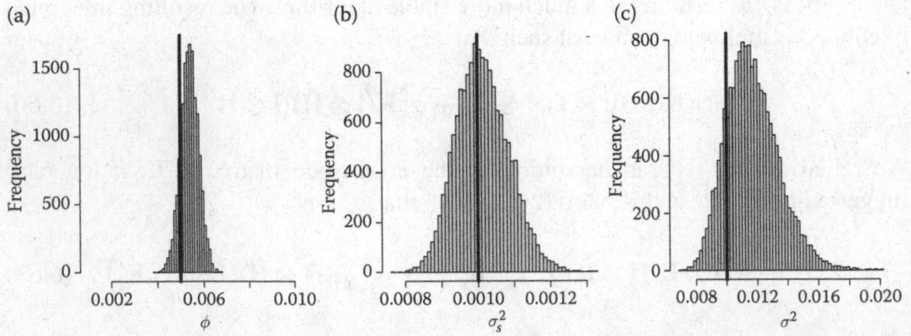

FIGURE 6.15 Marginal posterior distributions for FMM parameters resulting from a fit to simulated data arising from Equations 6.64 and 6.65: (a) ϕ, (b) σ_s^2, and (c) σ^2.

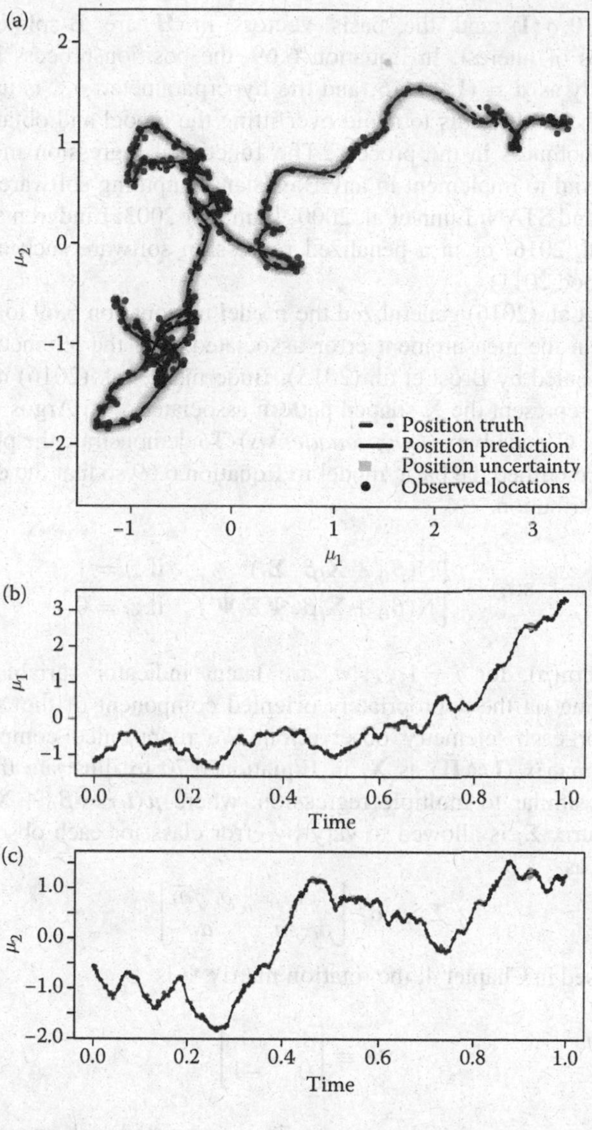

FIGURE 6.16 Panel (a) shows the simulated stochastic process (dashed line) and data (points) from the FMM in Equations 6.64 and 6.65 with posterior realizations of the position process (gray lines). Panels (b) and (c) show the marginal data and path as well as 95% credible interval (gray shaded region).

FMMs. A simplified first-order formulation, like that presented by Buderman et al. (2016), at the data level is

$$\mathbf{s} \sim \text{N}((\mathbf{I} \otimes \tilde{\mathbf{H}})\boldsymbol{\beta}, \sigma_s^2\mathbf{I}), \tag{6.69}$$

where $\boldsymbol{\beta} \sim \text{N}(\mathbf{0}, \sigma_\beta^2 \mathbf{I})$ and the basis vectors in $\tilde{\mathbf{H}}$ are B-splines at various temporal scales of interest. In Equation 6.69, the position process is represented deterministically as $\boldsymbol{\mu} \equiv (\mathbf{I} \otimes \tilde{\mathbf{H}})\boldsymbol{\beta}$, and the hyperparameter, σ_β^2, is used to impose shrinkage on the coefficients to avoid overfitting the model and obtain the optimal amount of smoothness in the process. The functional regression model in Equation 6.69 is trivial to implement in any Bayesian computing software (e.g., BUGS, JAGS, INLA, and STAN; Lunn et al. 2000; Plummer 2003; Lindgren and Rue 2015; Carpenter et al. 2016) or in a penalized regression software such as the "mgcv" R package (Wood 2011).

Buderman et al. (2016) generalized the model in Equation 6.69 to accommodate heterogeneity in the measurement error associated with the telemetry data. Similar to that presented by Brost et al. (2015), Buderman et al. (2016) used a mixture distribution to represent the X-shaped pattern associated with Argos telemetry data when modeling Canada lynx (*Lynx canadensis*). To demonstrate the phenomenological FMM, we modified the basic model in Equation 6.69 so that the data arise from the mixture distribution

$$\mathbf{s}(t_i) \sim \begin{cases} \text{N}(\boldsymbol{\beta}_0 + \mathbf{X}_i\boldsymbol{\beta}, \boldsymbol{\Sigma}_i) & \text{if } z_i = 1 \\ \text{N}(\boldsymbol{\beta}_0 + \mathbf{X}_i\boldsymbol{\beta}, \boldsymbol{\Psi}\boldsymbol{\Sigma}_i\boldsymbol{\Psi}') & \text{if } z_i = 0 \end{cases}, \tag{6.70}$$

where $z_i \sim \text{Bern}(p)$, for $i = 1, \ldots, n$, are latent indicator variables that act as switches, turning on the appropriately oriented component of the X-shaped error distribution for each telemetry observation. We represented components of the larger design matrix $(\mathbf{I} \otimes \tilde{\mathbf{H}})$ as $\mathbf{X}_i$ in Equation 6.70 to illustrate that the formulation is very similar to multiple regression, where $\boldsymbol{\mu}(t_i) = \boldsymbol{\beta}_0 + \mathbf{X}_i\boldsymbol{\beta}$. The error covariance matrix $\boldsymbol{\Sigma}_i$ is allowed to vary by error class for each observation and is parameterized as

$$\boldsymbol{\Sigma}_i \equiv \sigma_i^2 \begin{bmatrix} 1 & \rho_i\sqrt{a_i} \\ \rho_i\sqrt{a_i} & a_i \end{bmatrix}, \tag{6.71}$$

and, as discussed in Chapter 4, the rotation matrix $\boldsymbol{\Psi}$ is

$$\boldsymbol{\Psi} \equiv \begin{bmatrix} 1 & 0 \\ 0 & -1 \end{bmatrix}. \tag{6.72}$$

Each covariance parameter in Equation 6.71 is associated with an error class c (for $c = 1, \ldots, C$) such that $\sigma_i^2 = \sigma_c^2$, $\rho_i = \rho_c$, and $a_i = a_c$, for example, when the ith observation is designated as class c. Thus, the parameters for each of six error classes (i.e., 3, 2, 1, 0, A, B) associated with Argos telemetry data and a seventh for VHF telemetry data are specified with prior distributions and estimated while fitting the model. In the case of VHF telemetry data, the measurement error is much less than with Argos data and lacks the X-shaped pattern. Thus, for the VHF telemetry data, $z_i = 1$ and $\boldsymbol{\Sigma}_i = \sigma_i^2\mathbf{I}$ accommodate an error pattern with circular isopleths.

We used three sets of temporal basis vectors in our specification of $\tilde{\mathbf{H}}$ (and hence, $\mathbf{X}_i$) to describe the movement of an individual Canada lynx in Colorado (Figure 6.17). Following Buderman et al. (2016), we used B-splines at three different scales (i.e.,

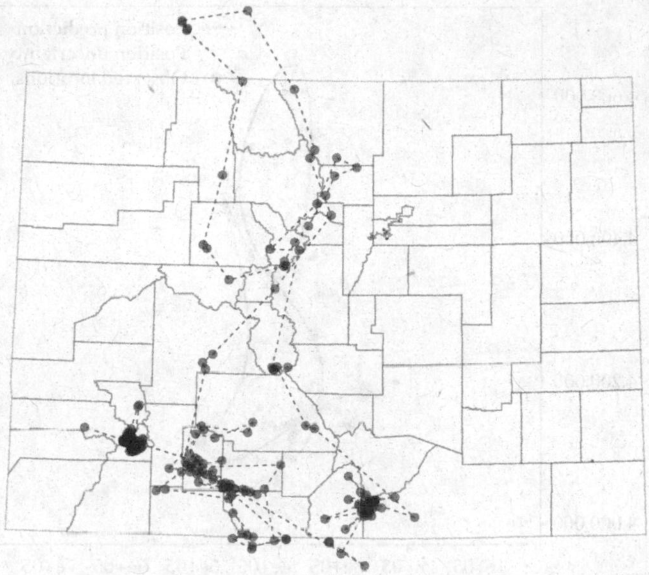

FIGURE 6.17 Observed Argos and VHF telemetry data (points) for an individual Canada lynx in Colorado (Colorado counties outlined in gray). Dashed lines are used to visualize the sequence of telemetry data only.

spanning the compact support of each B-spline basis function): 1 month, 3 months, and 1 year. Thus, the phenomenological FMM is capable of characterizing movement processes at the combination of those temporal scales representing the continuous-time trajectory that best explains the data.

Figure 6.18 shows the results of fitting the phenomenological FMM in Equation 6.70 to the Canada lynx telemetry data in Figure 6.17. While some of the Argos telemetry observations can be subject to extreme error, the VHF telemetry data provide consistently smaller errors and, thus, have a stronger influence on the model fit. Therefore, the northernmost portion of the position process in Figure 6.18 appears to show the predicted path missing the observed data. However, after incorporating uncertainty related to the telemetry data and the inherent smoothness in the remainder of the path, the predictions are optimal if they do not pass directly through the observed telemetry data. Finally, large time gaps in data collection (i.e., between 750 and 900 days) result in appropriately widened credible intervals for the predicted position process (Figure 6.18).

6.7.5 Velocity-Based Ornstein–Uhlenbeck Models

In the preceding sections, we covered the general framework for using continuous-time stochastic processes to model animal movement with varying levels of smoothness for the true individual position process. Thus, we can apply this same framework to more complicated processes than Brownian motion. For example, we saw before that a stationary Brownian motion model is also referred to as an OU model, and implies

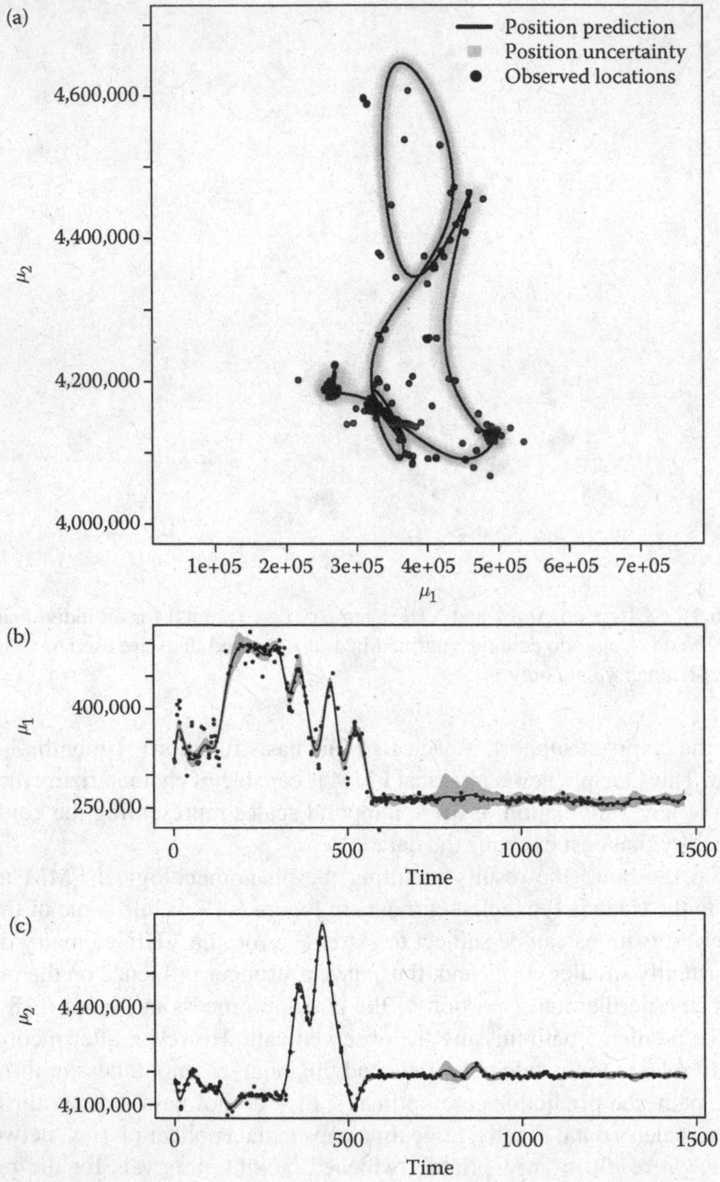

FIGURE 6.18 Observed Argos and VHF telemetry data (points) for an individual Canada lynx in Colorado. (a) Predicted position process (dark line) and position process realizations (gray lines) in 2-D geographic space. (b) Marginal position process (dark line) in easting, observed telemetry data (points), and 95% credible interval (gray). (c) Marginal position process (dark line) in northing, observed telemetry data (points), and 95% credible interval (gray).

that some force of attraction acts on the position process. In this case, recall that the basic SIE representation of a multivariate OU model is

$$\boldsymbol{\mu}(t) = \boldsymbol{\mu}(0) + \int_0^t (\mathbf{M} - \mathbf{I})(\boldsymbol{\mu}(\tau) - \boldsymbol{\mu}^*)\, d\tau + \mathbf{b}(t), \tag{6.73}$$

where $\boldsymbol{\mu}^*$ is the attracting point in geographic space and the matrix $\mathbf{M}$ is usually parameterized so that $\mathbf{M} \equiv \rho\mathbf{I}$. In a sense, the parameter ρ controls the attraction because, if $\rho = 1$, the process becomes nonstationary (and hence no effect of the attractor $\boldsymbol{\mu}^*$). On the other hand, if $0 < \rho < 1$ then the process is stationary. However, the parameter ρ (which is also called an autocorrelation parameter) also controls the smoothness of the process; at least up to a certain degree. As $\rho \to 1$, the process reverts to Brownian motion and assumes its inherent degree of smoothness, but as $\rho \to 0$, the position process in Equation 6.73 becomes less smooth than Brownian motion due to the fact that the autocorrelation approaches zero and the locations are independent and identically distributed realizations from $\text{N}(\boldsymbol{\mu}^*, \sigma^2\mathbf{I})$. Thus, to a certain extent, the parameter ρ is capable of smoothing the position process similar to the FMM approach described in the previous section. However, the OU model in Equation 6.73 can only smooth the process so much and still maintain attraction. Thus, we can combine the OU process with the FMM to achieve both smoothness and stationarity simultaneously.

One way to combine the OU and FMM models is to replace the Brownian motion component $\mathbf{b}(t)$ in Equation 6.73 with the smoothed process $\boldsymbol{\eta}(t)$, where

$$\boldsymbol{\eta}(t) = \int_0^T h(t,\tau)\mathbf{v}(\tau)\, d\tau, \tag{6.74}$$

where $\mathbf{v}(t)$ is a 2-D OU process instead of a Brownian motion. The benefit of this modification to the model is that, in the limits, the OU process ranges from a white noise process to a BM process. Therefore, with one additional parameter, we can control the smoothness from BM to an integrated BM model.

Combining these ideas with the exponential notation described in the previous sections, Johnson et al. (2008a) implicitly used the kernel function

$$h(t,\tau) = \begin{cases} 1 & \text{if } 0 < \tau \le t \\ 0 & \text{if } t < \tau \le T \end{cases}. \tag{6.75}$$

In doing so, they specify the OU process directly for the individual's velocity process in each direction (i.e., 1-D to simplify notation) and convolve with $h(t,\tau)$ to yield

$$\eta(t) = \int_0^T h(t,\tau)\left(\gamma + \frac{e^{-\theta\tau}}{\sqrt{2\theta}} b(e^{2\theta\tau})\right) d\tau, \tag{6.76}$$

where γ is the mean velocity and θ is an autocorrelation parameter. Then, the position process becomes

$$\mu(t) = \mu(0) + \eta(t). \tag{6.77}$$

To fit the model to a discrete and finite set of telemetry data, Johnson et al. (2008a) derived a discretization of the OU model as follows. First, they worked directly with the velocity process

$$v(t) = \gamma + \frac{e^{-\theta t}}{\sqrt{2\theta}} b(e^{2\theta t}), \tag{6.78}$$

which is another way to formulate the OU process. Then, for times $t_1, \ldots, t_i, \ldots, t_n$, we can write

$$v_i = \gamma + \frac{e^{-\theta t_i}}{\sqrt{2\theta}} b(e^{2\theta t_i}), \tag{6.79}$$

and conditioning on the result from the preceding section on OU models, results in

$$v_i | v_{i-1} \sim \text{N}\left(v_{i-1} e^{-\theta(t_i - t_{i-1})} + \gamma(1 - e^{-\theta(t_i - t_{i-1})}), \sigma^2 \frac{(1 - e^{-\theta(t_i - t_{i-1})})}{2\theta} \right). \tag{6.80}$$

To find the associated position process for μ_i, we start from Equation 6.77, but condition on the previous position μ_{i-1} and integrate from t_{i-1} to t_i (instead of 0 to t_i) so that

$$\mu_i = \mu_{i-1} + \int_{t_{i-1}}^{t_i} v_{i-1} e^{-\theta(\tau - t_{i-1})} + \gamma(1 - e^{-\theta(\tau - t_{i-1})}) + \xi_i \, d\tau \tag{6.81}$$

$$= \mu_{i-1} + v_{i-1} \left(\frac{1 - e^{-\theta(t_i - t_{i-1})}}{\theta} \right) + \gamma \left(t_i - t_{i-1} - \frac{1 - e^{-\theta(t_i - t_{i-1})}}{\theta} \right) + \xi_i, \tag{6.82}$$

where the additive error ξ_i has the following distribution:

$$\xi_i \sim \text{N}\left(0, \frac{\sigma^2}{\theta^2} \left(t_i - t_{i-1} - \frac{2}{\theta}(1 - e^{-\theta(t_i - t_{i-1})}) + \frac{1}{2\theta}(1 - e^{-2\theta(t_i - t_{i-1})}) \right) \right). \tag{6.83}$$

Together, the results from Equations 6.80 and 6.82 are valid for each coordinate axis and can be combined to yield a discretized smooth OU process for a bivariate movement process.*

The main reason for deriving the preceding results is so that we can use observed telemetry data to fit the CTCRW model. Thus, Johnson et al. (2008a) rely on the

* Johnson et al. (2008a) provides additional details on the derivations of the integrated OU model.

Gaussian state-space framework to relate the observations to the process

$$\mathbf{s}_i \sim \mathrm{N}(\mathbf{K}\mathbf{z}_i, \boldsymbol{\Sigma}_s), \tag{6.84}$$
$$\mathbf{z}_i \sim \mathrm{N}(\mathbf{L}_i\mathbf{z}_{i-1}, \boldsymbol{\Sigma}_{z,i}), \tag{6.85}$$

where $\mathbf{s}_i$ represents the observed position vector at time t_i, $\mathbf{K}$ is a 2×4 mapping matrix of zeros and ones that pulls out the first two elements of $\mathbf{z}_i$, the state vector $\mathbf{z}_i \equiv (\mu_{1,i}, \mu_{1,i}, v_{1,i}, v_{1,i})'$ is composed of both the position and velocity processes, and the 2×2 covariance matrix accounts for the telemetry error in the observations. The state Equation 6.85 serves as the real workhorse of the approach and arises as a result of the derivations above. While this smoothed OU model is not Markov in the position process alone, it is Markov in the joint position–velocity process.* This joint Markov result allows us to write the latent position–velocity process as a VAR(1) time series model.†

If we assume directional independence and homogeneity, the 4×4 propagator matrix $\mathbf{L}_i$, in Equation 6.85, can be written as

$$\mathbf{L_i} = \begin{pmatrix} 1 & (1 - e^{-\theta(t_i - t_{i-1})})/\theta \\ 0 & e^{-\theta(t_i - t_{i-1})} \end{pmatrix} \otimes \mathbf{I}, \tag{6.86}$$

where $\mathbf{I}$ is a 2×2 identity matrix. Similarly, the covariance matrix in Equation 6.85 can be written as $\boldsymbol{\Sigma}_{z,i} \equiv \mathbf{Q}_i \otimes \mathbf{I}$ such that

$$\mathbf{Q}_i = \begin{pmatrix} q_{1,1,i} & q_{1,2,i} \\ q_{1,2,i} & q_{2,2,i} \end{pmatrix}, \tag{6.87}$$

where

$$q_{1,1,i} = \frac{\sigma^2}{\theta^2}\left(t_i - t_{i-1} - \frac{2}{\theta}(1 - e^{-\theta(t_i - t_{i-1})}) + \frac{1}{2\theta}(1 - e^{-2\theta(t_i - t_{i-1})})\right), \tag{6.88}$$

$$q_{1,2,i} = \frac{\sigma^2}{2\theta^2}\left(1 - 2e^{-\theta(t_i - t_{i-1})} + e^{-2\theta(t_i - t_{i-1})}\right), \tag{6.89}$$

$$q_{2,2,i} = q_{2,2,i} = \frac{\sigma^2}{2\theta}\left(1 - e^{-2\theta(t_i - t_{i-1})}\right). \tag{6.90}$$

Using these expressions as a guideline, it is straightforward to generalize them by allowing the autocorrelation parameter θ and variance parameter σ^2 to vary by coordinate, and possibly over time.

A Gaussian state-space formulation such as that presented in Equation 6.85 allows for the use of fast computational approaches such as Kalman filtering methods when fitting the model (Chapter 3). In fact, the models described in this section can be fit

---

* Recall that the Markov property essentially says that a process is independent of all other time points when conditioned on its direct neighbors in time.

† Recall the VAR(1) from Chapter 3 on time series.

using the R package "crawl" (Johnson et al. 2008a). Kalman filtering methods provide a way to estimate the latent state vector $\mathbf{z}_i$ for all times $i = 1, \ldots, n$ when conditioning on the parameters in the model. Thus, we can numerically integrate out the latent state and maximize the resulting likelihood using standard optimization methods. For example, using the GPS telemetry data from an adult male mule deer (*Odocoileus hemionus*) in Figure 6.19, first analyzed by Hooten et al. (2010b), we were able to fit the CTCRW model in Equation 6.85 with the "crawl" R package. The maximum likelihood algorithm in "crawl" required only 1 s to fit, and the resulting MLEs for parameters associated with the OU process were $\widehat{\log(\theta)} = -3.61$ and $\widehat{\log(\sigma)} = 4.27$.

The state-space formulation presented by Johnson et al. (2008a) is also suited to Bayesian hierarchical modeling techniques and only needs priors for the unknown parameters to proceed. The fully Gaussian state-space model will result in conjugate full-conditional distributions (multivariate normal) for all $\mathbf{z}_i$, and thus, easy implementation in an MCMC algorithm.

While the level of smoothness in the OU velocity model can be controlled with the OU correlation parameter, it is still a nonstationary model. That is, a simulated

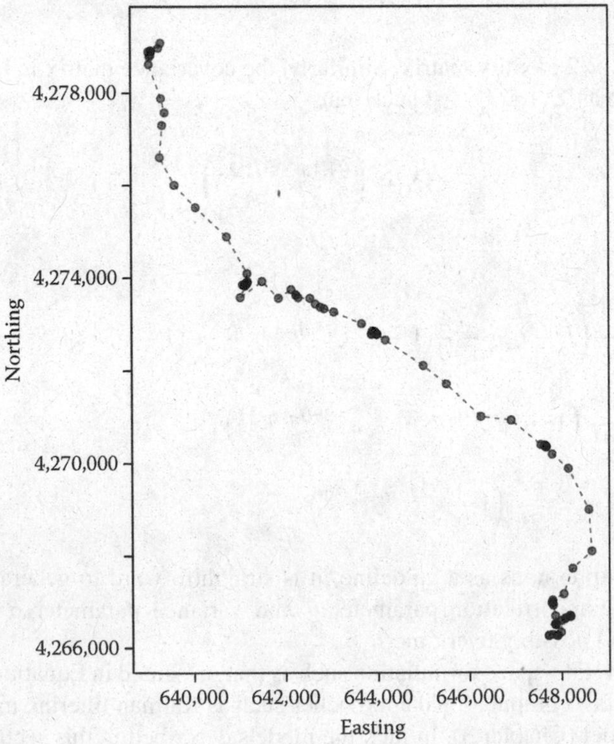

FIGURE 6.19 Observed GPS telemetry data ($n = 129$, points) from an adult male mule deer during autumn in southeastern Utah, USA. Dashed line is shown to connect the points in sequence only.

OU velocity realization will eventually wander away like a Brownian motion realization. In fact, the simulated OU velocity realization will wander away at a faster rate because it is smoother; for this reason, it is referred to as "superdiffusive." Superdiffusivity is not usually a problem when modeling animal movement because, when fitted to telemetry data (e.g., using the Kalman filter), the estimated state is constrained by the data. However, if there are extremely large time gaps in the data, this model can perform poorly because the estimated position process will tend to wander off in the direction of the last known velocity trajectory and will not begin to return until approximately half way to the time of the next observed location. To fix this, Fleming et al. (2014) proposed the Ornstein–Uhlenbeck foraging (OUF) model. The OUF model extends the OU velocity model by adding attraction to a central location in the OU velocity model. The OUF model can be characterized by the SIE

$$\boldsymbol{\mu}(t) = \int_0^t (\mathbf{M} - \mathbf{I})(\boldsymbol{\mu}(\tau) - \boldsymbol{\mu}^*)\, d\tau + \int_0^t \mathbf{v}(\tau)\, d\tau, \tag{6.91}$$

where $\mathbf{v}(\tau)$ is an OU velocity process (Fleming et al. 2015). The OUF model has the same SIE as an OU position model, with the integrated white noise, $d\mathbf{b}(t)$, replaced with a correlated OU process. By replacing the white noise with correlated noise, the OUF model produces a smooth position process in the short term, yet will not wander off as $t \to \infty$ as with the integrated OU velocity model.

6.7.6 Resource Selection and Ornstein–Uhlenbeck Models

We already introduced the concept of how auxiliary information about environmental variables (i.e., resources) can be used in both discrete- and continuous-time dynamic movement models. In the continuous-time case specifically, we discussed potential functions and their specification in animal movement models. This is one valid way to obtain inference about factors that affect animal movement; however, the question remains whether there is a direct connection to the spatial point process models and resource selection. Recall that most forms of point process models rely on a custom distribution for either geographic space or covariate space. In Chapter 4, when focusing on resource selection functions (RSFs), we saw that the point process distribution for the animal positions $\boldsymbol{\mu}_i$ can be written as

$$[\boldsymbol{\mu}_i|\boldsymbol{\beta},\boldsymbol{\theta}] \equiv \frac{g(\mathbf{x}(\boldsymbol{\mu}_i),\boldsymbol{\beta})f(\boldsymbol{\mu}_i,\boldsymbol{\theta})}{\int g(\mathbf{x}(\boldsymbol{\mu}),\boldsymbol{\beta})f(\boldsymbol{\mu},\boldsymbol{\theta})d\boldsymbol{\mu}}, \tag{6.92}$$

where $g(\mathbf{x}(\boldsymbol{\mu}_i),\boldsymbol{\beta})$ is the actual resource selection function and $f(\boldsymbol{\mu}_i,\boldsymbol{\theta})$ is often referred to as the availability distribution. The availability distribution represents locations in the spatial domain that are available in the time interval $(t_{i-1}, t_i]$. The function $f(\boldsymbol{\mu}_i,\boldsymbol{\theta})$ can differentially weight these locations based on a variety of things such as hard barriers to movement, physical limitations of the animal, territoriality, and so on. The most frequently chosen availability distribution in conventional RSF models is a uniform distribution on the spatial support of the point process (typically the study area or home range of the animal). The choice of availability distribution is often the largest factor affecting differences in resource selection inference using these methods (Hooten et al. 2013b). Thus, our specification of $f(\boldsymbol{\mu}_i, \boldsymbol{\theta})$ is a critical component of obtaining resource selection inference.

In a reconciliation of RSF and dynamic animal movement models, Johnson et al. (2008b) presented a general framework for considering these two approaches simultaneously. We described point process models in Chapters 2 through 4; however, we return to them now with a background in continuous-time stochastic models for movement. Johnson et al. (2008b) proposed that the availability distribution be linked to a dynamic animal movement model such that $f(\boldsymbol{\mu}_i, \boldsymbol{\theta}) = \exp((\boldsymbol{\mu}_i - \bar{\boldsymbol{\mu}}_i)'\boldsymbol{\Lambda}_i^{-1}(\boldsymbol{\mu}_i - \bar{\boldsymbol{\mu}}_i)/2)$, where $\bar{\boldsymbol{\mu}}_i = \boldsymbol{\mu}^* + \mathbf{B}_i(\boldsymbol{\mu}_{i-1} - \boldsymbol{\mu}^*)$ and $\mathbf{B}_i = \exp(-(t_i - t_{i-1})/\phi) \cdot \mathbf{I}$ is a 2×2 matrix with zeros on the off-diagonals, $\boldsymbol{\Lambda}_i = \boldsymbol{\Lambda} - \mathbf{B}_i\boldsymbol{\Lambda}\mathbf{B}_i'$, and $\boldsymbol{\Lambda}$ is a covariance matrix that controls the strength of attraction to the central place $\boldsymbol{\mu}^*$. Notice that this definition for the availability distribution $f(\boldsymbol{\mu}_i, \boldsymbol{\theta})$ is proportional to the multivariate OU process presented in the previous sections. The reason for the proportionality is that the normalizing constants in the rest of the Gaussian distribution cancel out in the numerator and denominator. To see this, we use an exponential selection function and an OU model for the availability distribution and update the point process model for $\boldsymbol{\mu}_i$

$$[\boldsymbol{\mu}_i|\boldsymbol{\beta}, \boldsymbol{\theta}] \equiv \frac{\exp(\mathbf{x}(\boldsymbol{\mu}_i)'\boldsymbol{\beta} + (\boldsymbol{\mu}_i - \bar{\boldsymbol{\mu}}_i)'\boldsymbol{\Lambda}_i^{-1}(\boldsymbol{\mu}_i - \bar{\boldsymbol{\mu}}_i)/2))}{\int \exp(\mathbf{x}(\boldsymbol{\mu})'\boldsymbol{\beta} + (\boldsymbol{\mu} - \bar{\boldsymbol{\mu}}_i)'\boldsymbol{\Lambda}_i^{-1}(\boldsymbol{\mu} - \bar{\boldsymbol{\mu}}_i)/2))d\boldsymbol{\mu}}. \quad (6.93)$$

Recall how similar Equation 6.93 is to the model in Equation 4.40, developed by Brost et al. (2015), for handling irregularly spaced telemetry data and constraints to movement. Thus, the OU model serves as a useful way to control for temporal autocorrelation based on the physics of movement in the standard resource selection framework. The two ways to approach fitting these types of point process models are either (1) jointly or (2) two-stage. Jointly, one would fit the point process model directly and estimate both the parameters in the selection and availability functions simultaneously. Brost et al. (2015) use the joint approach, and, while it is most rigorous statistically, it can also be computationally demanding, depending on how difficult it is to calculate the integral in the denominator of Equation 6.93. See Chapter 4 for details on that aspect of implementation.

The second approach to fitting this movement-constrained point process model (6.93) is to preestimate the availability distribution for all times of interest, $t_1, \ldots, t_n$, using the methods in the previous section and then use those estimates for availability parameters while fitting the point process model in a second step. This can be much more stable and less computationally demanding, allowing for things like parallelization of the first computational step across individuals in a population, for example. However, as with most two-stage modeling procedures, the validity of the final inference depends heavily on the appropriateness of the first step and requires minimal feedback from the second to the first step. That is, if statistical learning about resource selection significantly alters the future availability of resources to the individual, then some amount of feedback would be essential to fit the proper model. As usual, there is a trade-off in how important it is to fit the exact model versus how important it is to get at least tentative or preliminary results about the overall process. In an era of "big data," such trade-offs are being made every day because scientists need to fit approximate models that would otherwise be computationally intractable in their exact form. We return to these concepts of two-stage animal movement models (and the concept of multiple imputation) in Chapter 7.

6.7.7 Prediction Using Ornstein–Uhlenbeck Models

The methods for fitting stochastic differential or integral equation models described in the previous sections are particularly valuable in the statistical setting when the data (i.e., positions) are observed at irregular time intervals. Because the conditional distribution for a position at time t_i given the position at t_{i-1} depends on the time gap $t_i - t_{i-1}$ in Equation 6.82, for example, the time between telemetry fixes is inherently part of the statistical model. However, knowledge of the complete path on a larger time interval is often of interest as well. While we have shown that many of the OU statistical models can be fit by considering the process at a discrete and finite set of times, they still fundamentally rely on a continuous underlying process.

We cannot learn about the true continuous underlying process completely because the process must be discretized for computational reasons regardless. However, the discretization can be made sufficiently fine that we gain quasi-continuous inference. For example, it is often of interest to infer an animal's position during a time period between telemetry fixes, with the associated uncertainty. Another valuable use for inference about the complete path of an individual (or individuals) is to estimate the utilization distribution (UD). Recall from Chapter 4 that the UD has historically been used to learn about animal space use. The UD tells us where the individual spent most of its time, and can be broken up into inference in spatial regions or landscape/waterscape types. Classical methods for estimating UDs have depended on kernel density estimation (KDE) techniques that have been heavily scrutinized (e.g., Otis and White 1999; Fieberg 2007). KDE approaches have more recently been modified to better portray accurate space use patterns (e.g., Fleming et al. 2015), but most implementations still lack a fundamental connection to the process generating mechanism. In fact, nearly all KDE approaches for UD estimation are purely phenomenological and are a function of the data directly rather than the true path of the individual (which is unknown).

In addition to the UD, it is often of interest to infer various summary statistics or metrics as functions of the individual's complete position or velocity process (e.g., Buderman et al. 2016). Johnson et al. (2011) refers to these unknown quantities as "movement metrics" and describes a Bayesian approach for learning about them using CTCRW models. One good reason for the Bayesian approach in this setting is that the quantities of interest can be complicated functions of the unknown complete position process and it is challenging to obtain valid uncertainty inference for such quantities using other approaches, such as maximum likelihood.

In the Bayesian setting, a generic hierarchical model specification is

$$\mathbf{s} \sim [\mathbf{s}|\boldsymbol{\mu}], \tag{6.94}$$

$$\boldsymbol{\mu} \sim [\boldsymbol{\mu}|\boldsymbol{\theta}], \tag{6.95}$$

$$\boldsymbol{\theta} \sim [\boldsymbol{\theta}], \tag{6.96}$$

where $\mathbf{s}$ represents the set of telemetry observations (appropriately vectorized), $\boldsymbol{\mu}$ is the complete latent position process, and $\boldsymbol{\theta}$ represents the unknown process parameters. Note that we have omitted any parameters from the data model, assuming they are known for simplicity in presenting the basic strategy. This type of Bayesian hierarchical model assumes the same basic form as those described in the previous chapters; however, critically, the latent process $\boldsymbol{\mu}$ is continuous. Fitting the model involves finding the posterior distribution

$$[\boldsymbol{\mu}, \boldsymbol{\theta}|\mathbf{s}] \propto [\mathbf{s}|\boldsymbol{\mu}][\boldsymbol{\mu}|\boldsymbol{\theta}][\boldsymbol{\theta}]. \tag{6.97}$$

Computational methods (e.g., MCMC) can be used to sample from the posterior distribution in Equation 6.97. We obtain inference for the position process by integrating the parameters out of the joint posterior (6.97) to yield the posterior distribution

$$[\boldsymbol{\mu}|\mathbf{s}] = \int [\boldsymbol{\mu}, \boldsymbol{\theta}|\mathbf{s}]\, d\boldsymbol{\theta}. \tag{6.98}$$

Posterior inference for the position process, such as the posterior mean and variance of $\boldsymbol{\mu}$, is obtained easily using sample moments based on the resulting MCMC samples ($\boldsymbol{\mu}^{(k)}, k = 1, \ldots, K$) from the model fit. For example, the posterior mean of $\boldsymbol{\mu}$ is calculated as

$$E(\boldsymbol{\mu}|\mathbf{s}) = \int \boldsymbol{\mu}[\boldsymbol{\mu}|\mathbf{s}]\, d\boldsymbol{\mu} \tag{6.99}$$

$$= \int\int \boldsymbol{\mu}[\boldsymbol{\mu}, \boldsymbol{\theta}|\mathbf{s}]\, d\boldsymbol{\theta}\, d\boldsymbol{\mu} \tag{6.100}$$

$$\approx \frac{\sum_{k=1}^{K} \boldsymbol{\mu}^{(k)}}{K}. \tag{6.101}$$

using the MCMC samples $\boldsymbol{\mu}^{(k)}$. This procedure requires that we sample the complete position process ($\boldsymbol{\mu}$) in our MCMC algorithm.

In practice, we obtain MCMC samples for $\boldsymbol{\mu}^{(k)}$ at a finite set of prediction times. These times may or may not line up perfectly with the times for which telemetry data are available. Thus, consider two vectors; one vector containing the position process that lines up in time with the observations $\boldsymbol{\mu}$ and a second vector that contains the positions for all prediction times of interest $\tilde{\boldsymbol{\mu}}$. In this case, we can use composition sampling to obtain MCMC samples for $\tilde{\boldsymbol{\mu}}$ by first sampling from the full-conditional for the parameters $\boldsymbol{\theta}$, next sampling from the full-conditional distribution of $\boldsymbol{\mu}$ conditioned on $\boldsymbol{\theta}$, and finally sampling $\tilde{\boldsymbol{\mu}}$ from the conditional predictive distribution $[\tilde{\boldsymbol{\mu}}|\boldsymbol{\mu},\boldsymbol{\theta}]$.

We may also seek the posterior distribution for the movement metrics of interest. Given that these movement metrics (e.g., $f(\boldsymbol{\mu})$) are direct functions of the latent position process $\boldsymbol{\mu}$, they can be treated as derived quantities in the model. To obtain posterior inference for derived quantities that are functions of the complete position process, we often need to calculate posterior moments. An example derived quantity is the posterior mean of the movement metric itself

$$E(f(\boldsymbol{\mu})|\mathbf{s}) = \int\int\int f(\tilde{\boldsymbol{\mu}})[\tilde{\boldsymbol{\mu}}|\boldsymbol{\mu},\boldsymbol{\theta}][\boldsymbol{\mu},\boldsymbol{\theta}|\mathbf{s}]\, d\boldsymbol{\theta}\, d\boldsymbol{\mu}\, d\tilde{\boldsymbol{\mu}} \tag{6.102}$$

$$\approx \frac{\sum_{k=1}^{K} f(\tilde{\boldsymbol{\mu}}^{(k)})}{K}. \tag{6.103}$$

The ability to find posterior statistics (e.g., means, variance, credible intervals) using MCMC for functions of unknown quantities in Bayesian models arises as a result of the equivariance property (Hobbs and Hooten 2015).

An example of a useful movement metric might be the total amount of time an individual animal spent in geographic region $\mathcal{A}$; in practice, $\mathcal{A}$ could be an area of critical habitat, a national park, a highway buffer, or a city boundary. The associated movement metric is

$$f(\tilde{\boldsymbol{\mu}}) = \sum_{j=1}^{m} \Delta t I_{\{\tilde{\boldsymbol{\mu}}(t_j)\in\mathcal{A}\}}, \tag{6.104}$$

where the sum is over a set of m prediction times $(t_1, \ldots, t_j, \ldots, t_m)$ spaced Δt units apart. The movement metric in Equation 6.104 can be used to graphically portray the UD by calculating the posterior mean of it for a large set of grid cells in the study area, each represented by a different $\mathcal{A}$.

Another type of movement metric is total distance traveled by the individual. In this case, an appropriate movement metric can be defined as

$$f(\tilde{\boldsymbol{\mu}}) = \sum_{j=2}^{m} \sqrt{(\tilde{\boldsymbol{\mu}}(t_j) - \tilde{\boldsymbol{\mu}}(t_{j-1}))'(\tilde{\boldsymbol{\mu}}(t_j) - \tilde{\boldsymbol{\mu}}(t_{j-1}))}. \tag{6.105}$$

The metric in Equation 6.105 adds up the lengths of each of the steps to calculate the total path length. As with the first metric in Equation 6.104, the metric corresponding to total distance moved, Equation 6.105, will converge to the correct value as the time gap between prediction locations shrinks ($\Delta t \rightarrow 0$). From a computational storage perspective, one benefit of using these single-number summaries as metrics is that we can calculate running averages of them in the MCMC algorithm without having to save the entire position process at all prediction times for every iteration.

As an alternative to obtaining the posterior inference for the movement metrics concurrently with fitting the Bayesian model, Johnson et al. (2011) provided three methods for obtaining approximate inference using a two-stage approach. In each method, the first stage involves fitting the CTCRW model (i.e., fit using "crawl" R package) described in the previous sections. Recall that the CTCRW approach of Johnson et al. (2011) relies on maximum likelihood methods and uses the Kalman filter to estimate the latent state and is thus very computationally efficient.

For stage two, Johnson et al. (2011) suggest one of the following three approaches to sample realizations of the position process $\tilde{\boldsymbol{\mu}}(t_j)$ based on an implicit posterior distribution for $\boldsymbol{\mu}$.

1. *Plug-in*: Use the MLEs for the model parameters as a stand in for the posterior mode (under vague priors) in the full-conditional distribution $[\tilde{\boldsymbol{\mu}}(t_j)|\cdot]$ and sample from it to obtain realizations of the position process.
2. *Importance sampling*: Sample model parameters from a proposal distribution, weight them according to the implicit posterior, then sample $\tilde{\boldsymbol{\mu}}(t_j)$ from its full-conditional distribution given the model parameters resampled with probability proportional to the weights.
3. *Integrated nested Laplace approximation*: Deterministically sample model parameters from a distribution that mimics the posterior, construct weights based on the posterior at these sampled parameter locations, then sample $\tilde{\boldsymbol{\mu}}(t_j)$ from its full-conditional distribution.

All of these approaches assume that the MLE is a good representation of the posterior mode, and thus, make strong assumptions about the effect of the prior distribution (or lack thereof). However, the first approach will also be substantially faster to implement than fitting the full Bayesian model. Specifically, the downside to the first approach is that it will not properly accommodate the uncertainty in the parameters and may be a poor approximation in cases where the parameter uncertainty is relatively large. On the other hand, it is the fastest and easiest of the methods to implement. Approach two is more rigorous and will provide a good approximation to the true posterior when the proposal distribution is close to the target density. Otherwise, importance sampling methods are prone to degeneracy issues that result in posterior realizations that carry too much weight. Despite the additional complexity, Johnson et al. (2011) prefer the third approach because the first two were inadequate for their example.

Returning to the mule deer example, recall the GPS telemetry data for an autumn migration of a male mule deer (Figure 6.19). Based on the CTCRW model fit using maximum likelihood, we used the "crawl" R package to simulate 1000 realizations of the position process, $\boldsymbol{\mu}(t)$, by importance sampling (Johnson et al. 2011). Figure 6.20 shows the original telemetry data (points) and the distribution of the position process (gray shaded region).

Regardless of how the realizations of the position process are obtained, after they are in hand, they can be used for inference concerning the movement metrics of interest. The excellent properties of Monte Carlo integration and MCMC allow for a straightforward calculation of posterior summaries for derived quantities, regardless of whether they are linear functions of the position process or not.

For example, notice the uncertainty in the position process increases (i.e., gray regions widen) during periods where observations are spaced far apart in the mule deer example (Figure 6.20). We can properly account for the uncertainty in the

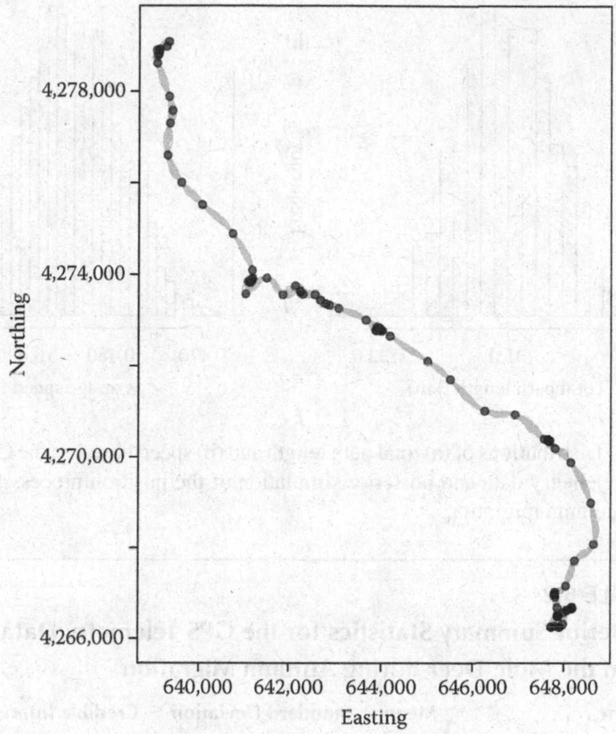

FIGURE 6.20 Observed GPS telemetry data ($\mathbf{s}(t)$, points) and predicted position process ($\boldsymbol{\mu}(t)$, gray shaded region) for an adult male mule deer during autumn in southeastern Utah, USA.

underlying position process when obtaining inference for movement metrics. For example, Figure 6.21 shows the distributions for the movement metrics based on the posterior simulation of the position process after fitting the CTCRW model to the GPS telemetry data from the mule deer during an autumn migration. Finally, the summary statistics in Table 6.2 indicates that the posterior mean path length during the fall migration for this individual mule deer was 31 km, and the average speed during the fall migration was 0.48 km/h.

6.8 CONNECTIONS AMONG DISCRETE AND CONTINUOUS MODELS

At the beginning of this chapter, we demonstrated how continuous-time SDE models for animal movement could be derived from discrete-time model formulations. This provides a natural link between the two different approaches for modeling movement. However, McClintock et al. (2014) point out that, while the true process of animal movement occurs in continuous time, it is often more intuitive for ecologists to think about the discrete-time setting. Thus, they set out to compare the most commonly used

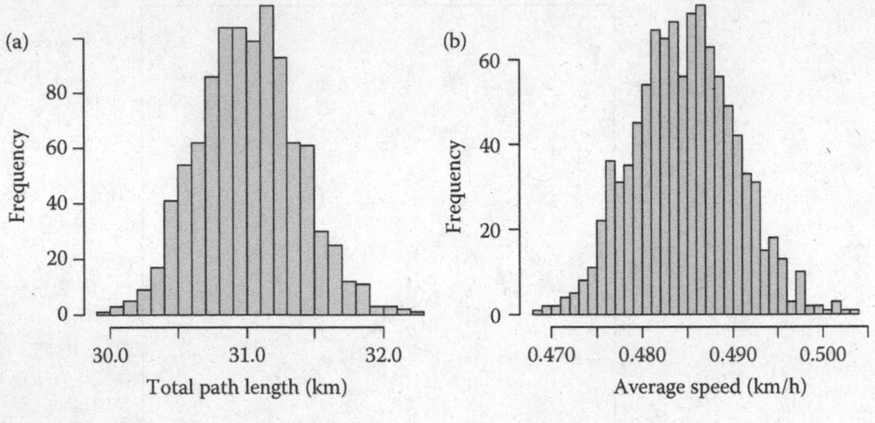

FIGURE 6.21 Distributions of (a) total path length and (b) speed based on the CTCRW model fit to the GPS telemetry data and posterior simulation of the position process from the mule deer during an autumn migration.

TABLE 6.2
Posterior Summary Statistics for the GPS Telemetry Data from the Mule Deer during Autumn Migration

Metric	Mean	Standard Deviation	Credible Interval
Total path length (km)	31.0	0.360	(30.34, 31.72)
Average speed (km/h)	0.48	0.006	(0.47, 0.50)

continuous- and discrete-time models both analytically and empirically. In doing so, they made several important points that we summarize in what follows.

One of the first points made by McClintock et al. (2014) is that the term "state-space model" refers to every hierarchical model that incorporates data and process model components. Both discrete-time and continuous-time animal movement models are state-space models if they accommodate measurement error. Thus, this term may not be the most appropriate way to distinguish among models forms. In fact, McClintock et al. (2014), in their Table 6.2, list 17 different forms of statistical movement models based on the following attributes: discrete/continuous time, discrete/continuous space, metric being modeled (e.g., position, velocity, turning angle, step length), directed or undirected movement, correlated or uncorrelated movement, and whether they are single-state or multistate models. Their synthesis suggests a huge variety in the type of movement models developed and used in practice and far from a consensus in their form.

As we have seen in this chapter, there has been a lengthy and sometimes parallel evolution of both discrete-time and continuous-time models for animal movement. Naturally, new developments are derived as generalizations of earlier models. For example, the discrete-time models described by Morales et al. (2004) and Jonsen et al. (2005) reduce to uncorrelated random walks under certain parameterizations (i.e., correlation parameters set to zero). Similarly, the continuous-time models of Dunn and Gipson (1977), Johnson et al. (2008a), and Harris and Blackwell (2013) reduce to Brownian motion when attractive forces are removed from the OU process.

McClintock et al. (2014) report one key difference in the mechanistic underpinnings implied in both models. That is, the CTCRW models (for both position and velocity) have extra correlation structure in speed, for example, that is not apparent in the discrete-time counterparts. Furthermore, in standard continuous-time models with attraction, the movement rate depends on distance from the point of attraction, implying that the movement of the individual will slow down as it approaches the central location. This type of behavior may not be realistic in all settings. Another key difference is that a discrete-time model will only be able to provide inference for movement at a fixed time interval between positions. This time interval is either controlled when setting the duty cycling for the telemetry device or after the fact based on subsampling of the original data. Continuous-time models, on the other hand, will yield the same results regardless of the temporal resolution of the data.

In our experience, one of the more complicated aspects of implementing contemporary animal movement models from scratch is allowing for state-switching behavior. McClintock et al. (2014) noted that, while change-point and hidden Markov modeling approaches have become common in discrete-time models, they are less used in the continuous-time framework (but see Blackwell et al. 2015), no doubt because of the inevitable increase in complexity of the mathematics involved. To get around this, Hanks et al. (2015b) accommodated changes in movement characteristics using temporally varying parameters. Much like the position process itself, model parameters are allowed to vary smoothly in continuous time. The degree of smoothness can be modeled, or tuned using predictive scoring approaches. This approach avoids the complications of classical state-switching models while still accounting for time-varying behavior. Even so, it may not be obvious how to use similar techniques in all animal movement models and, thus, more development of these approaches are needed.

Computational demands are also an important consideration in animal movement modeling (and any statistical modeling). McClintock et al. (2014) claimed that most continuous-time models are less computationally demanding than their discrete-time counterparts. However, substantial variability exists due to the computational platform (e.g., laptop vs. supercomputer) and software (e.g., C vs. R). It is certainly true that the complexity of a model strongly correlates with an increase in required computing time. For comparison, basic continuous-time model fits may take only seconds or minutes using the "crawl" software of Johnson et al. (2008a), while fits of more complicated discrete-time models (e.g., McClintock et al. 2012) are expected to take hours or days. However, in the absence of measurement error, hidden Markov model machinery (e.g., Franke et al. 2006; Holzmann et al. 2006; Patterson et al. 2009; Langrock et al. 2012) can fit the discrete-time movement process models described by McClintock et al. (2012) in considerably less time. It is worth noting that newer computational tools such as Rcpp and parallel computing, as well as new model reparameterizations (e.g., Hanks et al. 2015b), are leading to vast improvements in speed for algorithms associated with animal movement models. Also, we are now obtaining substantially more variety in inference than with the simpler and faster models used previously. Thus, some would argue the time for information trade-off is worth it.

From our perspective, when considering the speed of obtaining animal movement inference, one should consider the time it takes to develop the model, the code, and the actual computational time together. More time spent on optimizing computer code leads to increases in speed. Thus, when an algorithm needs to be used repetitively (e.g., for several hundred individuals in a larger population), it can be worth the extra programming time up front. Likewise, although models already exist that could be used to analyze telemetry data, they can always be improved upon to yield faster algorithms. Thus, ongoing development of both discrete- and continuous-time animal movement models is essential. However, we need not always focus on extending animal movement models to more complicated settings; we should continue to pursue important ways to facilitate the use of existing model forms.

6.9 ADDITIONAL READING

The topic of continuous-time stochastic processes can be quite technical. However, actual animal movement trajectories occur in continuous time; thus, concepts like stochastic differential equations, Brownian motion, and potential functions are essential tools that can be used to model animal movement. Formal references on stochastic processes and calculus are Durrett (1996), Karatzas and Shreven (2012), and Grimmett and Stirzaker (2001), and for a solid reference on potential functions, see Taylor (2005). Despite the use of stochastic process models for trajectories in multiple fields (e.g., human movement, iceberg motion, ocean drifter devices), much of the relevant literature applied to animal ecology was written by David Brillinger and colleagues (see Brillinger 2010 for an overview).

For other recent references on continuous-time stochastic process models for animal movement, see Russell et al. (2016b) and Hooten and Johnson (2016). Russell et al. (2016b) extend potential function models to include "friction" or "motility" surfaces that multiplicatively operate on the potential-based movement by slowing it down or speeding it up depending on where it is in space or time. Hooten and Johnson (2016) extended the basis function approaches presented in this chapter to accommodate heterogeneous dynamics in smoothed Brownian motion processes. They temporally warp the time domain to allow the smoothness to vary throughout the process. The approach proposed by Hooten and Johnson (2016) allows the analyst to use a single algorithm to fit continuous-time animal movement models in parallel and then recombine them using Bayesian model averaging for final inference.

Finally, Turchin (1998) discussed the concept of scaling from the individual to the population level (i.e., Lagrangian vs. Eulerian) mathematically, as we summarized in the first section of this chapter. Garlick et al. (2011) and Garlick et al. (2014) presented a computationally efficient mathematical scaling approach that leads to optimally fast algorithms for solving Eulerian PDE models based on Lagrangian animal movement processes. Hooten et al. (2013a) and Hefley et al. (2016b) showed how to use the same computationally efficient approach (i.e., homogenization) in a statistical framework for fitting PDE models to aggregate animal movement data.

7 Secondary Models and Inference

In the previous chapter, we showed how one can build additional complexity and realism into CTCRW movement models through the use of potential functions. However, owing to the readily available and user-friendly software provided by Johnson et al. (2008a) that fits smooth velocity-based OU models to irregularly spaced data while accounting for measurement error, there are many reasons to rely on the resulting model output for further inference. Using "crawl" to obtain posterior realizations of the position process $\tilde{\boldsymbol{\mu}}(t_j)$ allows for much more complicated inference than that proposed by Johnson et al. (2011). In fact, entirely new movement models can be fit using the output from "crawl" (or similar first stage models) as data. In what follows, we describe several approaches for using first-stage posterior realizations of $\tilde{\boldsymbol{\mu}}(t_j)$ in secondary statistical models to learn about additional factors influencing movement.

The basic concept is to think of the types of statistical models you might fit if you could have perfect knowledge about the true position process $\boldsymbol{\mu}$ (i.e., $\boldsymbol{\mu}(t), \forall t \in \mathcal{T}$, for the compact time period of interest $\mathcal{T}$).* In this case, we can build models that rely on the entire continuous position process (i.e., a line on a map) and we can characterize the path using the methods in the preceding section to obtain inference. We can build population-level models that pool or cluster similar behaviors among individuals. We can also obtain inference that improves the understanding of how animals choose to move among resources and interact with each other at any temporal scale of interest.

7.1 MULTIPLE IMPUTATION

Lacking an exact measurement of the true position process, the simplest approach is to use the posterior predictive mean for the process and pretend that it is the truth. Then we condition on the position process as data in a secondary statistical model that provides the desired inference; this procedure is called "imputation." The imputation concept of "doing statistics on statistics" may not accommodate the proper uncertainty pertaining to knowledge of the process. However, a technique referred to as "multiple imputation" can help account for the uncertainty associated with the modeled process we intend to use as data in a secondary model.

The heuristic for multiple imputation is to use an imputation distribution $[\tilde{\boldsymbol{\mu}}|\mathbf{s}]$ that closely resembles the true posterior predictive distribution of interest $[\boldsymbol{\mu}|\mathbf{s}]$ and then fit a secondary model while conditioning on the imputation distribution, allowing the uncertainty to propagate into the secondary inference. Multiple imputation provides

* With telemetry technology rapidly improving, semicontinuous data may not be far away, but we will always have historical data sets for which inference is desired.

more accurate uncertainty estimates for the secondary model parameters than only conditioning on the posterior mean for $\boldsymbol{\mu}$.

Traditionally, multiple imputation treats $\tilde{\boldsymbol{\mu}}$ as missing data and $[\boldsymbol{\beta}|\boldsymbol{\mu}]$ is assumed to be asymptotically Gaussian (Rubin 1987, 1996). Furthermore, if we condition on the augmented $\tilde{\boldsymbol{\mu}}$ and fit the Bayesian model, the posterior distribution $[\boldsymbol{\beta}|\tilde{\mathbf{s}}]$ will converge to the distribution of the MLE for $\boldsymbol{\beta}$ conditioned on $\tilde{\boldsymbol{\mu}}$. These ideas allow us to use maximum likelihood methods to obtain the point estimate for $\boldsymbol{\beta}$ (i.e., $E(\boldsymbol{\beta}|\tilde{\boldsymbol{\mu}})$) and associated variance (i.e., $\text{Var}(\boldsymbol{\beta}|\tilde{\boldsymbol{\mu}})$), which can then be averaged to arrive at inference for $\boldsymbol{\beta}$ conditioned on $\boldsymbol{\mu}$ using the following conditional mean and variance relationships:

$$E(\boldsymbol{\beta}|\boldsymbol{\mu}) \approx E_{\tilde{\boldsymbol{\mu}}}(E(\boldsymbol{\beta}|\tilde{\boldsymbol{\mu}})), \tag{7.1}$$

and

$$\text{Var}(\boldsymbol{\beta}|\boldsymbol{\mu}) \approx E_{\tilde{\boldsymbol{\mu}}}(\text{Var}(\boldsymbol{\beta}|\tilde{\boldsymbol{\mu}})) + \text{Var}_{\tilde{\boldsymbol{\mu}}}(E(\boldsymbol{\beta}|\tilde{\boldsymbol{\mu}})). \tag{7.2}$$

In practice, we fit individual models using maximum likelihood methods and $\tilde{\boldsymbol{\mu}}^{(k)}$ as data to yield $\hat{\boldsymbol{\beta}}^{(k)}$ and $\text{Var}(\hat{\boldsymbol{\beta}}^{(k)})$ for $k = 1, \ldots, K$ realizations from a first-stage model fit. We approximate the required integrals in the conditional mean and variance relationships using Monte Carlo integration, essentially computing sample averages and variances using $\hat{\boldsymbol{\beta}}^{(k)}$ and $\text{Var}(\hat{\boldsymbol{\beta}}^{(k)})$ for the K imputation samples. We have found that only a relatively small number of imputation samples provide stable inference (i.e., on the order of 10s rather than 100s or 1000s). This approach to multiple imputation is well known and performs well in most cases, but also requires stronger assumptions and provides only approximate inference.

An alternative approach to multiple imputation used by Hooten et al. (2010b), Hanks et al. (2011), and Hanks et al. (2015a) can be formulated as

$$[\boldsymbol{\beta}|\mathbf{s}] = \int [\boldsymbol{\beta}, \boldsymbol{\mu}|\mathbf{s}]d\boldsymbol{\mu} \tag{7.3}$$

$$= \int [\boldsymbol{\beta}|\boldsymbol{\mu}, \mathbf{s}][\boldsymbol{\mu}|\mathbf{s}]d\boldsymbol{\mu} \tag{7.4}$$

$$\approx \int [\boldsymbol{\beta}|\tilde{\boldsymbol{\mu}}][\tilde{\boldsymbol{\mu}}|\mathbf{s}]d\boldsymbol{\mu}, \tag{7.5}$$

where, as long as the distribution of secondary model parameters $\boldsymbol{\beta}$ is nearly conditionally independent of $\mathbf{s}$ given $\tilde{\boldsymbol{\mu}}$ (i.e., $[\boldsymbol{\beta}|\boldsymbol{\mu}, \mathbf{s}]$ is close to $[\boldsymbol{\beta}|\tilde{\boldsymbol{\mu}}]$), the approximation is adequate. In the context of continuous-time animal movement modeling, the approximation implies that we can predict the true position process (i.e., $\tilde{\boldsymbol{\mu}}$; path) well enough using a CTCRW model and account for the inherent uncertainty in the predicted path, that the final inference for $\boldsymbol{\beta}$ will be close to accurate, or if not, then at least conservative.

To implement the multiple imputation procedure in a Bayesian model using MCMC is trivial, which highlights one of the advantages. The necessary steps in the Bayesian multiple imputation procedure are

1. Fit the CTCRW model proposed by Johnson et al. (2008a) to original telemetry data set. The R package "crawl" can be used.
2. Use the methods described by Johnson et al. (2011) to sample K posterior predictive realizations of the position process $\tilde{\boldsymbol{\mu}}^{(k)}$ at the desired temporal resolution for MCMC samples $k = 1, \ldots, K$.
3. Fit a secondary model using a modified MCMC algorithm. Instead of conditioning on a fixed data set, on the kth iteration of the MCMC algorithm, use $\tilde{\boldsymbol{\mu}}^{(k)}$ as the data.
4. Obtain posterior summaries for the model parameters (i.e., $\boldsymbol{\beta}$) as usual.

This type of modified MCMC algorithm, which is related to data augmentation and composition sampling algorithms, will integrate over the uncertainty in the true position process and incorporate the uncertainty in the inference for the model parameters.

Multiple imputation works well when the imputation distribution accurately represents the true distribution, but, as with any two-stage statistical method, certain pathological situations can arise. Given that the type of imputation approach described here can be useful for fitting models that would otherwise be computationally intractable, they certainly need to be considered as part of the broader toolbox. However, further research is needed to develop procedures for identifying problematic situations. We provide examples where secondary modeling techniques can be useful in what follows.

7.2 TRANSITIONS IN DISCRETE SPACE

Hooten et al. (2010b) proposed a secondary modeling approach that utilizes posterior predictive output from "crawl" in a discretized form that matches up with spatial covariate data on a grid. This modeling approach is useful when there is a set of environmental covariate data already available on a grid of prespecified resolution. Often, remotely sensed and digital elevation data products are created at a 30×30 m spatial resolution and consist of a lattice of pixels in a rectangular arrangement, each with one or more quantitative or qualitative attributes. Given that a finer resolution of these data is often not available, Hooten et al. (2010b) proposed to transform the continuous-space data to discrete-space form to match the grid. Using posterior realizations from "crawl" based on the methods of Johnson et al. (2011), each interpolated position $\tilde{\boldsymbol{\mu}}_j^{(k)}$ in a given realization of the individual's path can be transformed to a binary vector $\mathbf{y}_j^{(k)} \equiv (y_{1,j}^{(k)}, y_{2,j}^{(k)}, y_{3,j}^{(k)}, y_{4,j}^{(k)}, y_{5,j}^{(k)})'$ representing the center pixel and nearest neighboring four pixels (i.e., in each of the cardinal directions if the grid is oriented north and south; Figure 7.1). The vector $\mathbf{y}_j^{(k)}$ represents the "move" that occurs between times t_{j-1} and t_j. The elements of the vector can be arranged in any sensible way as long as they consistently refer to the same directions. For example, using the third position as the center pixel in the neighborhood, we could assign the rest in a clockwise fashion starting with north (Figure 7.2). Thus, the vector $\mathbf{y}_j = (1, 0, 0, 0, 0)'$ indicates the individual moved north at time t_j (Figure 7.2a). Similarly, the vector $\mathbf{y}_j = (0, 0, 1, 0, 0)'$ indicates the individual stayed in the center pixel (Figure 7.2c).

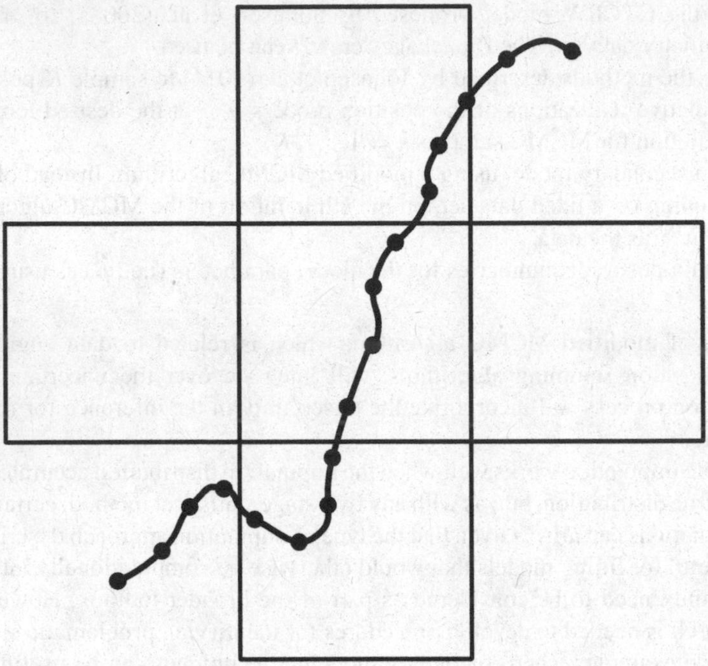

FIGURE 7.1 Schematic of telemetry observations (points) and interpolated position process (line) passing through a neighborhood of the center pixel in a larger lattice.

The transformed posterior realizations can be viewed as multinomial data and a model can be constructed to learn about how the moves correlate with the underlying landscape (Hooten et al. 2010b). Thus, we specify the data model so that $\mathbf{y}_j \sim \text{MN}(1, \mathbf{p}_j)$ for $j = 1, \ldots, m$, where the 1 indicates the individual can only move to single pixel at each time step and $\mathbf{p}_j$ is a vector (dimension 5×1) of probabilities corresponding to the move.* We can now use a generalized linear modeling framework to link the move probabilities $\mathbf{p}_j$ to a linear combination of environmental conditions.

The multinomial vectors $\mathbf{y}_j$ can be thought of as a discretized version of a velocity vector, and thus, can be modeled in similar ways. That is, because we are now modeling the moves rather than the positions directly, we can use the potential function ideas described in the previous sections to incorporate covariate information. Hooten et al. (2010b) proposed linear predictors for a transformation of movement probabilities $\mathbf{p}_j$ that are based on several different possible drivers of animal movement. The

* The chosen dimension of 5×1 is not arbitrary. It arises from the fact that a truly continuous path must first pass to one of the first-order neighbors in a square lattice of pixels before moving into other pixels (i.e., it cannot pass directly through the corner). Our temporal discretization can always be made fine enough so that successive points will be no further away than a single pixel. In practice, for large data sets, this could be computationally demanding. In such cases, the methods of Hanks et al. (2015a) may be necessary.

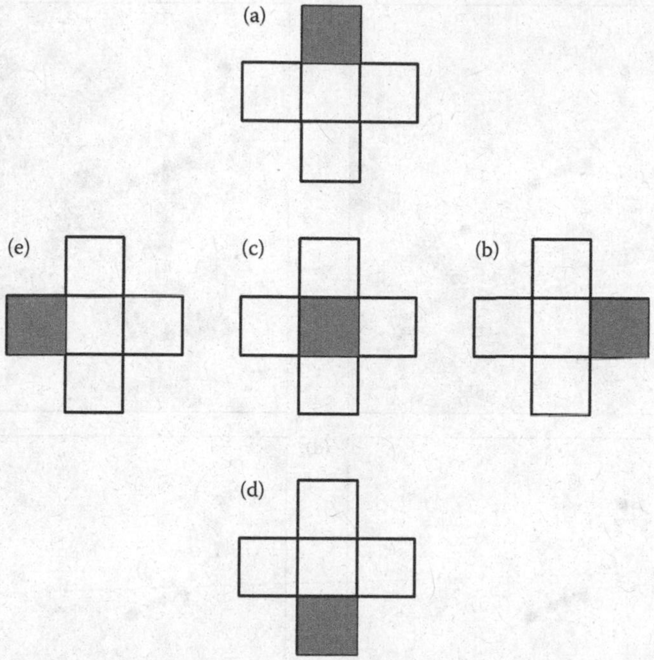

FIGURE 7.2 Possible first-order moves on a regular lattice with square cells (i.e., pixels). (a) Move north, $\mathbf{y}_j = (1,0,0,0,0)'$, (b) move east, $\mathbf{y}_j = (0,1,0,0,0)'$, (c) stay, $\mathbf{y}_j = (0,0,1,0,0)'$, (d) move south, $\mathbf{y}_j = (0,0,0,1,0)'$, and (e) move west, $\mathbf{y}_j = (0,0,0,0,1)'$.

simplest of these drivers is based on the concept that an animal might move in a certain direction given the landscape it was in at time t_{j-1}. For example, possible drivers of mule deer movement in southeastern Utah are shown in Figure 7.3. We can link the movement probabilities with covariates so that $g(p_{j,l}) = \mathbf{x}_j' = \mathbf{x}(\tilde{\boldsymbol{\mu}}_j)'\boldsymbol{\beta}_1$ for the lth neighbor of the cell corresponding to position $\tilde{\boldsymbol{\mu}}_{j-1}$.

Similarly, moves based on changes in the landscape can also be modeled. Recall from the previous section on potential functions that we can model changes in position as a function of the gradient associated with a potential function. In this context, the movement probability is modeled as a function of the difference in covariates, $\boldsymbol{\delta}_j = \mathbf{x}(\tilde{\boldsymbol{\mu}}_j) - \mathbf{x}(\tilde{\boldsymbol{\mu}}_{j-1})$, at the center pixel and the neighboring pixel such that $g(p_{j,l}) = \boldsymbol{\delta}_j'\boldsymbol{\beta}_2$. This is the same general approach described in the spatio-temporal modeling literature (Hooten and Wikle 2010; Hooten et al. 2010a; Broms et al. 2016).

Furthermore, when the individual does not move to a new pixel between successive prediction times (t_{j-1} and t_j), we can model the residence probability as a function of covariates in the residing pixel and neighboring pixels.* For example, Hooten et al. (2010b) describe two possible residence models:

* For a temporally fine set of prediction times, there will be many more "stays" than "moves." Again, Hanks et al. (2015a) generalize these "stays" to a residence time, which reduces the computational demand significantly.

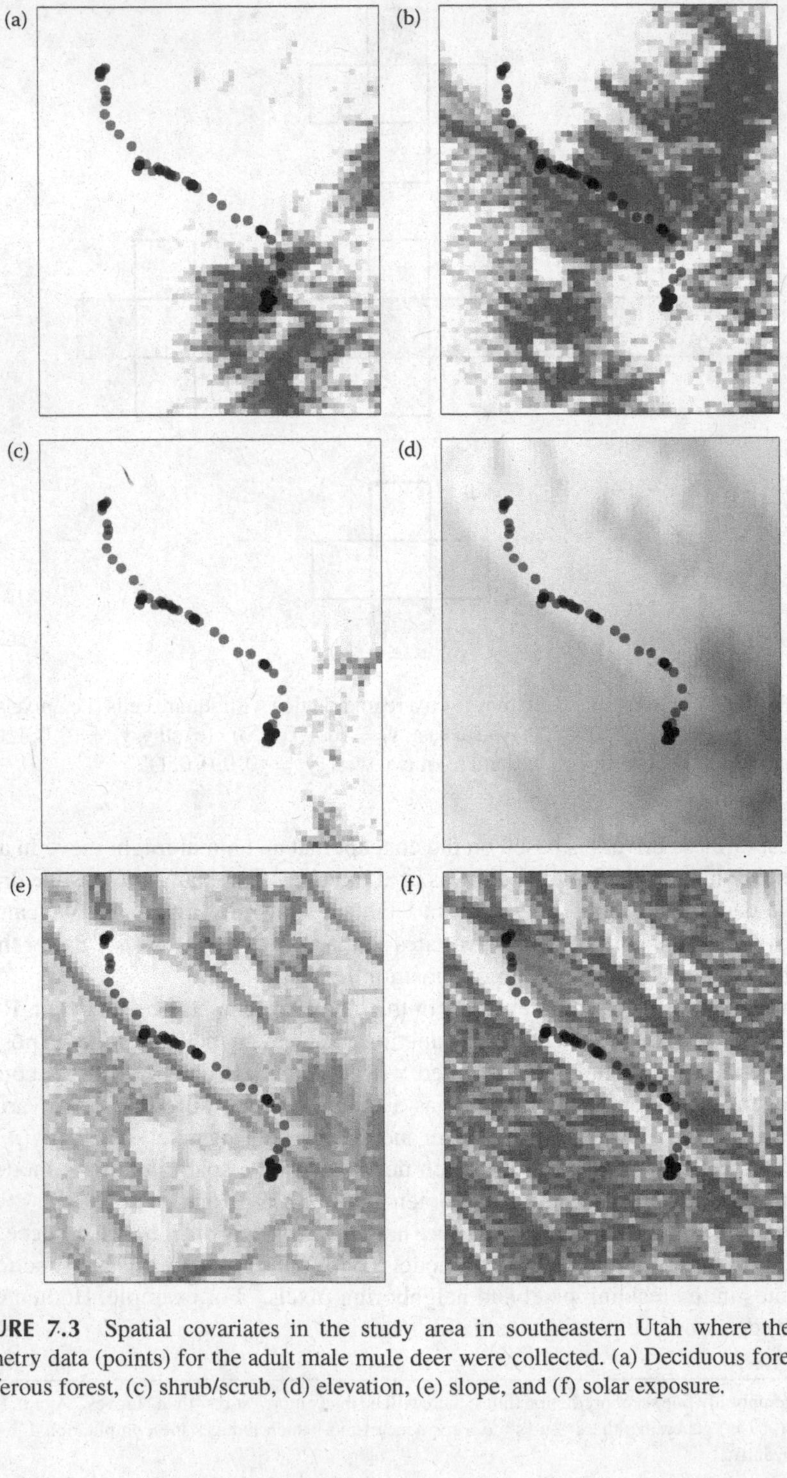

FIGURE 7.3 Spatial covariates in the study area in southeastern Utah where the GPS telemetry data (points) for the adult male mule deer were collected. (a) Deciduous forest, (b) coniferous forest, (c) shrub/scrub, (d) elevation, (e) slope, and (f) solar exposure.

1. Stays are based on current environmental factors: $g(p_{j,3}) = \mathbf{x}'_{j-1,3}\boldsymbol{\beta}_3$.
2. Stays are based on the surrounding environmental factors: $g(p_{j,3}) = \sum_{l \neq 3} \mathbf{x}'_{j-1,l}\boldsymbol{\beta}_4$, where $\mathbf{x}'_{j-1,l}$ denotes the covariates associated with the lth neighboring landscape of the pixel where the individual stayed.

To implement the model, various types of link functions could be used for $g(p)$. Hooten et al. (2010b) employed a hierarchical Bayesian framework that involves the use of latent auxiliary variables $z_{j,l}$. Combining all previously mentioned drivers of animal movement together, we can write a model for the continuous latent movement variable as

$$z_{j,l} = \begin{cases} \beta_{0,1} + \mathbf{x}'_j\boldsymbol{\beta}_1 + \boldsymbol{\delta}'_j\boldsymbol{\beta}_2 + \varepsilon_{j,l} & \text{if move on step } j \\ \beta_{0,2} + \mathbf{x}'_{j-1,3}\boldsymbol{\beta}_3 + \sum_{l \neq 3} \mathbf{x}'_{j-1,l}\boldsymbol{\beta}_4 + \varepsilon_{j,l} & \text{if stay on step } j \end{cases}, \tag{7.6}$$

where $\varepsilon_{j,l} \sim \text{N}(0, 1)$. Following Albert and Chib (1993), if the data model is specified such that $p_{j,l} \equiv P(z_{j,l} > z_{j,\tilde{l}}, \forall \tilde{l} \neq l)$, we are implicitly assuming a probit link function in this model.* This particular specification for a multinomial response model also yields an MCMC algorithm that is fully conjugate, meaning that no tuning of the algorithm is necessary. This is the primary advantage to using the popular auxiliary variable approach of Albert and Chib (1993).

When implementing this model, we need to use the multiple imputation approach; thus, there are K sets of data and covariates that we cycle through on each MCMC iteration when sampling the sets of parameters $\boldsymbol{\beta}_1$, $\boldsymbol{\beta}_2$, $\boldsymbol{\beta}_3$, and $\boldsymbol{\beta}_4$. We suppressed the k notation for each $\tilde{\boldsymbol{\mu}}_j$ in the model statements of this section for simplicity. However, the reason why we need K different sets of corresponding covariates is because the covariates will change as the position $\tilde{\boldsymbol{\mu}}_j^{(k)}$ changes. Thus, despite its utility for providing new inference, this approach can also be computationally demanding.

We fit the hierarchical discrete-space continuous-time movement model described in this section to the mule deer GPS telemetry data in Figure 6.19 using Bayesian multiple imputation based on the position process predictions in Figure 6.20. Focusing on the marginal posterior distributions for the coefficients associated with moves based on the gradient of a potential function ($\boldsymbol{\beta}_4$) in Equation 7.6, Figure 7.4 shows violin plots for each coefficient. The most striking effect in Figure 7.4 is that of elevation (d) on autumn movement of the mule deer individual. The strong negative coefficient indicates that increasing elevation has a negative effect on movement because the individual is descending as temperatures decrease in the autumn and forage becomes scarce.

An alternative way to view the inference of the environmental covariates is to visualize them spatially. Figure 7.5 shows the posterior mean potential function and resulting directional derivative functions associated with the term $\boldsymbol{\delta}'_j\boldsymbol{\beta}_2$ from Equation 7.6. Hooten et al. (2010b) did not use a negative in the gradient function

---

* The probit link function is the standard normal cumulative distribution function. It transforms variables on real support to the compact support of (0,1). The probit link is an alternative to the logit link.

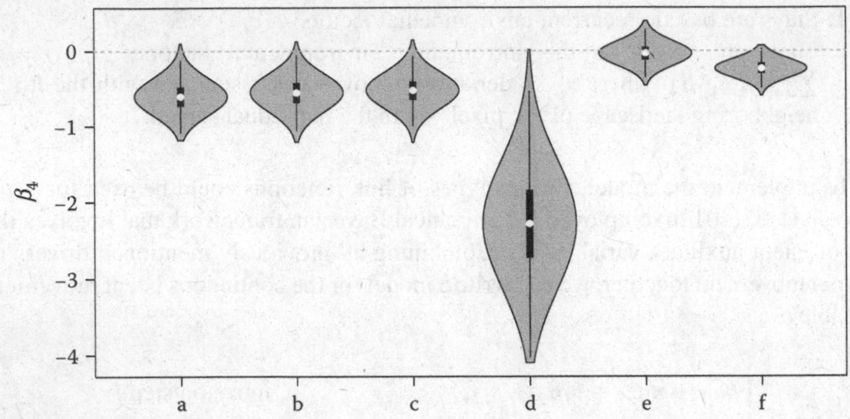

FIGURE 7.4 Marginal posterior distributions (shown as violin plots representing the shape of the posterior density functions) resulting from fitting the discrete-space continuous-time movement model to the mule deer GPS telemetry data. Each coefficient corresponds to the covariates in Figure 7.3: (a) Deciduous forest, (b) coniferous forest, (c) shrub/scrub, (d) elevation, (e) slope, and (f) solar exposure. Internal dark bars represent typical boxplots and white points represent the median for each coefficient.

in their model specification (7.6); thus, Figure 7.5c shows the spatial potential function increasing (darker shading) toward high potential.* In this case, the potential function (Figure 7.5c) is controlled mostly by elevation, as we discussed in relation to the parameter estimates in Figure 7.4.

7.3 TRANSITIONS IN CONTINUOUS SPACE

Hooten et al. (2010b) provided the basic framework for using posterior realizations of the position process for further inference in a secondary model and focused on the discrete space setting in which transitions among areal units on a landscape can be represented as multinomial moves, or discretized velocities. Hanks et al. (2011) considered velocities as response variables directly so that the models could be used with both continuous and discrete spatial covariate data and more information can be retained in the response variable.†

Hanks et al. (2011) define the velocity vector as $\mathbf{y}_j = \tilde{\boldsymbol{\mu}}_j - \tilde{\boldsymbol{\mu}}_{j-1}$, where we have dropped the k superscript notation again for simplicity, but a $\mathbf{y}_j$ is calculated for each k realization of the position process (for $k = 1, \ldots, K$ and $j = 2, \ldots, m$) resulting from the posterior predictive inference as described in the previous section. Then, rather than discretize the velocity as in Hooten et al. (2010b), the velocity is now modeled

* Recall that we defined the gradient function with a negative sign in Section 6.6 to be consistent with the notation used by Brillinger (2010).

† Generally, a discretized response variable will carry less information than the continuous response variable it is based on. A binary response variable y, where $y = I_{\{z>0\}}$ for $z \sim \text{N}(0, \sigma^2)$, contains much less information than z.

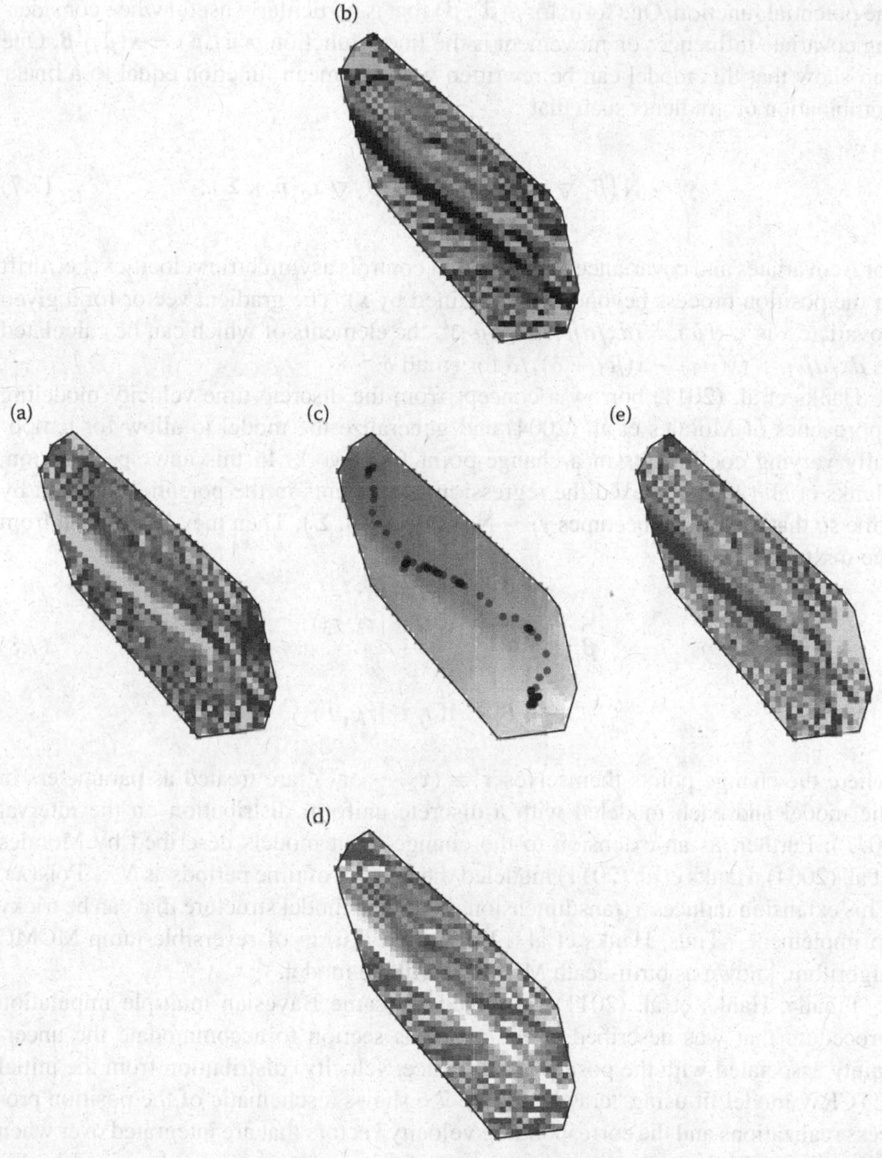

FIGURE 7.5 Posterior mean potential function (c) and directional derivatives: (a) west, (b) north, (d) south, (e) east. Dark regions indicate large values. Telemetry data are shown in panel (c) as dark points.

directly as a function of the gradient of the underlying potential function. This specification results in a simpler model form than the multinomial, where the velocity vectors are modeled as $\mathbf{y}_j \sim \mathrm{N}\left(\nabla p(\tilde{\boldsymbol{\mu}}_j, \boldsymbol{\beta}), \boldsymbol{\Sigma}\right)$. Recall, from the previous section on potential functions, that the term $\nabla p(\tilde{\boldsymbol{\mu}}_j, \boldsymbol{\beta})$ represents the gradient operator of the spatially explicit function $p(\tilde{\boldsymbol{\mu}}_j, \boldsymbol{\beta})$. As noted previously, there are several options for the potential function. One form for $p(\tilde{\boldsymbol{\mu}}_j, \boldsymbol{\beta})$ that is particularly useful when considering covariate influences on movement is the linear function $p(\tilde{\boldsymbol{\mu}}_j, \boldsymbol{\beta}) = \mathbf{x}(\tilde{\boldsymbol{\mu}}_j)'\boldsymbol{\beta}$. One can show that this model can be rewritten with the mean function equal to a linear combination of gradients such that

$$\mathbf{y}_j \sim \text{N}\left(\beta_1 \nabla x_1(\tilde{\boldsymbol{\mu}}_j) + \cdots + \beta_q \nabla x_q(\tilde{\boldsymbol{\mu}}_j), \boldsymbol{\Sigma}\right), \tag{7.7}$$

for q covariates and covariance matrix $\boldsymbol{\Sigma}$ that controls asymmetric velocities (i.e., drift in the position process beyond that explained by $\mathbf{x}$). The gradient vector for a given covariate x is $\nabla x(\tilde{\boldsymbol{\mu}}) = (dx/d\tilde{\mu}_1, dx/d\tilde{\mu}_2)'$, the elements of which can be calculated as $dx/d\tilde{\mu}_1 \approx (x(\tilde{\mu}_1) - x(\tilde{\mu}_1 + \delta))/\delta$ for small δ.

Hanks et al. (2011) borrow a concept from the discrete-time velocity modeling approaches of Morales et al. (2004) and generalize the model to allow for temporally varying coefficients in a change-point framework. In this new specification, Hanks et al. (2011) indexed the regression coefficients in the potential function by time so that the model becomes $\mathbf{y}_j \sim \text{N}\left(\nabla p(\tilde{\boldsymbol{\mu}}_j, \boldsymbol{\beta}_j), \boldsymbol{\Sigma}\right)$. Then they let $\boldsymbol{\beta}_j$ arise from the mixture

$$\boldsymbol{\beta}_j = \begin{cases} \boldsymbol{\beta}_1 & \text{if } t_j \in (0, \tau_2) \\ \boldsymbol{\beta}_2 & \text{if } t_j \in [\tau_2, \tau_3) \\ \vdots & \\ \boldsymbol{\beta}_N & \text{if } t_j \in [\tau_N, T) \end{cases}, \tag{7.8}$$

where the change points themselves $\boldsymbol{\tau} \equiv (\tau_2, \ldots, \tau_N)'$ are treated as parameters in the model and each modeled with a discrete uniform distribution on the interval $(0, T)$. Further, as an extension to the change-point models described by Morales et al. (2004), Hanks et al. (2011) modeled the number of time periods as $N \sim \text{Pois}(\lambda)$. This extension induces a transdimensionality to the model structure that can be tricky to implement.* Thus, Hanks et al. (2011) used a form of reversible-jump MCMC algorithm, known as birth-death MCMC, to fit the model.

Finally, Hanks et al. (2011) employed the same Bayesian multiple imputation procedure that was described in the previous section to accommodate the uncertainty associated with the position (and hence, velocity) distribution from the initial CTCRW model fit using "crawl." Figure 7.6 shows a schematic of the position process realizations and the corresponding velocity vectors that are integrated over when using the multiple imputation procedure to fit the velocity model described in this section.

We fit the model described in Equations 7.7 and 7.8 to observed telemetry data from an adult male northern fur seal (*Callorhinus ursinus*) on an 18-day foraging trip during the summer in the Bering Sea near the Pribilof Islands (Figure 7.7). The

* Transdimensionality means that the parameter space changes on every iteration of a statistical algorithm for fitting the model, like MCMC. These changes in the parameter space require modifications to the MCMC algorithm so that the models with different numbers of change points can be fairly visited by the algorithm.

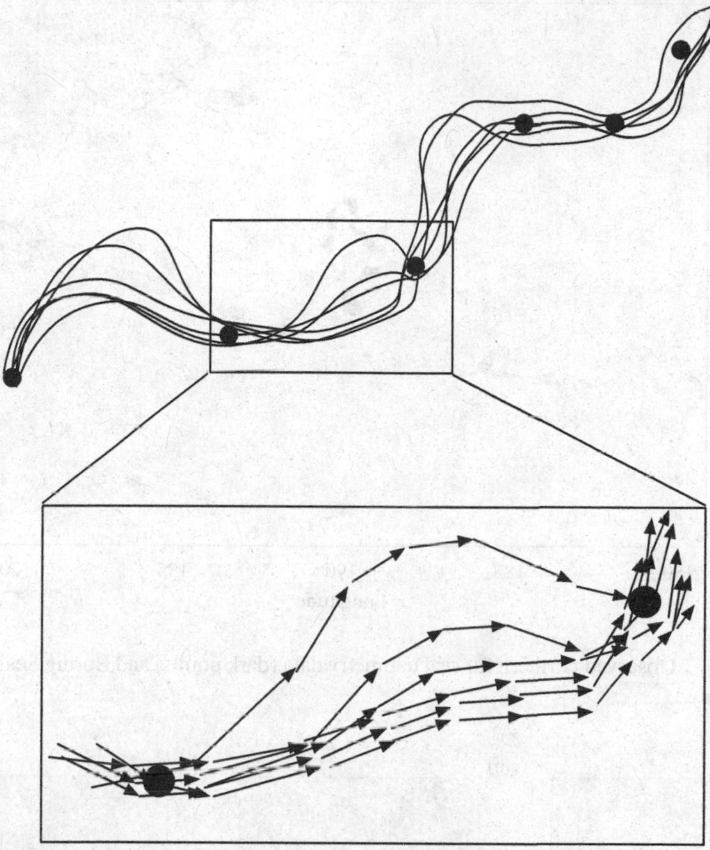

FIGURE 7.6 Position process realizations (top lines), corresponding velocity vectors (bottom arrows), and telemetry observations (points).

largest northern fur seal rookeries exist at the Pribilof (i.e., Saint Paul and Saint George) and Commander Islands (i.e., Bering Island and Medney Island) in the summer. Male northern fur seals establish territories and breed with large groups of females early in the summer. Generally, northern fur seals are pelagic foragers, feeding on fish in the open ocean. During summer months, most northern fur seals behave like central place foragers and respond to various environmental covariates during their foraging trips. We used distance to rookery, sea surface temperature, and primary productivity as covariates in the model (Figure 7.8). Figure 7.9 shows the inference pertaining to the time-varying coefficients induced by the change-point model in Equation 7.8. For this adult male northern fur seal, the credible intervals for the coefficients indicate that the individual traveled away from the rookery (i.e., the coefficient for the gradient of distance to rookery was positive) during the early part of the trip (up to day 12, approximately), then switched to respond negatively to the gradient (Figure 7.9a). Figure 7.9b shows a similar temporal effect for sea surface

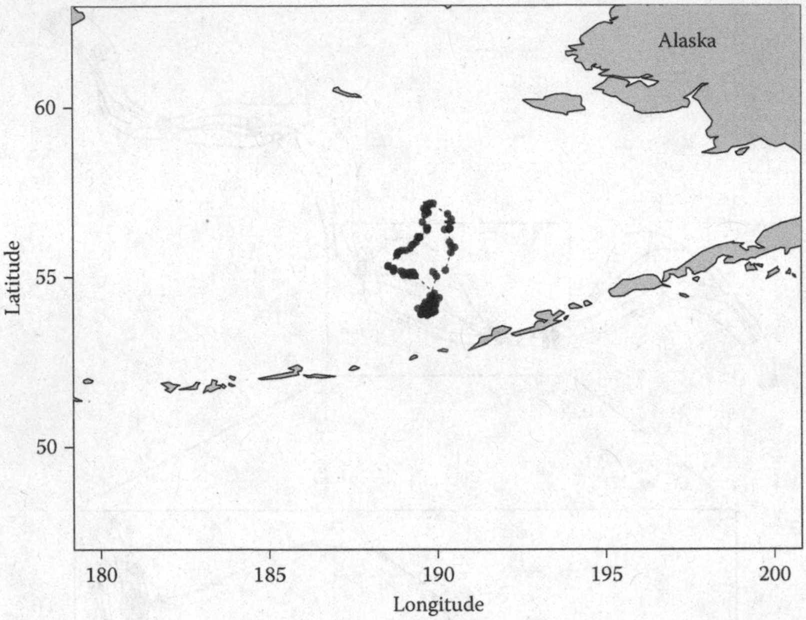

FIGURE 7.7 Observed northern fur seal telemetry data (dark points) and Bering Sea. Alaska shown in gray.

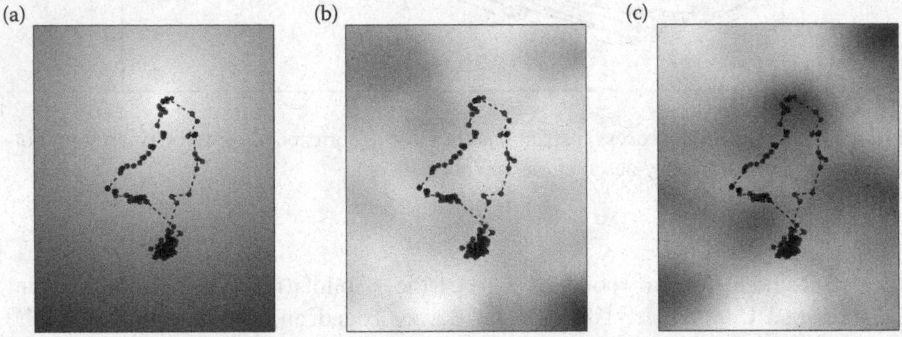

FIGURE 7.8 Bering sea environmental covariates: (a) distance to rookery, (b) sea surface temperature, and (c) primary productivity. Observed telemetry data are shown as points.

temperature, whereas Figure 7.9c indicates a lack of response to primary productivity given the other covariates in the model.

The change-point model also allows us to examine the overall posterior mean gradient field as a function of time as well. Figure 7.10 shows a sequence of eight regularly spaced posterior mean gradient fields resulting from fitting the velocity model in Equations 7.7 and 7.8 to the northern fur seal data. At the beginning of the trip, the gradient field is indicating movement away from the rookery in a transiting behavior (i.e., between 0 and 126 h). However, the individual northern fur seal

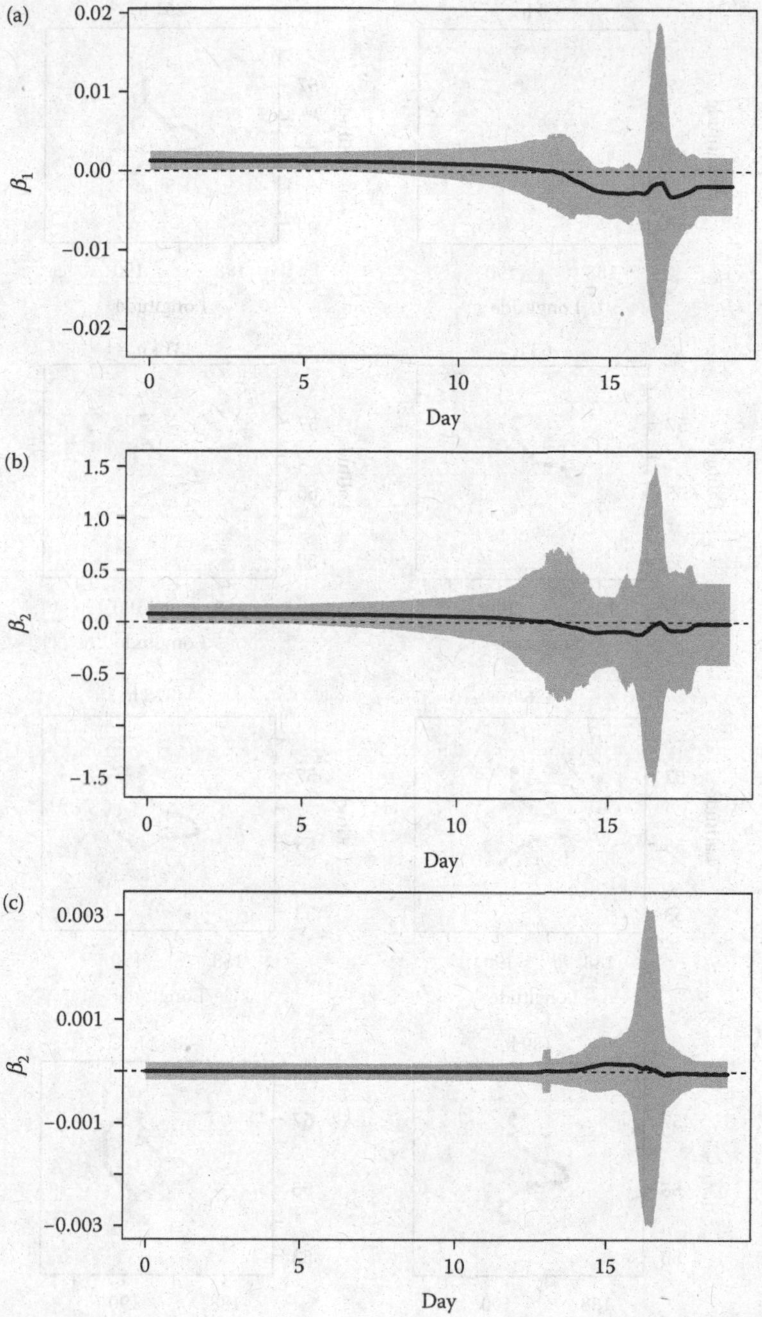

FIGURE 7.9 Posterior 95% credible intervals (gray region) and posterior mean (dark line) for the coefficients associated with covariates: (a) distance to rookery, (b) sea surface temperature, and (c) primary productivity.

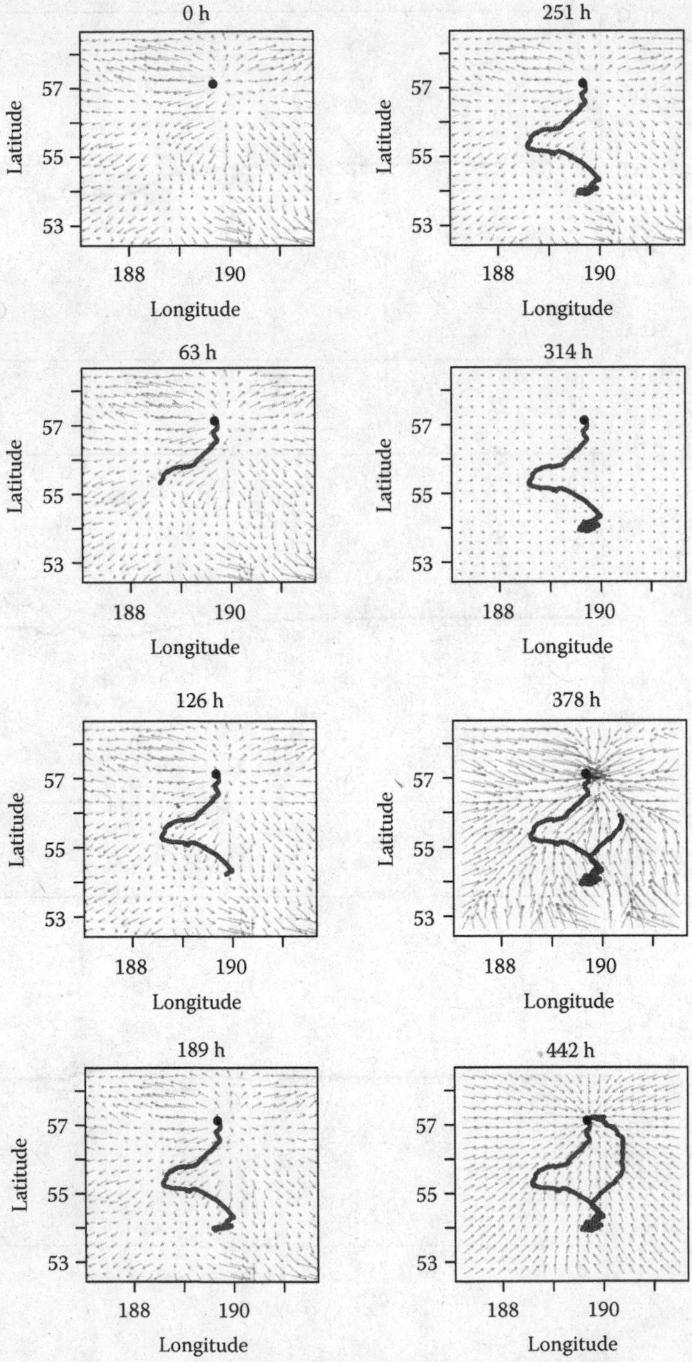

FIGURE 7.10 Posterior mean gradient surface shown as arrows pointing in the direction of largest gradient at a subset of time points during the 18-day foraging trip for the northern fur seal. The position process is shown as a dark line.

changes its behavior between 189 and 314 h, exhibiting more of a foraging pattern (Figure 7.10). Finally, after 378 h, the individual returns to the rookery in a transiting behavior again, indicated by the strong gradient field pointing toward the rookery.

7.4 GENERALIZED MODELS FOR TRANSITIONS IN DISCRETE SPACE

The secondary modeling approaches presented thus far are powerful in that they allow for additional inference that would be difficult to obtain using the correlated random walk models conditioned directly on the original telemetry data. However, both the discrete-space and continuous-space secondary modeling approaches presented by Hooten et al. (2010b) and Hanks et al. (2011) can be computationally intensive for large data sets. Hanks et al. (2015a) extended the methods presented in the previous sections by reparameterizing the discrete-time approach of Hooten et al. (2010b) for faster implementation and more flexibly modeling the relationship between transitions and environmental conditions in discrete areal units.

Hooten et al. (2010b) relied on a transformation of the position process ($\tilde{\mu}_j$) as discrete transitions among grid cells in the geographic space of interest. The variables indicating the transitions ($\mathbf{y}_j$) represent moves to neighboring grid cells (or stays in the same grid cell) during a time period of length Δt. An alternative approach developed by Hanks et al. (2015a) views the process in two parts: (1) transition rates and (2) residence times. There are distinct computational advantages to this approach. Heuristically, the approach of Hanks et al. (2015a) can result in a dimension reduction of the discrete-space model proposed by Hooten et al. (2010b) because, as $\Delta t \rightarrow 0$, the number of time periods that the individual remains in the same cell increases, causing computational problems. However, if we instead model residence time in the grid cell of interest, then there is only a single quantity representing the stays. Thus, Hanks et al. (2015a) developed a combined approach to model residence time and cell transitions using a continuous-time Markov chain formulation. In what follows, we show how this new model arises from the original discrete-space formulation of Hooten et al. (2010b).

Recall that Hooten et al. (2010b) proposed a multinomial model for transitions to neighboring cells (including possible stays in the same cell during period Δt). First, we assume the same five-pixel neighborhood ($l = 1, \ldots, 5$) for each pixel described previously, where pixel $l = 3$ represents the middle pixel, and the probability of a transition to the lth neighboring cell between times t_{j-1} and t_j can be written as $p_{j,l}$. The probability of remaining in the current pixel for period of time τ_j is $p_{j,3}^{\tau_j/\Delta t}$. Alternatively, $p_{j,3}^{\tau_j/\Delta t}$ can be written as $(1 - p_{j,\text{move}})^{\tau_j/\Delta t}$, where $p_{j,\text{move}}$ represents the probability of a move. If we let $p_{j,\text{move}} = \Delta t \cdot \lambda_{j,\text{move}}$ and we decrease the gap between prediction times ($\Delta t \rightarrow 0$), we arrive at an asymptotic result relating to residence time

$$\lim_{\Delta t \rightarrow 0} (1 - p_{j,\text{move}})^{\tau_j/\Delta t} = e^{-\tau_j \lambda_{j,\text{move}}}. \tag{7.9}$$

The move probability ($p_{j,\text{move}}$) in Equation 7.9 can be thought of as a movement rate scaled by a decreasing unit of time ($\Delta t \cdot \lambda_{j,\text{move}}$). Thus, if we properly normalize Equation 7.9 so that it integrates to one, we have

$$\frac{e^{-\tau_j \lambda_{j,\text{move}}}}{\int_0^\infty e^{-\tau \lambda_{j,\text{move}}} d\tau}, \tag{7.10}$$

which results in the model for residence time

$$\tau_j \sim \text{Exp}(\lambda_{j,\text{move}}) \equiv \lambda_{j,\text{move}} e^{-\tau_j \lambda_{j,\text{move}}}. \tag{7.11}$$

Thus, the asymptotic residence time model is exponentially distributed with parameter $\lambda_{j,\text{move}}$.

Returning to the multinomial model for moves, we arrive at a similar asymptotic result for transitions to new pixels. Given that the individual is moving to a new pixel, the probability of it moving to the lth neighboring cell is $p_{j,l}/p_{j,\text{move}}$. As before, if we replace the transition probabilities with the associated rates scaled by Δt and take the limit, we have

$$\begin{aligned} \lim_{\Delta t \to 0} \frac{p_{j,l}}{p_{j,\text{move}}} &= \lim_{\Delta t \to 0} \frac{\Delta t \cdot \lambda_{j,l}}{\Delta t \cdot \lambda_{j,\text{move}}} \\ &= \frac{\lambda_{j,l}}{\lambda_{j,\text{move}}}. \end{aligned} \tag{7.12}$$

The limit is not necessary in Equation 7.12 because the Δt cancels in the numerator and denominator, but we retain it to remain consistent with the derivation for residence time.

We now have a model for residence time (7.11) and for movement (7.12). If we assume conditional independence, a model for the joint process of residence and movement arises as a product of Equations 7.11 and 7.12

$$\frac{\lambda_{j,l}}{\lambda_{j,\text{move}}} \lambda_{j,\text{move}} e^{-\tau_j \lambda_{j,\text{move}}} = \lambda_{j,l} e^{-\tau_j \lambda_{j,\text{move}}}. \tag{7.13}$$

Based on Equation 7.13, Hanks et al. (2015a) noticed that, for all pairs of sequential stays and moves, the resulting likelihood is equivalent to a Poisson regression with a temporally heterogeneous offset. To show this, note that we can always expand the transition rate for the lth neighboring pixel in a product as $\lambda_{j,l} = \prod_{\tilde{l} \neq 3} \lambda_{j,\tilde{l}}^{y_{j,\tilde{l}}}$, where the $y_{j,\tilde{l}}$ are

$$y_{j,\tilde{l}} = \begin{cases} 1 & \text{if } \tilde{l} = l \\ 0 & \text{otherwise} \end{cases}, \tag{7.14}$$

as defined in Hooten et al. (2010b). Also, recall that the overall movement rate is a sum of pixel movement rates $\lambda_{j,\text{move}} = \sum_{\tilde{l} \neq 3} \lambda_{j,\tilde{l}}$. Substituting these quantities into

Equation 7.13 yields

$$\prod_{\tilde{l}\neq 3} \lambda_{j,\tilde{l}}^{y_{j,\tilde{l}}} e^{-\tau_j \lambda_{j,\tilde{l}}}, \tag{7.15}$$

which is proportional to a product of Poisson probability mass functions for the random variables $y_{j,\tilde{l}}$ with offsets τ_j. Thus, for a sequence of stay/move pairs that occur at the subset of prediction times $\mathcal{J}$, we arrive at the likelihood

$$\prod_{j\in\mathcal{J}} \prod_{\tilde{l}\neq 3} \lambda_{j,\tilde{l}}^{y_{j,\tilde{l}}} e^{-\tau_j \lambda_{j,\tilde{l}}}. \tag{7.16}$$

One beneficial consequence of the model developed by Hanks et al. (2015a) is that a reparameterization of the multinomial model of Hooten et al. (2010b) leads to a secondary statistical model that is computationally efficient. There are two reasons for the computational improvement: (1) The original set of prediction times needs to approach infinity, but this model depends only on the total number of moves, which is a function of pixel size, and (2) by using the sufficient statistics $(y_{j,1}, y_{j,2}, y_{j,4}, y_{j,5}, \tau_j)$ for $j \in \mathcal{J}$ of the data structure used by Hooten et al. (2010b), the reparameterized model of Hanks et al. (2015a) is a Poisson GLM and can be fit with any statistical software.*

The last step in setting up a useful model framework is to link the movement rates $\lambda_{j,l}$ with covariates. Thus, consider the standard log-linear regression model $\log(\lambda_{j,l}) = \mathbf{x}'_{j,l}\boldsymbol{\beta}$, where $\mathbf{x}_{j,l}$ are the covariates associated with the lth neighbor of the pixel in which $\tilde{\boldsymbol{\mu}}_{j-1}$ falls, and $\boldsymbol{\beta}$ are the usual regression coefficients to be estimated.

As with any regression model, this one (7.16) can be generalized further to allow for varying coefficients. In the animal movement context, it is sensible to allow for time-varying coefficients, which could account for the individual's residence time and movement probabilities that may change during the period of time for which data are collected. The resulting semiparametric model has the same form as Equation 7.16 but with link function modified so that

$$\begin{aligned} \log(\lambda_{j,l}) &= \mathbf{x}'_{j,l}\boldsymbol{\beta}_j \\ &= \mathbf{x}'_{j,l}\mathbf{W}_j\boldsymbol{\alpha}, \end{aligned} \tag{7.17}$$

where $\mathbf{W}_j$ is a matrix of basis functions indexed in time and $\boldsymbol{\alpha}$ is a new set of coefficients to be estimated, instead of estimating $\boldsymbol{\beta}$ directly. The implementation of this new model (7.17) only requires the creation of a modified set of covariates $\mathbf{x}'_{j,l}\mathbf{W}_j$ and then the estimated coefficients can be recombined with the matrices of basis functions to recover $\boldsymbol{\beta}_j = \mathbf{W}_j\boldsymbol{\alpha}$ after the model has been fit to data. This procedure allows us to view the $\boldsymbol{\beta}_j$ as they vary over time. For example, based on a telemetry

* Notice that $y_{j,3}$ is missing from the list of sufficient statistics because it originally represented a stay, but now stays are represented by τ_j and moves are represented by the remaining multinomial zeros and ones $(y_{j,1}, y_{j,2}, y_{j,4}, y_{j,5})$.

data set spanning an entire year, we can obtain explicit statistical inference to assess whether residence time is influenced more by forest cover in the winter or summer. The choice of basis functions, $\mathbf{W}_j$, should match the goals of the study, and various forms of regularization or model selection can be used to assess which coefficients in $\boldsymbol{\alpha}$ are helpful for prediction. By shrinking $\boldsymbol{\alpha}$ toward zero with a penalized likelihood approach or a Bayesian prior, one can essentially identify the optimal level of smoothness in the $\boldsymbol{\beta}_j$ over time. We would expect smoother $\boldsymbol{\beta}_j$ over time in cases with limited data.

Hanks et al. (2015a) examined various approaches for regularization (Hooten and Hobbs 2015) of the parameters $\boldsymbol{\alpha}$ and made a strong case for the use of a lasso penalty (based on an L_1 norm). Regularization can be used in Bayesian and non-Bayesian contexts and the amount of shrinkage can be chosen via cross-validation. Hanks et al. (2015a) employed both approaches to multiple imputation described earlier (i.e., approximate and fully Bayesian) and found strong agreement among inference using as little as 50 imputation samples for $\tilde{\boldsymbol{\mu}}$.

We fit the reparameterized continuous-time discrete-space model developed by Hanks et al. (2015a) to a subset of GPS telemetry data arising from an individual female mountain lion (*Puma concolor*) in Colorado, USA. Based on the covariates in Figure 7.11, we used the forest versus non-forest covariate (Figure 7.11a) for a "static driver" of movement and the distance to potential kill site (Figure 7.11b) as a "dynamic driver" of movement.

Using a semiparametric specification as in Equation 7.17 with hour of day represented in the basis function ($\mathbf{W}_j$), we fit the movement model to the data from the adult female mountain lion. Figure 7.12 shows the inference obtained for the effects of forest versus non-forest and distance to nearest kill site as a function of time of day (in hours). The results in Figure 7.12a suggest a lack of evidence for an effect of forest presence on the individual mountain lion. However, Figure 7.12b provides some evidence that distance to nearest kill site temporally affects the potential function that could influence movement. We also fit a temporally homogeneous Poisson GLM to the same data and found strong evidence for an effect of distance to nearest kill site on the potential function ($p < 0.001$).

7.5 CONNECTIONS WITH POINT PROCESS MODELS

7.5.1 Continuous-Time Models

Despite the increase in sophistication and accessibility of dynamic animal movement models (in both discrete- and continuous-time), the most popular approach to analyze telemetry data is using resource selection functions (RSFs) based on point process models. However, as discussed in Chapter 4, it is absolutely critical to account for movement in point process models when the telemetry data are temporally close together relative to the movement dynamics. Thus, the point process models of Johnson et al. (2008b), Forester et al. (2009), and Brost et al. (2015) represent rigorous approaches that account for inherent autocorrelation in the telemetry data beyond resource selection. These approaches all incorporate an animal movement mechanism in the point process modeling framework in the form of an availability distribution

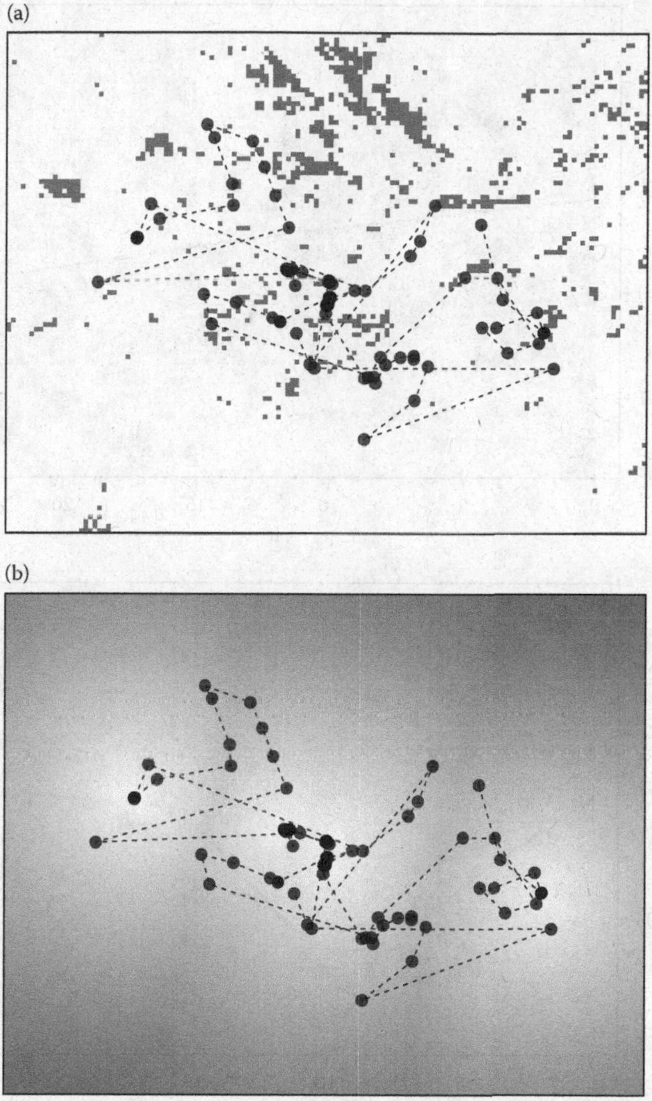

FIGURE 7.11 GPS telemetry data (points connected in sequence by dashed lines) for an adult female mountain lion (*Puma concolor*) in Colorado and two spatial covariates: (a) presence of non-forest (dark) versus forest (light) and (b) distance to nearest potential kill site (dark is far, light is near).

$f(\boldsymbol{\mu}_i|\boldsymbol{\mu}_{i-1}, \boldsymbol{\theta})$ such that the model for the position process at time t_i is

$$[\boldsymbol{\mu}_i|\boldsymbol{\mu}_{i-1}, \boldsymbol{\beta}, \boldsymbol{\theta}] \equiv \frac{g(\mathbf{x}(\boldsymbol{\mu}_i), \boldsymbol{\beta})f(\boldsymbol{\mu}_i|\boldsymbol{\mu}_{i-1}, \boldsymbol{\theta})}{\int g(\mathbf{x}(\boldsymbol{\mu}), \boldsymbol{\beta})f(\boldsymbol{\mu}|\boldsymbol{\mu}_{i-1}, \boldsymbol{\theta})d\boldsymbol{\mu}}. \tag{7.18}$$

In Chapter 4, we mentioned that these types of point process models are sometimes referred to as step selection functions (Fortin et al. 2005; Avgar et al. 2016).

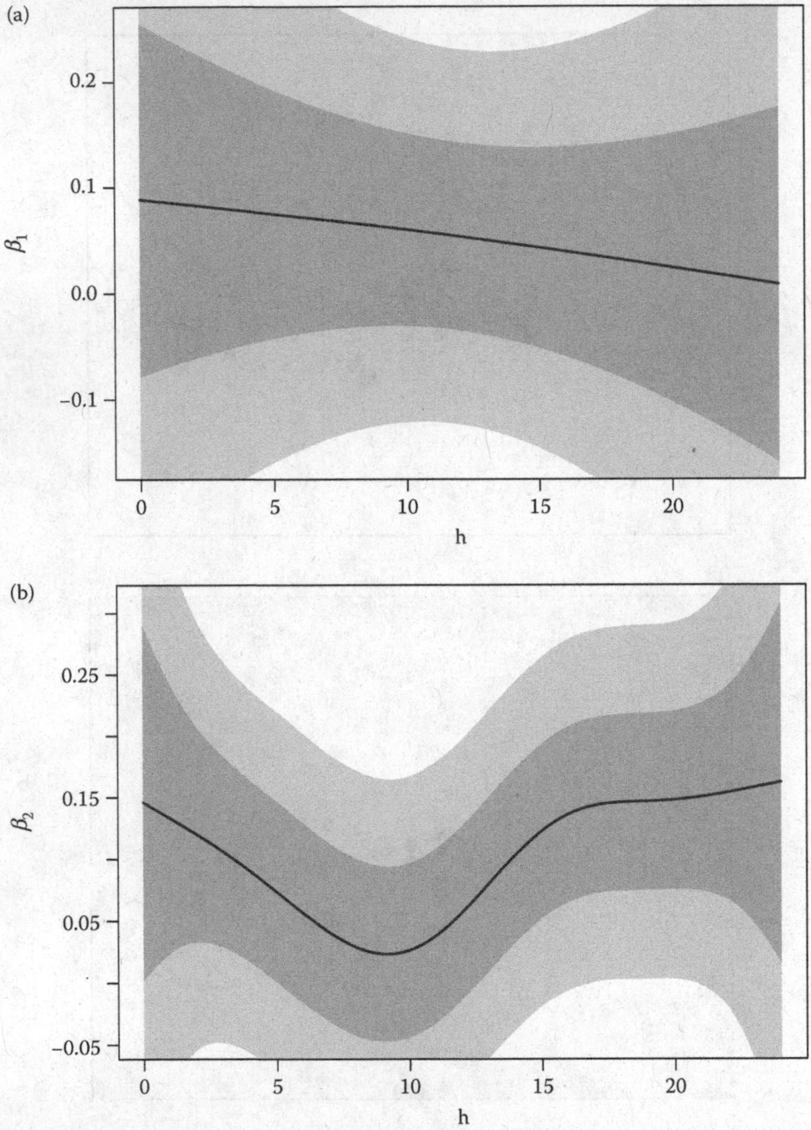

FIGURE 7.12 Inference for $\boldsymbol{\beta}$ resulting from fitting the reparameterized continuous-time discrete-space model to an adult female mountain lion (*Puma concolor*) in Colorado using the two spatial covariates: (a) presence of non-forest versus forest and (b) distance to nearest potential kill site. Light shading represents a 95% confidence interval for the temporally varying coefficient and dark shading represents a 67% confidence interval. The temporally varying point estimate is shown as the dark line.

These methods definitely account for temporal scale while providing resource selection inference, but they do not allow you to choose the scale for inference. To put the choice of scale back in the hands of the analyst, Hooten et al. (2014) developed an approach for combining continuous-time movement models with resource selection functions. Their approach relied on the OU models of Johnson et al. (2008a) to characterize the use and availability distributions (i.e., $[\boldsymbol{\mu}_i|\boldsymbol{\mu}_{i-1}, \boldsymbol{\beta}, \boldsymbol{\theta}]$ and $f(\boldsymbol{\mu}_i|\boldsymbol{\mu}_{i-1}, \boldsymbol{\theta})$ from Equation 7.18). They reconciled the two distributions to obtain resource selection inference (i.e., inference for $\boldsymbol{\beta}$).

To characterize use and availability, Hooten et al. (2014) proposed to use the smoother and predictor distributions resulting from a hierarchical model for the true position process $\boldsymbol{\mu}(t)$ (Figure 7.13). As we discussed in Chapter 3, the Kalman filter, smoother, and predictor distributions all pertain to our understanding of the latent temporal process. These distributions are useful for estimating state variables in hierarchical time series models and are often paired with maximum likelihood or EM

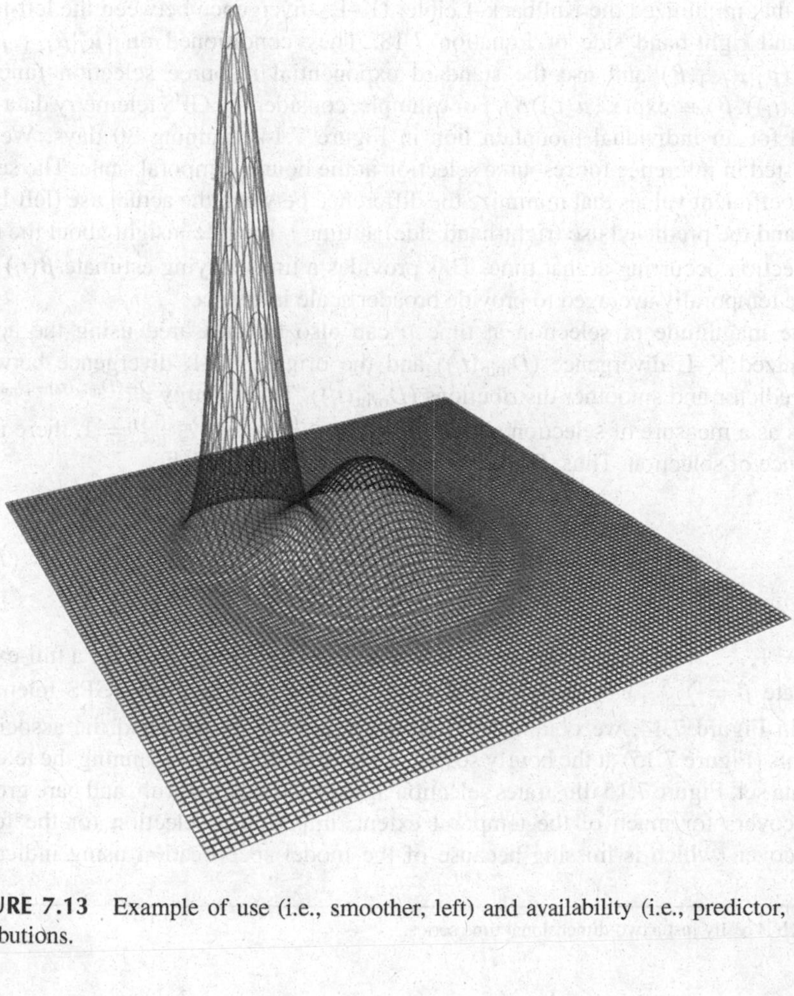

FIGURE 7.13 Example of use (i.e., smoother, left) and availability (i.e., predictor, right) distributions.

algorithms to fit non-Bayesian models. In the animal movement context,* the predictor distribution is the distribution of $\boldsymbol{\mu}(t_i)$ given everything up to, but not including, time t_i. The filter is the distribution of $\boldsymbol{\mu}(t_i)$ given everything up to and including time t_i. Finally, the smoother distribution is the distribution of $\boldsymbol{\mu}(t_i)$ given everything before and after time t_i. Recall, from Chapter 3, that the predictor distribution is the most diffuse, with the filter and smoother each more precise. In fact, the smoother distribution is our best estimate of $\boldsymbol{\mu}(t_i)$ using all information about the individual's path. The predictor distribution tells us about the likely location of the individual given only past movement. Thus, the predictor serves as a good estimator of availability, informing us about where the individual is likely to be based on previous movement alone. By contrast, the smoother serves as a good estimator for actual space use.

Hooten et al. (2014) define $[\boldsymbol{\mu}_i|\boldsymbol{\mu}_{i-1}, \boldsymbol{\beta}, \boldsymbol{\theta}]$ from Equation 7.18 as the smoother distribution, and $f(\boldsymbol{\mu}_i|\boldsymbol{\mu}_{i-1}, \boldsymbol{\theta})$ as the predictor distribution. Because Kalman methods are used to implement the CTCRW model of Johnson et al. (2008a), the smoother and predictor distributions can be easily obtained using the "crawl" R package. To estimate the selection coefficients $\boldsymbol{\beta}$, Hooten et al. (2014) used the point estimate for $\boldsymbol{\beta}$ that minimized the Kullback–Leibler (K–L) divergence between the left-hand side and right-hand side of Equation 7.18. They conditioned on $[\boldsymbol{\mu}_i|\boldsymbol{\mu}_{i-1}, \boldsymbol{\beta}, \boldsymbol{\theta}]$ and $f(\boldsymbol{\mu}_i|\boldsymbol{\mu}_{i-1}, \boldsymbol{\theta})$ and use the standard exponential resource selection function $g(\mathbf{x}(\boldsymbol{\mu}(t_i)), \boldsymbol{\beta}) \equiv \exp(\mathbf{x}'(\boldsymbol{\mu}(t_i))\boldsymbol{\beta})$. For example, consider the GPS telemetry data collected for an individual mountain lion in Figure 7.14 spanning 30 days. We are interested in inference for resource selection at the hourly temporal scale. The selection coefficient values that minimize the difference between the actual use (left-hand side) and the predicted use (right-hand side) at time t_i provide insight about the type of selection occurring at that time. This provides a time-varying estimate $\hat{\boldsymbol{\beta}}(t_i)$ that can be temporally averaged to provide broader scale inference.

The magnitude of selection at time t_i can also be measured using the actual minimized K–L divergence ($D_{\text{min}}(t_i)$) and the original K–L divergence between the predictor and smoother distributions ($D_{\text{orig}}(t_i)$). The quantity $e^{-(D_{\text{orig}}(t_i)-D_{\text{min}}(t_i))}$ serves as a measure of selection at time t_i; when $e^{-(D_{\text{orig}}(t_i)-D_{\text{min}}(t_i))} = 1$, there is no evidence of selection. Thus, Hooten et al. (2014) used the weights

$$w(t_i) = \frac{1 - e^{-(D_{\text{orig}}(t_i)-D_{\text{min}}(t_i))}}{\sum_{i=1}^{n} 1 - e^{-(D_{\text{orig}}(t_i)-D_{\text{min}}(t_i))}}, \tag{7.19}$$

for $i = 1, \ldots, n$ to average the selection coefficients over time to obtain a full-extent estimate $\bar{\boldsymbol{\beta}} = \sum_{i=1}^{n} w(t_i)\hat{\boldsymbol{\beta}}(t_i)$ for selection. For the mountain lion GPS telemetry data in Figure 7.14, we obtained the optimal coefficients $\hat{\boldsymbol{\beta}}(t)$ and the associated weights (Figure 7.15) at the hourly scale for a period of 30 days spanning the telemetry data set. Figure 7.15 illustrates selection against the urban, shrub, and bare ground land covers for much of the temporal extent, implying a selection for the forest land cover (which is missing because of the model specification using indicators

* Which is really just a two-dimensional time series.

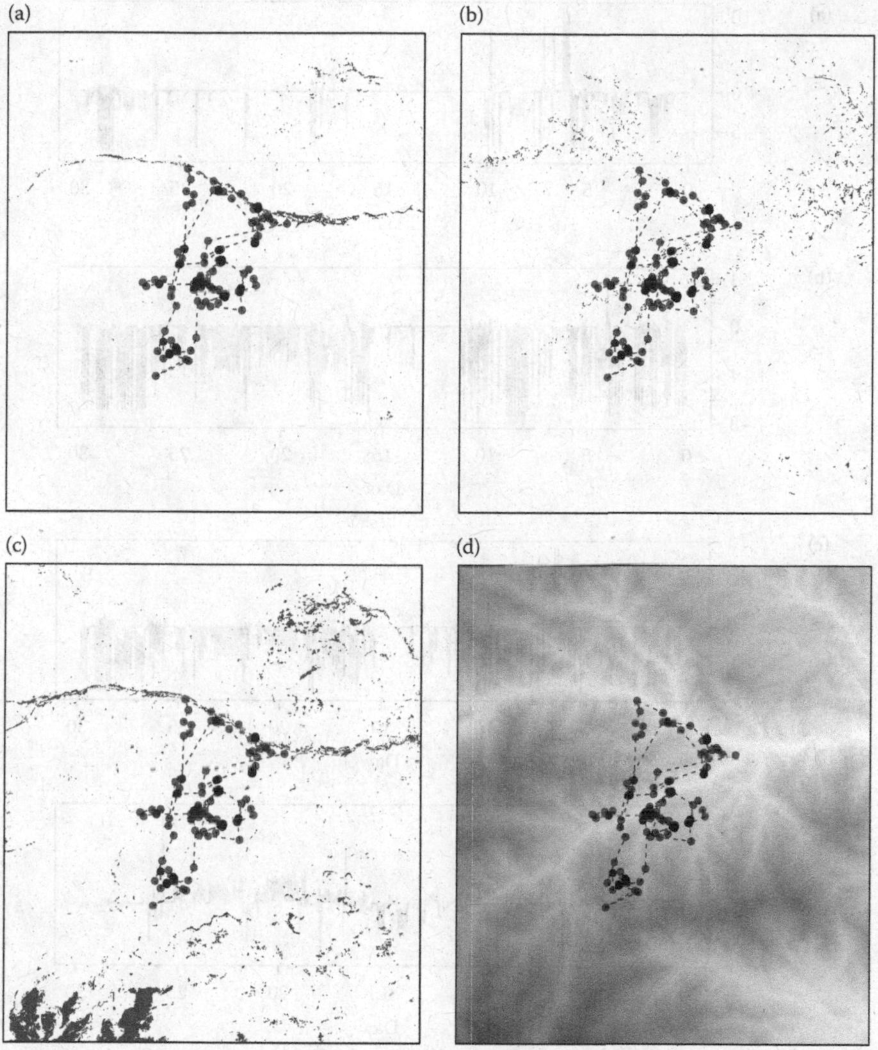

FIGURE 7.14 GPS telemetry data (dark points connected by dashed lines) and spatial covariates (background image) for resource selection inference: (a) Urban, (b) shrub, (c) bare ground, and (d) elevation.

for the other covariates). Furthermore, the selection for higher elevations tends to vary throughout the month-long period, but is somewhat temporally clustered (e.g., approximately day 20). Finally, the weights $w(t)$ shown in Figure 7.15e indicate that certain periods of time (e.g., day 20–25) are near zero, indicating a lack of evidence for selection during that period. The temporally averaged coefficients were $\bar{\boldsymbol{\beta}} = (-1.64, -3.51, -3.04, -0.01)'$. The fact that the averaged coefficient for elevation ($\bar{\beta}_4$) is close to zero suggests that, over the period of a month, elevation is not consistently selected for or against.

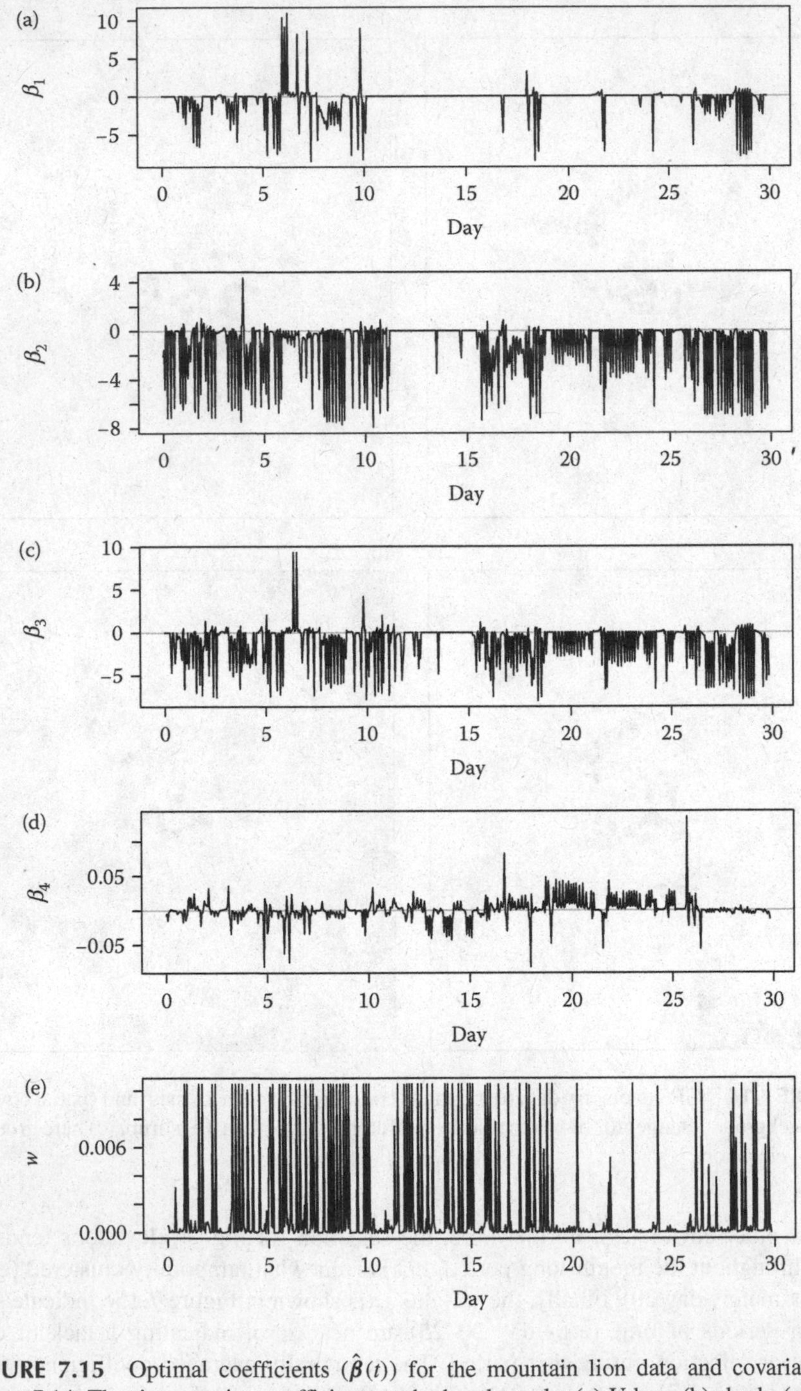

FIGURE 7.15 Optimal coefficients ($\hat{\boldsymbol{\beta}}(t)$) for the mountain lion data and covariates in Figure 7.14. The time-varying coefficients at the hourly scale: (a) Urban, (b) shrub, (c) bare ground, and (d) elevation. The time-varying weights, calculated using Equation 7.19 are shown in panel (e).

While this approach is not considered to be fully model-based, it does rely on the point process model formulation and the continuous-time movement modeling approaches to estimate use and availability distributions (i.e., smoother and predictor). It also provides a way to control the scale of inference directly through the predictor distribution. If the prediction is farther ahead in time, the predictor distribution widens. This yields different inference because the availability for the animal changes as the time interval increases. Thus, the analyst can quickly examine the temporally averaged selection coefficients $\bar{\boldsymbol{\beta}}$ for a range of time scales to better understand how the individual might be responding at different temporal scales.

7.5.2 Discrete-Time Models

In Chapter 4, we discussed how point process models are useful for learning about space use, availability, and resource selection by relating animal locations to their environment. In Chapter 5, we demonstrated several discrete-time multistate movement models that can be used to identify different movement behaviors. For certain applications, it may be desirable to combine these two approaches. Suppose one were to fit a resource selection model to the elk data from Morales et al. (2004) (Section 5.2.1) using the different habitat types (e.g., open habitat, dense deciduous forest) as covariates. This could be problematic because the elk seem to exhibit distinct modes of movement behavior (Figure 5.9), "encamped" (which could include resting or foraging within a patch) and "exploratory" (for movement between patches), which have different time allocations (i.e., activity budgets) and are inherently likely to have different relationships with habitat type. For example, if one were interested in grazing habitat selection, it may be unwise to include resting locations in the response variable for the resource selection model. In the elk data, it would be difficult to identify and remove resting locations without auxiliary information (e.g., from head tilt sensors).

Sometimes it is straightforward to remove locations related to particular behaviors prior to fitting point process models. In their northern fur seal example, Johnson et al. (2013) limited their resource selection analysis to foraging trip locations with reasonable confidence by excluding periods where their biotelemetry wet/dry sensors indicated the animals were hauled out on land. When particular behaviors cannot be identified and excluded easily, then multistate movement models such as those described in Chapter 5 might be required. As with continuous-time movement models, multiple imputation can be used to fit point process models to the output from discrete-time multistate movement models. We provide a brief example below based on recent work by Cameron et al. (2016), where posterior samples from a discrete-time multistate movement model were utilized to better understand the role of resource selection in bearded seal foraging ecology.

McClintock et al. (2016) extended the approach of McClintock et al. (2013) and McClintock et al. (2015) to identify six movement behavior states (hauled out on land, hauled out on ice, resting at sea, mid-water foraging, benthic foraging, and transit) for seven bearded seals captured and deployed with Argos tags off the coast of Alaska, USA. They accomplished this by combining biotelemetry data (location, time-at-depth, number of dives to depth, dry time) and environmental data (sea ice cover concentration, sea floor depth) in a Bayesian model that was fit using MCMC. As part of an interdisciplinary collaboration on bearded seal ecology, Cameron et al. (2016) were able to examine benthic foraging resource selection by drawing from the posterior output of McClintock et al. (2016) and using multiple imputation to account for location and state assignment uncertainty.

For their analysis, Cameron et al. (2016) synthesized trawl survey data for dozens of benthic taxa sampled along the Chukchi corridor of Alaska, including most of those known to be prey species of bearded seals based on stomach content data (e.g., bivalves). Using prey species biomass, sediment type, and sea floor depth as predictors partitioned into a fine set of grid cells, Cameron et al. (2016) fit a resource selection model similar to the space-only Poisson point process model described by Johnson et al. (2013) to each draw from the posterior output of McClintock et al. (2016) using the readily available R package "INLA" (Lindgren and Rue 2015). This is the exact same model as (4.51), but, in this case, the response variable, $y_l^{(k)}$, is the number of locations $\boldsymbol{\mu}^{(k)}$ in grid cell l that were assigned to the benthic foraging state for the kth draw from the posterior. Because benthic species distribution

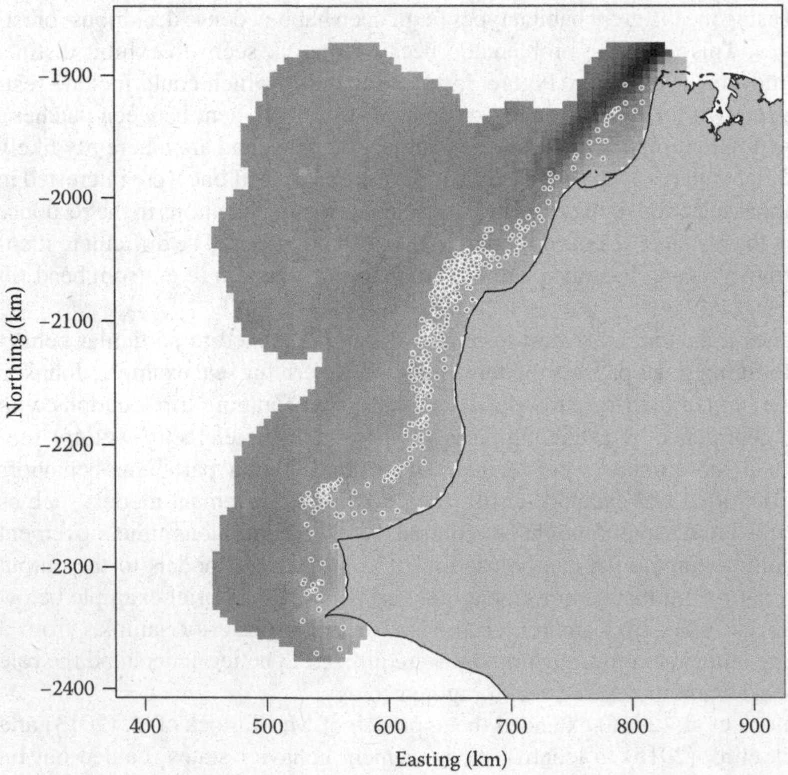

FIGURE 7.16 Bearded seal benthic foraging locations identified within the Chukchi corridor study area near Alaska, USA, from a discrete-time multistate movement model. Study area grid is shaded by sea floor depth, where darker shades indicate deeper waters.

and community composition are interrelated and the product of complicated ecological relationships that are spatially correlated, Cameron et al. (2016) dealt with the problem of multicollinearity by using principal component regression techniques and singular value decomposition of the (standardized) design matrix $\mathbf{X} = \mathbf{UDV}'$, where the columns of $\mathbf{U}$ (i.e., the left singular vectors) form an orthonormal basis that were used in lieu of $\mathbf{X}$ in Equation 4.51. After model fitting, the regression coefficients can be easily back-transformed for inference on the original scale of the predictors.

For demonstration, we replicated the analysis of Cameron et al. (2016) for a single bearded seal (Figure 7.16). We fit (4.51) to $K = 4000$ posterior samples of the locations and state assignments from the output of the multistate model fit by McClintock et al. (2016). This particular seal exhibited several benthic foraging resource selection "hotspots" off the coast of Alaska along the Chukchi corridor (Figure 7.17). Bivalves, sculpins (family Cottidae), sea urchins (class Echinoidea), and shrimp (infraorder Caridea) represented a subset of the prey taxa that exhibited positive selection coefficients for this particular seal (Figure 7.18).

While it is theoretically possible to explicitly incorporate dozens of environmental covariates related to specific behaviors into the multistate movement models described in Chapter 5, it is more computationally efficient to perform a two-stage analysis using multiple imputation as in Cameron et al. (2016). While multiple imputation allowed Cameron et al. (2016) to account for location and state assignment uncertainty, a disadvantage of their two-stage approach is that the prey biomass data were not used to inform the movement process itself (and hence the estimated locations of benthic foraging activity). This could be particularly important when trying

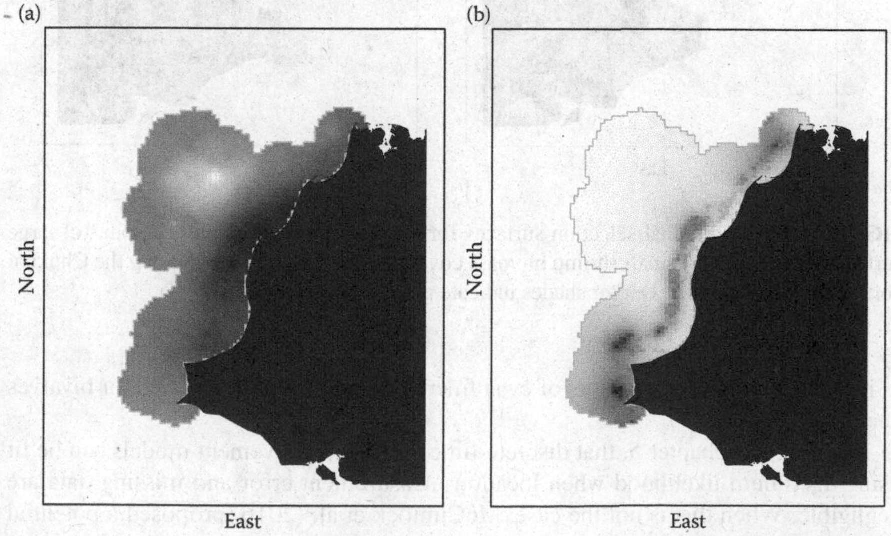

FIGURE 7.17 Overall fitted (a) selection and (b) availability surfaces for a bearded seal along the Chukchi corridor of Alaska, USA. Selection covariates included sea floor depth, sediment type, and dozens of benthic taxa (e.g., bivalves, fish). Darker shades indicate greater intensity.

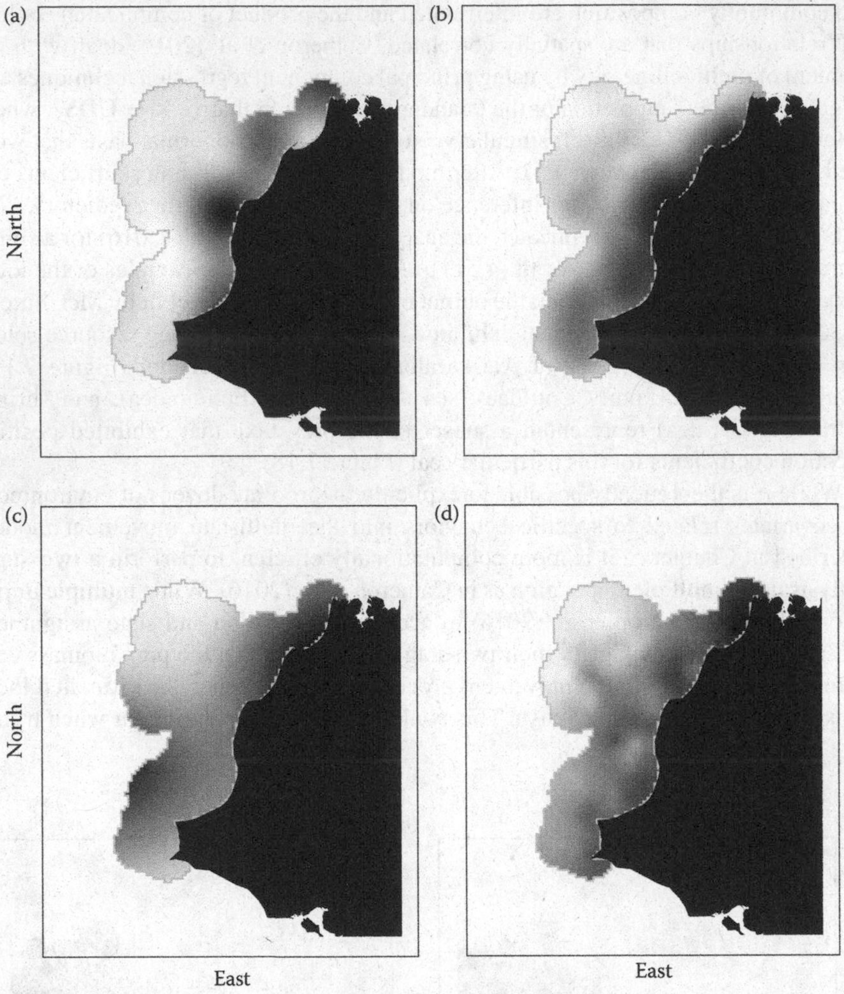

FIGURE 7.18 Individual selection surfaces for (a) sea urchin, (b) small sculpin, (c) large surface bivalve, and (d) small shrimp biomass covariates for a bearded seal along the Chukchi corridor of Alaska, USA. Darker shades indicate greater intensity.

to identify movement behaviors of even finer detail (e.g., foraging dives for bivalves versus cod).

Recall, from Chapter 5, that discrete-time multistate movement models can be fit using maximum likelihood when location measurement error and missing data are negligible. When this is not the case, McClintock et al. (2016) proposed a potential alternative using multiple imputation; instead of using computationally demanding MCMC methods to fit discrete-time multistate movement models that account for location measurement error (e.g., Jonsen et al. 2005; McClintock et al. 2012), realizations of the movement path obtained from "crawl" can be used as the data for hidden Markov models such as those implemented in the R package "moveHMM" (Michelot et al. 2015). Because both steps can be performed using maximum likelihood, this is a fast and easy method for practitioners to obtain behavior state assignments from a movement model that accounts for both location and state assignment uncertainty. More work is needed to assess the potential advantages and disadvantages of this approach, but it remains a promising avenue for further research.

7.6 ADDITIONAL READING

Secondary models can be useful in situations where it is not computationally feasible to fit a larger hierarchical model. Thus, a number of secondary models described in the literature utilize output (i.e., estimated quantities) from an initial model fit to data. However, many of them are considered *ad hoc* because they do not allow for a propagation of uncertainty from the first model into the secondary inference. As an example, the RUF models described in Section 4.3 involve secondary modeling because a KDE is used to first estimate the UD, which is then used in a geostatistical model to provide inference for resource selection. Multiple imputation facilitates the propagation of uncertainty but will be conservative (i.e., propagate excessive uncertainty) if the imputation distribution does not approximate the true distribution of the response variable being used in the secondary model well.

Another useful type of secondary modeling approach for obtaining population-level inference based on a set of individual-based model output was described by Hooten et al. (2016). For certain hierarchical animal movement model specifications, Hooten et al. (2016) showed that a meta-analytic computing strategy (Lunn et al. 2013) can be used to obtain exact inference about population-level parameters by resampling individual-level parameters in a secondary MCMC. This two-stage approach can be advantageous in big data situations where the likelihood is time consuming to compute because the first-stage individual-level models can be fit in parallel.

Glossary

Activity budget: Proportion of time an individual spends performing activities.

Autocorrelation: Dependence among random variables, often used in a spatial or temporal context.

Basis vector: Discrete version of basis function; vectors in a matrix (resembling a design matrix) spanning the space of functions in a model.

Bayesian model: A statistical model that treats all unobserved variables as random and is specified using conditional probability.

Behavioral state (or mode): An activity that is sustained for a period of time.

Brownian motion: Continuous-time stochastic process that is a sum of white noise (mean zero and constant variance continuous, random numbers). Also called a Weiner process.

Deviance information criterion (DIC): A function of data, model, and parameters used to score different models for Bayesian model selection.

Diffusion: Spreading at the population level. Typically described by a mathematical model.

Eulerian: Perspective of movement that is focused on space, involving densities of individuals (i.e., large-scale).

Fix: The acquisition of a telemetry observation.

Fixed effect: Parameters in a statistical model that do not arise from a distribution with unknown parameters.

Frequentist: Nonparametric statistical inference concerned with the estimation of fixed, but unknown, population characteristics.

Gaussian: Most commonly referring to a probability distribution for a continuous random variable that is mound shaped, characterized by a mean and variance parameter (i.e., a normal distribution).

Hidden Markov model: A latent process component, in a broader hierarchical model, that has Markovian dynamics (if temporal).

Hierarchical model: A joint probability model for many variables that is often specified as a sequence of simpler conditional distributions. Most commonly has three levels: data, process, and parameters (Berliner 1996).

Integral equation: An equation for a continuous process involving an integral.

Integro-difference equation: A discrete-time equation that uses an integral (often in the form of a convolution) to get from one time to the next.

Kernel: A function in a space of interest (e.g., time or geographic space) with a finite integral, usually having mass concentrated in some region of the space.

Lagrangian: Perspective of movement that is focused on the individual (i.e., individual- or agent-based; small-scale).

Likelihood: A function that describes the shape and position of the probability or density of the response variable given the parameters.

Moment: A characteristic of a random variable or data set that is calculated by integration of the probability distribution or summation of the data set.

Markov: A random variable that is dependent on the rest of a process only through its neighbors.

Markov process: A set of random variables that depend on each other only through their neighbors.

Mixed model: A statistical model that contains both fixed and random effects.

Monte Carlo: Obtaining realizations of random variables by drawing them from a probability distribution.

Multistate model: In animal movement ecology, a clustering model allowing for the data or process to arise from a discrete set of probability regimes.

Nonparametric: A statistical model that does not fully specify a specific function as a probability distribution for a random variable.

Norm: A distance function (not necessarily Euclidean). For example, $|\mathbf{a}-\mathbf{b}|$ is the L_1 norm between vectors **a** and **b** (i.e., Manhattan distance).

OU process (Ornstein–Uhlenbeck): A Brownian motion process that has attraction to a point.

Parametric: A statistical model that involves a specific probability distribution for the random variable whose functional form depends on a set of parameters that are often unknown.

Point process: A stochastic process where the positions of the events are the random quantity of interest. In movement ecology, the events are typically either the observed or true locations of the individual.

Posterior distribution: Probability distribution of parameters given observed data.

Posterior predictive distribution: Probability distribution of future data given the observed data.

Precision: The inverse of variance (e.g., $1/\sigma^2$, or $\boldsymbol{\Sigma}^{-1}$ if $\boldsymbol{\Sigma}$ is a covariance matrix).

Prior distribution: A probability distribution useful in Bayesian modeling containing known information about the model parameters before the current data are analyzed.

Probability density (or mass) function, PDF or PMF: A function expressing the stochastic nature of a continuous or discrete random variable (usually denoted as $f(y)$ or $[y]$ for random variable y).

Random effect: Parameters in a statistical model that are allowed to arise from a distribution with unknown parameters.

Random field: A continuous stochastic process over space or time that is usually correlated in some way.

Random walk: A dynamic temporal stochastic process that is not necessarily to a central location.

Redistribution kernel: A function that describes the probability of moving from one location to another in a period of time.

Seasonality: Periodicity in temporal processes, a commonly used term in time series.

Singular value decomposition: The decomposition of a matrix ($\mathbf{X}$) into a product of three matrices (i.e., $\mathbf{X}=\mathbf{UDV}'$), the left singular vectors $\mathbf{U}$, a diagonal matrix with singular values on the diagonal, and the right singular vectors $\mathbf{V}$.

Spectral decomposition: An Eigen decomposition of a matrix (e.g., $\boldsymbol{\Sigma}=\mathbf{Q}\boldsymbol{\Lambda}\mathbf{Q}'$).

Stationary process: A process with covariance structure that does not vary with location (in space or time).

State-space model: See "hierarchical model."

Support: The set of values that a random variable can assume (sometimes referred to as the sample space).

Type 1 error: Incorrectly rejecting a hypothesis when it is actually true (a "false positive").

Type 2 error: Failing to reject a hypothesis when it is actually false (a "false negative").

References

Aarts, G., J. Fieberg, and J. Matthiopoulos. 2012. Comparative interpretation of count, presence-absence, and point methods for species distribution models. *Methods in Ecology and Evolution*, 3:177–187.

Albert, J. and S. Chib. 1993. Bayesian analysis of binary and polychotomous response data. *Journal of the American Statistical Association*, 88:669–679.

Altman, R. 2007. Mixed hidden Markov models: An extension of the hidden Markov model to the longitudinal data setting. *Journal of the American Statistical Association*, 102:201–210.

Andow, D., P. Kareiva, S. Levin, and A. Okubo. 1990. Spread of invading organisms. *Landscape Ecology*, 4:177–188.

Arthur, S., B. Manly, L. McDonald, and G. Garner. 1996. Assessing habitat selection when availability changes. *Ecology*, 77:215–227.

Austin, D., W. Bowen, J. McMillan, and D. Boness. 2006. Stomach temperature telemetry reveals temporal patterns of foraging success in a free-ranging marine mammal. *Journal of Animal Ecology*, 75:408–420.

Avgar, T., J. Baker, G. Brown, J. Hagens, A. Kittle, E. Mallon, M. McGreer et al. 2015. Space-use behaviour of woodland caribou based on a cognitive movement model. *Journal of Animal Ecology*, 84:1059–1070.

Avgar, T., R. Deardon, and J. Fryxell. 2013. An empirically parameterized individual based model of animal movement, perception, and memory. *Ecological Modelling*, 251:158–172.

Avgar, T., J. Potts, M. Lewis, and M. Boyce. 2016. Integrated step selection analysis: Bridging the gap between resource selection and animal movement. *Methods in Ecology and Evolution*, 7:619–630.

Baddeley, A., E. Rubak, and R. Turner. 2016. *Spatial Point Patterns: Methodology and Applications with R*. Chapman & Hall/CRC, Boca Raton, Florida, USA.

Baddeley, A. and R. Turner. 2000. Practical maximum pseudolikelihood for spatial point patterns. *Australian & New Zealand Journal of Statistics*, 42:283–322.

Banerjee, S., B. P. Carlin, and A. E. Gelfand. 2014. *Hierarchical Modeling and Analysis for Spatial Data*. CRC Press, Boca Raton, Florida, USA.

Banerjee, S., A. Gelfand, A. Finley, and H. Sang. 2008. Gaussian predictive process models for large spatial datasets. *Journal of the Royal Statistical Society, Series B*, 70:825–848.

Barnett, A. and P. Moorcroft. 2008. Analytic steady-state space use patterns and rapid computations in mechanistic home range analysis. *Journal of Mathematical Biology*, 57:139–159.

Barry, R. and J. Ver Hoef. 1996. Blackbox kriging: Spatial prediction without specifying variogram models. *Journal of Agricultural, Biological and Environmental Statistics*, 1:297–322.

Berliner, L. 1996. Hierarchical Bayesian time series models. In Hanson, K. and R. Silver, editors, *Maximum Entropy and Bayesian Methods*, pages 15–22. Kluwer Academic Publishers, Dordrecht, The Netherlands.

Berman, M. and T. Turner. 1992. Approximating point process likelihoods with GLIM. *Applied Statistics*, 41(1):31–38.

Besag, J. 1974. Spatial interaction and the statistical analysis of lattice systems. *Journal of the Royal Statistical Society, Series B*, 36:192–225.

Beyer, H., J. Morales, D. Murray, and M.-J. Fortin. 2013. Estimating behavioural states from movement paths using Bayesian state-space models: A proof of concept. *Methods in Ecology and Evolution*, 4:433–441.

Bidder, O., J. Walker, M. Jones, M. Holton, P. Urge, D. Scantlebury, N. Marks, E. Magowan, I. Maguire, and R. Wilson. 2015. Step by step: Reconstruction of terrestrial animal movement paths by dead-reckoning. *Movement Ecology*, 3:1–16.

Biuw, M., B. McConnell, C. Bradshaw, H. Burton, and M. Fedak. 2003. Blubber and buoyancy: Monitoring the body condition of free-ranging seals using simple dive characteristics. *Journal of Experimental Biology*, 206:3405–3423.

Blackwell, P. 1997. Random diffusion models for animal movement. *Ecological Modelling*, 100:87–102.

Blackwell, P. 2003. Bayesian inference for Markov processes with diffusion and discrete components. *Biometrika*, 90:613–627.

Blackwell, P., M. Niu, M. Lambert, and S. LaPoint. 2015. Exact Bayesian inference for animal movement in continuous time. *Methods in Ecology and Evolution*, 7:184–195.

Boersma, P. and G. Rebstock. 2009. Foraging distance affects reproductive success in Magellanic penguins. *Marine Ecology Progress Series*, 375:263–275.

Bolker, B. M. 2008. *Ecological Models and Data in R*. Princeton University Press, Princeton, New Jersey, USA.

Borger, L., B. Dalziel, and J. Fryxell. 2008. Are there general mechanisms of animal home range behaviour? A review and prospects for future research. *Ecology Letters*, 11:637–650.

Bowler, D. and T. Benton. 2005. Causes and consequences of animal dispersal strategies: Relating individual behaviour to spatial dynamics. *Biological Reviews*, 80:205–225.

Boyce, M., J. Mao, E. Merrill, D. Fortin, M. Turner, J. Fryxell, and P. Turchin. 2003. Scale and heterogeneity in habitat selection by elk in Yellowstone National Park. *Ecoscience*, 10:321–332.

Boyd, J. and D. Brightsmith. 2013. Error properties of Argos satellite telemetry locations using least squares and Kalman filtering. *PLoS One*, 8:e63051.

Breed, G., D. Costa, M. Goebel, and P. Robinson. 2011. Electronic tracking tag programming is critical to data collection for behavioral time-series analysis. *Ecosphere*, 2:1–12.

Breed, G. A., I. D. Jonsen, R. A. Myers, W. D. Bowen, and M. L. Leonard. 2009. Sex-specific, seasonal foraging tactics of adult grey seals (Halichoerus grypus) revealed by state–space analysis. *Ecology*, 90:3209–3221.

Bridge, E., K. Thorup, M. Bowlin, P. Chilson, R. Diehl, R. Fléron, P. Hartl et al. 2011. Technology on the move: Recent and forthcoming innovations for tracking migratory birds. *BioScience*, 61:689–698.

Brillinger, D. 2010. Modeling spatial trajectories. In Gelfand, A., P. Diggle, M. Fuentes, and P. Guttorp, editors, *Handbook of Spatial Statistics*, pages 463–475. Chapman & Hall/CRC, Boca Raton, Florida, USA.

Brillinger, D., H. Preisler, A. Ager, and J. Kie. 2001. The use of potential functions in modeling animal movement. In Saleh, E., editor, *Data Analysis from Statistical Foundations*, pages 369–386. Nova Science Publishers, Huntington, New York, USA.

Brockwell, P. and R. Davis. 2013. *Time Series: Theory and Methods*. Springer Science & Business Media, New York, New York, USA.

Broms, K., M. Hooten, R. Altwegg, and L. Conquest. 2016. Dynamic occupancy models for explicit colonization processes. *Ecology*, 97:194–204.

Brost, B., M. Hooten, E. Hanks, and R. Small. 2015. Animal movement constraints improve resource selection inference in the presence of telemetry error. *Ecology*, 96:2590–2597.

Brown, J. 1969. Territorial behavior and population regulation in birds: A review and re-evaluation. *The Wilson Bulletin*, 81:293–329.

Buderman, F., M. Hooten, J. Ivan, and T. Shenk. 2016. A functional model for characterizing long distance movement behavior. *Methods in Ecology and Evolution*, 7:264–273.

Burt, W. 1943. Territoriality and home range concepts as applied to mammals. *Journal of Mammalogy*, 24:346–352.

Cagnacci, F., L. Boitani, R. A. Powell, and M. S. Boyce. 2010. Animal ecology meets GPS-based radiotelemetry: A perfect storm of opportunities and challenges. *Philosophical Transactions of the Royal Society of London B: Biological Sciences*, 365:2157–2162.

Calder, C. 2007. Dynamic factor process convolution models for multivariate space-time data with application to air quality assessment. *Environmental and Ecological Statistics*, 14:229–247.

Cameron, M., B. McClintock, A. Blanchard, S. Jewett, B. Norcross, R. Lauth, J. Grebmeier, J. Lovvorn, and P. Boveng. 2016. Bearded seal foraging resource selection related to benthic communities and environmental characteristics of the Chukchi Sea. In Review.

Carbone, C., G. Cowlishaw, N. Isaac, and J. Rowcliffe. 2005. How far do animals go? Determinants of day range in mammals. *The American Naturalist*, 165:290–297.

Carpenter, B., A. Gelman, M. Hoffman, D. Lee, B. Goodrich, M. Betancourt, M. Brubaker, J. Guo, P. Li, and A. Riddell. 2016. Stan: A probabilistic programming language. *Journal of Statistical Software*.

Caswell, H. 2001. *Matrix Population Models*. Wiley Online Library, Sunderland, Massachusetts, USA.

Christ, A., J. Ver Hoef, and D. Zimmerman. 2008. An animal movement model incorporating home range and habitat selection. *Environmental and Ecological Statistics*, 15:27–38.

Clark, J. 1998. Why trees migrate so fast: Confronting theory with dispersal biology and the paleorecord. *The American Naturalist*, 152:204–224.

Clark, J. 2007. *Models for Ecological Data: An Introduction*. Princeton University Press, Princeton, New Jersey, USA.

Clark, J., M. Lewis, J. McLachlan, and J. HilleRisLambers. 2003. Estimating population spread: What can we forecast and how well? *Ecology*, 84:1979–1988.

Clobert, J. 2000. *Dispersal*. Oxford University Press, New York, USA.

Clobert, J., L. Galliard, J. Cote, S. Meylan, and M. Massot. 2009. Informed dispersal, heterogeneity in animal dispersal syndromes and the dynamics of spatially structured populations. *Ecology Letters*, 12:197–209.

Codling, E., M. Plank, and S. Benhamou. 2008. Random walk models in biology. *Journal of the Royal Society Interface*, 5:813–834.

Cooke, S., S. Hinch, M. Wikelski, R. Andrews, L. Kuchel, T. Wolcott, and P. Butler. 2004. Biotelemetry: A mechanistic approach to ecology. *Trends in Ecology and Evolution*, 19:334–343.

Costa, D., P. Robinson, J. Arnould, A.-L. Harrison, S. E. Simmons, J. L. Hassrick, A. J. Hoskins et al. 2010. Accuracy of Argos locations of pinnipeds at-sea estimated using Fastloc GPS. *PLoS One*, 5:e8677.

Cote, J. and J. Clobert. 2007. Social information and emigration: Lessons from immigrants. *Ecology Letters*, 10:411–417.

Coulson, T., E. Catchpole, S. Albon, B. Morgan, J. Pemberton, T. Clutton-Brock, M. Crawley, and B. Grenfell. 2001. Age, sex, density, winter weather, and population crashes in soay sheep. *Science*, 292:1528–1531.

Couzin, I., J. Krause, N. Franks, and S. Levin. 2005. Effective leadership and decision-making in animal groups on the move. *Nature*, 433:513–516.

Couzin, I. D., J. Krause, R. James, G. D. Ruxton, and N. R. Franks. 2002. Collective memory and spatial sorting in animal groups. *Journal of Theoretical Biology*, 218(1):1–11.

Cox, D. and D. Oakes. 1984. *Analysis of Survival Data*, volume 21. CRC Press, Boca Raton, Florida, USA.

Craighead, F. and J. Craighead. 1972. Grizzly bear prehibernation and denning activities as determined by radiotracking. *Wildlife Monographs*, (32):3–35.

Cressie, N. 1990. The origins of Kriging. *Mathematical Geology*, 22:239–252.

Cressie, N. 1993. *Statistics for Spatial Data: Revised Edition*. John Wiley and Sons, New York, New York, USA.

Cressie, N. and C. Wikle. 2011. *Statistics for Spatio-Temporal Data*. John Wiley and Sons, New York, New York, USA.

Dall, S., L.-A. Giraldeau, O. Olsson, J. McNamara, and D. Stephens. 2005. Information and its use by animals in evolutionary ecology. *Trends in Ecology & Evolution*, 20:187–193.

Dall, S., A. Houston, and J. McNamara. 2004. The behavioural ecology of personality: Consistent individual differences from an adaptive perspective. *Ecology Letters*, 7:734–739.

Dalziel, B., J. Morales, and J. Fryxell. 2008. Fitting probability distributions to animal movement trajectories: Using artificial neural networks to link distance, resources, and memory. *The American Naturalist*, 172:248–258.

Danchin, E., L.-A. Giraldeau, T. Valone, and R. Wagner. 2004. Public information: From nosy neighbors to cultural evolution. *Science*, 305:487–491.

Datta, A., S. Banerjee, A. O. Finley, and A. E. Gelfand. 2016. Hierarchical nearest-neighbor Gaussian process models for large geostatistical datasets. *Journal of the American Statistical Association*, 111:800–812.

Davis, R. A., S. H. Holan, R. Lund, and N. Ravishanker. 2016. *Handbook of Discrete-Valued Time Series*. CRC Press, Boca Raton, Florida, USA.

Delgado, M. and V. Penteriani. 2008. Behavioral states help translate dispersal movements into spatial distribution patterns of floaters. *The American Naturalist*, 172:475–485.

Delgado, M., V. Penteriani, J. Morales, E. Gurarie, and O. Ovaskainen. 2014. A statistical framework for inferring the influence of conspecifics on movement behaviour. *Methods in Ecology and Evolution*, 5:183–189.

Deneubourg, J.-L., S. Goss, N. Franks, and J. Pasteels. 1989. The blind leading the blind: Modeling chemically mediated army ant raid patterns. *Journal of Insect Behavior*, 2:719–725.

deSolla, S., R. Shane, R. Bonduriansky, and R. Brooks. 1999. Eliminating autocorrelation reduces biological relevance of home range estimates. *Journal of Animal Ecology*, 68:221–234.

Diggle, P. 1985. A kernel method for smoothing point process data. *Applied Statistics*, 34:138–147.

Diggle, P., R. Menezes, and T. Su. 2010a. Geostatistical inference under preferential sampling. *Journal of the Royal Statistical Society: Series C (Applied Statistics)*, 59:191–232.

Diggle, P. and P. Ribeiro. 2002. Bayesian inference in Gaussian model-based geostatistics. *Geographical and Environmental Modelling*, 6:129–146.

Diggle, P. and P. Ribeiro. 2007. *Model-Based Geostatistics*. Springer, New York, New York, USA.

Diggle, P., J. Tawn, and R. Moyeed. 1998. Model-based geostatistics. *Journal of the Royal Statistical Society: Series C (Applied Statistics)*, 47(3):299–350.

Diggle, P. J., I. Kaimi, and R. Abellana. 2010b. Partial-likelihood analysis of spatio-temporal point-process data. *Biometrics*, 66:347–354.

Dorazio, R. M. 2012. Predicting the geographic distribution of a species from presence-only data subject to detection errors. *Biometrics*, 68:1303–1312.

Douglas, D., R. Weinzierl, S. Davidson, R. Kays, M. Wikelski, and G. Bohrer. 2012. Moderating Argos location errors in animal tracking data. *Methods in Ecology and Evolution*, 3:999–1007.

Duchesne, T., D. Fortin, and L.-P. Rivest. 2015. Equivalence between step selection functions and biased correlated random walks for statistical inference on animal movement. *PloS One*, 10:e0122947.

Dunn, J. and P. Gipson. 1977. Analysis of radio-telemetry data in studies of home range. *Biometrics*, 33:85–101.

Durbin, J. and G. Watson. 1950. Testing for serial correlation in least squares regression, i. *Biometrika*, 37:409–428.

Durrett, R. 1996. *Stochastic Calculus: A Practical Introduction*. CRC Press, Boca Raton, Florida, USA.

Durrett, R. and S. Levin. 1994. The importance of being discrete (and spatial). *Theoretical Population Biology*, 46:363–394.

Eckert, S., J. Moore, D. Dunn, R. van Buiten, K. Eckert, and P. Halpin. 2008. Modeling loggerhead turtle movement in the Mediterranean: Importance of body size and oceanography. *Ecological Applications*, 18:290–308.

Eftimie, R., G. De Vries, and M. Lewis. 2007. Complex spatial group patterns result from different animal communication mechanisms. *Proceedings of the National Academy of Sciences*, 104:6974–6979.

Ellner, S. and M. Rees. 2006. Integral projection models for species with complex demography. *The American Naturalist*, 167:410–428.

Fagan, W., M. Lewis, M. Auger-Méthé, T. Avgar, S. Benhamou, G. Breed, L. LaDage et al. 2013. Spatial memory and animal movement. *Ecology Letters*, 16:1316–1329.

Fahrig, L. 2001. How much habitat is enough? *Biological Conservation*, 100:65–74.

Fieberg, J. 2007. Kernel density estimators of home range: Smoothing and the autocorrelation red herring. *Ecology*, 88:1059–1066.

Fieberg, J. and M. Ditmer. 2012. Understanding the causes and consequences of animal movement: A cautionary note on fitting and interpreting regression models with time-dependent covariates. *Methods in Ecology and Evolution*, 3:983–991.

Fieberg, J., J. Matthiopoulos, M. Hebblewhite, M. Boyce, and J. Frair. 2010. Correlation and studies of habitat selection: Problem, red herring or opportunity? *Philosophical Transactions of the Royal Society, B*, 365:2233–2244.

Fisher, R. 1937. The wave of advance of advantageous genes. *Annals of Eugenics*, 7:355–369.

Fleming, C., J. Calabrese, T. Mueller, K. Olson, P. Leimgruber, and W. Fagan. 2014. From fine-scale foraging to home ranges: A semivariance approach to identifying movement modes across spatiotemporal scales. *The American Naturalist*, 183:154–167.

Fleming, C., W. Fagan, T. Mueller, K. Olson, P. Leimgruber, and J. Calabrese. 2015. Rigorous home range estimation with movement data: A new autocorrelated kernel density estimator. *Ecology*, 96(5):1182–1188.

Flierl, G., D. Grünbaum, S. Levins, and D. Olson. 1999. From individuals to aggregations: The interplay between behavior and physics. *Journal of Theoretical biology*, 196: 397–454.

Forester, J., H. Im, and P. Rathouz. 2009. Accounting for animal movement in estimation of resource selection functions: Sampling and data analysis. *Ecology*, 90:3554–3565.

Forester, J., A. Ives, M. Turner, D. Anderson, D. Fortin, H. Beyer, D. Smith, and M. Boyce. 2007. State-space models link elk movement patterns to landscape characteristics in Yellowstone National Park. *Ecological Monographs*, 77:285–299.

Fortin, D., H. Beyer, M. Boyce, D. Smith, T. Duchesne, and J. Mao. 2005. Wolves influence elk movements: Behavior shapes a trophic cascade in Yellowstone National Park. *Ecology*, 86:1320–1330.

Frair, J., E. Merrill, J. Allen, and M. Boyce. 2007. Know thy enemy: Experience affects elk translocation success in risky landscapes. *The Journal of Wildlife Management*, 71: 541–554.

Franke, A., T. Caelli, G. Kuzyk, and R. Hudson. 2006. Prediction of wolf (*Canis lupus*) kill-sites using hidden Markov models. *Ecological Modelling*, 197(1):237–246.

Fraser, D., J. Gilliam, M. Daley, A. Le, and G. Skalski. 2001. Explaining leptokurtic movement distributions: Intrapopulation variation in boldness and exploration. *The American Naturalist*, 158:124–135.

Fryxell, J., A. Mosser, A. Sinclair, and C. Packer. 2007. Group formation stabilizes predator–prey dynamics. *Nature*, 449:1041–1043.

Garlick, M., J. Powell, M. Hooten, and L. McFarlane. 2011. Homogenization of large-scale movement models in ecology. *Bulletin of Mathematical Biology*, 73:2088–2108.

Garlick, M., J. Powell, M. Hooten, and L. McFarlane. 2014. Homogenization, sex, and differential motility predict spread of chronic wasting disease in mule deer in Southern Utah. *Journal of Mathematical Biology*, 69:369–399.

Gaspar, P., J.-Y. Georges, S. Fossette, A. Lenoble, S. Ferraroli, and Y. Le Maho. 2006. Marine animal behaviour: Neglecting ocean currents can lead us up the wrong track. *Proceedings of the Royal Society of London B: Biological Sciences*, 273(1602): 2697–2702.

Gelfand, A. and A. Smith. 1990. Sampling-based approaches to calculating marginal densities. *Journal of the American Statistical Association*, 85:398–409.

Gelfand, A. E., P. Diggle, P. Guttorp, and M. Fuentes. 2010. *Handbook of Spatial Statistics*. CRC Press, Boca Raton, Florida, USA.

Gelman, A., J. B. Carlin, H. S. Stern, and D. B. Rubin. 2014. *Bayesian Data Analysis*. Taylor & Francis, Boca Raton, Florida, USA.

Gelman, A. and J. Hill. 2006. *Data Analysis Using Regression and Multilevel Hierarchical Models*. Cambridge University Press, Cambridge, United Kingdom.

Getz, W., S. Fortman-Roe, P. Cross, A. Lyons, S. Ryan, and C. Wilmers. 2007. LoCoH: Nonparameteric kernel methods for constructing home ranges and utilization distributions. *PLoS One*, 2:e207.

Giuggioli, L. and V. Kenkre. 2014. Consequences of animal interactions on their dynamics: Emergence of home ranges and territoriality. *Movement Ecology*, 2:20.

Giuggioli, L., J. Potts, and S. Harris. 2012. Predicting oscillatory dynamics in the movement of territorial animals. *Journal of The Royal Society Interface*, 9:1529–1543.

Grimmett, G. and D. Stirzaker. 2001. *Probability and Random Processes*. Oxford University Press, New York, New York, USA.

Gurarie, E., C. Bracis, M. Delgado, T. Meckley, I. Kojola, and C. Wagner. 2016. What is the animal doing? Tools for exploring behavioural structure in animal movements. *Journal of Animal Ecology*, 85(1):69–84.

Gurarie, E. and O. Ovaskainen. 2011. Characteristic spatial and temporal scales unify models of animal movement. *The American Naturalist*, 178(1):113–123.

Gurarie, E. and O. Ovaskainen. 2013. Towards a general formalization of encounter rates in ecology. *Theoretical Ecology*, 6:189–202.

Hanks, E. and M. Hooten. 2013. Circuit theory and model-based inference for landscape connectivity. *Journal of the American Statistical Association*, 108:22–33.

Hanks, E., M. Hooten, and M. Alldredge. 2015a. Continuous-time discrete-space models for animal movement. *Annals of Applied Statistics*, 9:145–165.

Hanks, E., M. Hooten, D. Johnson, and J. Sterling. 2011. Velocity-based movement modeling for individual and population level inference. *PLoS One*, 6:e22795.

Hanks, E., E. Schliep, M. Hooten, and J. Hoeting. 2015b. Restricted spatial regression in practice: Geostatistical models, confounding, and robustness under model misspecification. *Environmetrics*, 26:243–254.

Hanski, I. and O. Gaggiotti. 2004. *Ecology, Genetics, and Evolution of Metapopulations*. Academic Press, Burlington, Massachusetts, USA.

Harris, K. and P. Blackwell. 2013. Flexible continuous-time modeling for heterogeneous animal movement. *Ecological Modelling*, 255:29–37.

Harrison, X., J. Blount, R. Inger, D. Norris, and S. Bearhop. 2011. Carry-over effects as drivers of fitness differences in animals. *Journal of Animal Ecology*, 80:4–18.

Haydon, D., J. Morales, A. Yott, D. Jenkins, R. Rosatte, and J. Fryxell. 2008. Socially informed random walks: Incorporating group dynamics into models of population spread and growth. *Proceedings of the Royal Society of London B: Biological Sciences*, 275:1101–1109.

Haynes, K. and J. Cronin. 2006. Interpatch movement and edge effects: The role of behavioral responses to the landscape matrix. *Oikos*, 113:43–54.

Hefley, T., K. Broms, B. Brost, F. Buderman, S. Kay, H. Scharf, J. Tipton, P. Williams, and M. Hooten. 2016a. The basis function approach to modeling dependent ecological data. *Ecology*, In Press.

Hefley, T., M. Hooten, R. Russell, D. Walsh, and J. Powell. 2016b. Ecological diffusion models for large data sets and fine-scale inference. In Review.

Higdon, D. 1998. A process-convolution approach to modeling temperatures in the North Atlantic Ocean. *Environmental and Ecological Statistics*, 5:173–190.

Higdon, D. 2002. Space and space-time modeling using process convolutions. In Anderson, C., V. Barnett, P. Chatwin, and A. El-Shaarawi, editors, *Quantitative Methods for Current Environmental Issues*, pages 37–56. Springer-Verlag, London, UK.

Higgs, M. and J. V. Hoef. 2012. Discretized and aggregated: Modeling dive depth of harbor seals from ordered categorical data with temporal autocorrelation. *Biometrics*, 68:965–974.

Hobbs, N., C. Geremia, J. Treanor, R. Wallen, P. White, M. Hooten, and J. Rhyan. 2015. State-space modeling to support adaptive management of brucellosis in the Yellowstone bison population. *Ecological Monographs*, 85:525–556.

Hobbs, N. and M. Hooten. 2015. *Bayesian Models: A Statistical Primer for Ecologists*. Princeton University Press, Princeton, New Jersey, USA.

Hodges, J. and B. Reich. 2010. Adding spatially-correlated errors can mess up the fixed effect you love. *The American Statistician*, 64:325–334.

Holford, T. 1980. The analysis of rates and of survivorship using log-linear models. *Biometrics*, 36:299–305.

Holling, C. 1959a. Some characteristics of simple types of predation and parasitism. *The Canadian Entomologist*, 91:385–398.

Holling, C. 1959b. The components of predation as revealed by a study of small-mammal predation of the European pine sawfly. *The Canadian Entomologist*, 91:293–320.

Holzmann, H., A. Munk, M. Suster, and W. Zucchini. 2006. Hidden Markov models for circular and linear-circular time series. *Environmental and Ecological Statistics*, 13(3):325–347.

Hooker, S., S. Heaslip, J. Matthiopoulos, O. Cox, and I. Boyd. 2008. Data sampling options for animal-borne video cameras: Considerations based on deployments with Antarctic fur seals. *Marine Technology Society Journal*, 42:65–75.

Hooten, M., J. Anderson, and L. Waller. 2010a. Assessing North American influenza dynamics with a statistical SIRS model. *Spatial and Spatio-Temporal Epidemiology*, 1:177–185.

Hooten, M., F. Buderman, B. Brost, E. Hanks, and J. Ivan. 2016. Hierarchical animal movement models for population-level inference. *Environmetrics*, 27:322–333.

Hooten, M., M. Garlick, and J. Powell. 2013a. Computationally efficient statistical differential equation modeling using homogenization. *Journal of Agricultural, Biological and Environmental Statistics*, 18:405–428.

Hooten, M., E. Hanks, D. Johnson, and M. Aldredge. 2013b. Reconciling resource utilization and resource selection functions. *Journal of Animal Ecology*, 82:1146–1154.

Hooten, M., E. Hanks, D. Johnson, and M. Aldredge. 2014. Temporal variation and scale in movement-based resource selection functions. *Statistical Methodology*, 17:82–98.

Hooten, M. and N. Hobbs. 2015. A guide to Bayesian model selection for ecologists. *Ecological Monographs*, 85:3–28.

Hooten, M. and D. Johnson. 2016. Basis function models for animal movement. *Journal of the American Statistical Association*, In Press.

Hooten, M., D. Johnson, E. Hanks, and J. Lowry. 2010b. Agent-based inference for animal movement and selection. *Journal of Agricultural, Biological and Environmental Statistics*, 15:523–538.

Hooten, M., D. Larsen, and C. Wikle. 2003. Predicting the spatial distribution of ground flora on large domains using a hierarchical Bayesian model. *Landscape Ecology*, 18:487–502.

Hooten, M. and C. Wikle. 2008. A hierarchical Bayesian non-linear spatio-temporal model for the spread of invasive species with application to the Eurasian collared-dove. *Environmental and Ecological Statistics*, 15:59–70.

Hooten, M. and C. Wikle. 2010. Statistical agent-based models for discrete spatio-temporal systems. *Journal of the American Statistical Association*, 105:236–248.

Horne, J., E. Garton, S. Krone, and J. Lewis. 2007. Analyzing animal movements using Brownian bridges. *Ecology*, 88:2354–2363.

Horning, M. and R. Hill. 2005. Designing an archival satellite transmitter for life-long deployments on oceanic vertebrates: The life history transmitter. *IEEE Journal of Oceanic Engineering*, 30:807–817.

Hughes, J. and M. Haran. 2013. Dimension reduction and alleviation of confounding for spatial generalized linear mixed models. *Journal of the Royal Statistical Society, Series B*, 75:139–159.

Hutchinson, J. and P. Waser. 2007. Use, misuse and extensions of "ideal gas" models of animal encounter. *Biological Reviews*, 82:335–359.

Illian, J., S. Martino, S. Sørbye, J. Gallego-Fernández, M. Zunzunegui, M. Esquivias, and J. Travis. 2013. Fitting complex ecological point process models with integrated nested Laplace approximation. *Methods in Ecology and Evolution*, 4:305–315.

Illian, J., A. Penttinen, H. Stoyan, and D. Stoyan. 2008. *Statistical Analysis and Modelling of Spatial Point Patterns*. Wiley-Interscience, West Sussex, England.

Illian, J., S. Sorbye, H. Rue, and D. Hendrichsen. 2012. Using INLA to fit a complex point process model with temporally varying effects—A case study. *Journal of Environmental Statistics*, 3:1–25.

Iranpour, R., P. Chacon, and M. Kac. 1988. *Basic Stochastic Processes: The Mark Kac Lectures*. Macmillan, New York.

Isojunno, S. and P. Miller. 2015. Sperm whale response to tag boat presence: Biologically informed hidden state models quantify lost feeding opportunities. *Ecosphere*, 6(1):1–46.

Jetz, W., C. Carbone, J. Fulford, and J. Brown. 2004. The scaling of animal space use. *Science*, 306:266–268.

Ji, W., P. White, and M. Clout. 2005. Contact rates between possums revealed by proximity data loggers. *Journal of Applied Ecology*, 42:595–604.

Johnson, A., J. Wiens, B. Milne, and T. Crist. 1992. Animal movements and population dynamics in heterogeneous landscapes. *Landscape Ecology*, 7:63–75.

Johnson, D. 1980. The comparison of usage and availability measurements for evaluating resource preference. *Ecology*, 61:65–71.

Johnson, D., M. Hooten, and C. Kuhn. 2013. Estimating animal resource selection from telemetry data using point process models. *Journal of Animal Ecology*, 82:1155–1164.

Johnson, D., J. London, and C. Kuhn. 2011. Bayesian inference for animal space use and other movement metrics. *Journal of Agricultural, Biological and Environmental Statistics*, 16:357–370.

Johnson, D., J. London, M. Lea, and J. Durban. 2008a. Continuous-time correlated random walk model for animal telemetry data. *Ecology*, 89:1208–1215.

Johnson, D., D. Thomas, J. Ver Hoef, and A. Christ. 2008b. A general framework for the analysis of animal resource selection from telemetry data. *Biometrics*, 64:968–976.

Jonsen, I. 2016. Joint estimation over multiple individuals improves behavioural state inference from animal movement data. *Scientific Reports*, 6:20625.

Jonsen, I., J. Flemming, and R. Myers. 2005. Robust state-space modeling of animal movement data. *Ecology*, 45:589–598.

Jonsen, I., R. Myers, and J. Flemming. 2003. Meta-analysis of animal movement using state-space models. *Ecology*, 84:3055–3063.

Jonsen, I., R. Myers, and M. James. 2006. Robust hierarchical state-space models reveal diel variation in travel rates of migrating leatherback turtles. *Journal of Animal Ecology*, 75:1046–1057.

Jonsen, I., R. Myers, and M. James. 2007. Identifying leatherback turtle foraging behaviour from satellite telemetry using a switching state-space model. *Marine Ecology Progress Series*, 337:255–264.

Jønsson, K., A. Tøttrup, M. Borregaard, S. Keith, C. Rahbek, and K. Thorup. 2016. Tracking animal dispersal: From individual movement to community assembly and global range dynamics. *Trends in Ecology & Evolution*, 31(3):204–214.

Kalman, R. 1960. A new approach to linear filtering and prediction problems. *Transactions of the ASME—Journal of Basic Engineering*, 82:35–45.

Karatzas, I. and S. Shreven. 2012. *Brownian Motion and Stochastic Calculus*, volume 113. Springer Science & Business Media, New York, New York, USA.

Katzfuss, M. 2016. A multi-resolution approximation for massive spatial datasets. *Journal of the American Statistical Association*, In Press.

Kays, R., M. Crofoot, W. Jetz, and M. Wikelski. 2015. Terrestrial animal tracking as an eye on life and planet. *Science*, 348(6240):aaa2478.

Keating, K. A. and S. Cherry. 2009. Modeling utilization distributions in space and time. *Ecology*, 90:1971–1980.

Kendall, D. 1974. Pole-seeking Brownian motion and bird navigation. *Journal of the Royal Statistical Society, Series B*, 36:365–417.

Kenward, R. 2000. *A Manual for Wildlife Radio Tagging*. Academic Press, San Diego, California, USA.

Kery, M. and J. Royle. 2008. Hierarchical Bayes estimation of species richness and occupancy in spatially replicated surveys. *Journal of Applied Ecology*, 45:589–598.

Kot, M., M. Lewis, and P. van den Driessche. 1996. Dispersal data and the spread of invading organisms. *Ecology*, 77:2027–2042.

Langrock, R., J. Hopcraft, P. Blackwell, V. Goodall, R. King, M. Niu, T. Patterson, M. Pedersen, A. Skarin, and R. Schick. 2014. Modelling group dynamic animal movement. *Methods in Ecology and Evolution*, 5:190–199.

Langrock, R., R. King, J. Matthiopoulos, L. Thomas, D. Fortin, and J. Morales. 2012. Flexible and practical modeling of animal telemetry data: Hidden Markov models and extensions. *Ecology*, 93:2336–2342.

Lapanche, C., T. Marques, and L. Thomas. 2015. Tracking marine mammals in 3d using electronic tag data. *Methods in Ecology and Evolution*, 6:987–996.

Laver, P. and M. Kelly. 2008. A critical review of home range studies. *The Journal of Wildlife Management*, 72:290–298.

Le, N. D. and J. V. Zidek. 2006. *Statistical Analysis of Environmental Space-Time Processes*. Springer Science & Business Media, New York, New York.

LeBoeuf, B., D. Crocker, D. Costa, S. Blackwell, P. Webb, and D. Houser. 2000. Foraging ecology of northern fur seals. *Ecological Monographs*, 70:353–382.

Lee, H., D. Higdon, C. Calder, and C. Holloman. 2005. Efficient models for correlated data via convolutions of intrinsic processes. *Statistical Modelling*, 5:53–74.

Lele, S. and J. Keim. 2006. Weighted distributions and estimation of resource selection probability functions. *Ecology*, 87:3021–3028.

LeSage, J. and R. Pace. 2009. *Introduction to Spatial Econometrics*. Chapman & Hall/CRC, Boca Raton, Florida, USA.

Levey, D., B. Bolker, J. Tewksbury, S. Sargent, and N. Haddad. 2005. Effects of landscape corridors on seed dispersal by birds. *Science*, 309:146–148.

Lima, S. and P. Zollner. 1996. Towards a behavioral ecology of ecological landscapes. *Trends in Ecology and Evolution*, 11:131–135.

Lindgren, F. and H. Rue. 2015. Bayesian spatial modelling with R-INLA. *Journal of Statistical Software*, 63(19):1–25.

Lindgren, F., H. Rue, and J. Lindstrom. 2011. An explicit link between Gaussian fields and Gaussian Markov random fields: The SPDE approach (with discussion). *Journal of the Royal Statistical Society, Series B*, 73:423–498.

Liu, Y., B. Battaile, J. Zidek, and A. Trites. 2014. Bayesian melding of the dead-reckoned path and GPS measurements for an accurate and high-resolution path of marine mammals. *arXiv preprint: 1411.6683*.

Liu, Y., B. Battaile, J. Zidek, and A. Trites. 2015. Bias correction and uncertainty characterization of dead-reckoned paths of marine mammals. *Animal Biotelemetry*, 3(51): 1–11.

Lloyd, M. 1967. Mean crowding. *The Journal of Animal Ecology*, 36:1–30.

Long, R., J. Kie, T. Bowyer, and M. Hurley. 2009. Resource selection and movements by female mule deer *Odocoileus hemionus*: Effects of reproductive stage. *Wildlife Biology*, 15:288–298.

Lundberg, J. and F. Moberg. 2003. Mobile link organisms and ecosystem functioning: Implications for ecosystem resilience and management. *Ecosystems*, 6: 87–98.

Lunn, D., J. Barrett, M. Sweeting, and S. Thompson. 2013. Fully Bayesian hierarchical modelling in two stages, with application to meta-analysis. *Journal of the Royal Statistical Society: Series C (Applied Statistics)*, 62:551–572.

Lunn, D., A. Thomas, N. Best, and D. Spiegelhalter. 2000. WinBUGS—A Bayesian modelling framework: Concepts, structure, and extensibility. *Statistics and Computing*, 10(4):325–337.

Lyons, A., W. Turner, and W. Getz. 2013. Home range plus: A space-time characterization of movement over real landscapes. *Movement Ecology*, 1:2.

Manly, B., L. McDonald, D. Thomas, T. McDonald, and W. Erickson. 2007. *Resource Selection by Animals: Statistical Design and Analysis for Field Studies*. Springer Science & Business Media, Dordrecht, The Netherlands.

Marzluff, J., J. Millspaugh, P. Hurvitz, and M. Handcock. 2004. Relating resources to a probabilistic measure of space use: Forest fragments and Stellar's jays. *Ecology*, 85:1411–1427.

Matheron, G. 1963. Principles of geostatistics. *Economic Geology*, 58:1246–1266.

Matthiopoulos, J., J. Fieberg, G. Aarts, H. Beyer, J. Morales, and D. Haydon. 2015. Establishing the link between habitat selection and animal population dynamics. *Ecological Monographs*, 85:413–436.

Maxwell, J. 1860. V. Illustrations of the dynamical theory of gases. Part I. On the motions and collisions of perfectly elastic spheres. *The London, Edinburgh, and Dublin Philosophical Magazine and Journal of Science*, 19:19–32.

McClintock, B., D. Johnson, M. Hooten, J. Ver Hoef, and J. Morales. 2014. When to be discrete: The importance of time formulation in understanding animal movement. *Movement Ecology*, 2:21.

McClintock, B., R. King, L. Thomas, J. Matthiopoulos, B. McConnell, and J. Morales. 2012. A general discrete-time modeling framework for animal movement using multistate random walks. *Ecological Monographs*, 82:335–349.

McClintock, B., J. London, M. Cameron, and P. Boveng. 2015. Modelling animal movement using the argos satellite telemetry location error ellipse. *Methods in Ecology and Evolution*, 6:266–277.

McClintock, B., J. London, M. Cameron, and P. Boveng. 2016. Bridging the gaps in animal movement: hidden behaviors and ecological relationships revealed by integrated data streams. In Review.

McClintock, B., D. Russell, J. Matthiopoulos, and R. King. 2013. Combining individual animal movement and ancillary biotelemetry data to investigate population-level activity budgets. *Ecology*, 94(4):838–849.

McIntyre, N. and J. Wiens. 1999. How does habitat patch size affect animal movement? An experiment with darkling beetles. *Ecology*, 80:2261–2270.

McMichael, G., M. Eppard, T. Carlson, J. Carter, B. Ebberts, R. Brown, M. Weiland, G. Ploskey, R. Harnish, and Z. Deng. 2010. The juvenile salmon acoustic telemetry system: A new tool. *Fisheries*, 35:9–22.

Merkle, J., D. Fortin, and J. Morales. 2014. A memory-based foraging tactic reveals an adaptive mechanism for restricted space use. *Ecology Letters*, 17:924–931.

Merrill, E., H. Sand, B. Zimmermann, H. McPhee, N. Webb, M. Hebblewhite, P. Wabakken, and J. Frair. 2010. Building a mechanistic understanding of predation with GPS-based movement data. *Philosophical Transactions of the Royal Society of London B: Biological Sciences*, 365:2279–2288.

Metz, J. and O. Diekmann. 2014. *The Dynamics of Physiologically Structured Populations*. Springer, Berlin, Germany.

Michelot, T., R. Langrock, T. Patterson, and E. Rexstad. 2015. *moveHMM: Animal Movement Modelling Using Hidden Markov Models*, 2015. R package version 1.1.

Millspaugh, J. and J. M. Marzluff. 2001. *Radio Tracking and Animal Populations*. Academic Press, San Diego, California, USA.

Millspaugh, J., R. Nielson, L. MacDonald, J. Marzluff, R. Gitzen, C. Rittenhouse, M. Hubbard, and S. Sheriff. 2006. Analysis of resource selection using utilization distributions. *Journal of Wildlife Management*, 70:384–395.

Moll, R., J. Millspaugh, J. Beringer, J. Sartwell, Z. He, J. Eggert, and X. Zhao. 2009. A terrestrial animal-borne video system for large mammals. *Computers and Electronics in Agriculture*, 66:133–139.

Møller, J. and R. Waagepetersen. 2004. *Statistical Inference and Simulation for Spatial Point Processes*. Chapman & Hall/CRC, Boca Raton, Florida, USA.

Moorcroft, P. and A. Barnett. 2008. Mechanistic home range models and resource selection analysis: A reconciliation and unification. *Ecology*, 89:1112–1119.

Moorcroft, P. and M. Lewis. 2013. *Mechanistic Home Range Analysis.* Princeton University Press, Princeton, New Jersey, USA.

Moorcroft, P., M. Lewis, and R. Crabtree. 1999. Home range analysis using a mechanistic home range model. *Ecology*, 80:1656–1665.

Moorcroft, P., M. Lewis, and R. Crabtree. 2006. Mechanistic home range model capture spatial patterns and dynamics of coyote territories in Yellowstone. *Proceedings of the Royal Society B*, 273:1651–1659.

Morales, J. 2002. Behavior at habitat boundaries can produce leptokurtic movement distributions. *The American Naturalist*, 160:531–538.

Morales, J. and S. Ellner. 2002. Scaling up animal movements in heterogeneous landscapes: The importance of behavior. *Ecology*, 83:2240–2247.

Morales, J., D. Fortin, J. Frair, and E. Merrill. 2005. Adaptive models for large herbivore movements in heterogeneous landscapes. *Landscape Ecology*, 20:301–316.

Morales, J., J. Frair, E. Merrill, H. Beyer, and D. Haydon. 2016. Patch use of reintroduced elk in the Canadian Rockies: Memory effects and home range development. Unpublished Manuscript.

Morales, J., D. Haydon, J. Friar, K. Holsinger, and J. Fryxell. 2004. Extracting more out of relocation data: Building movement models as mixtures of random walks. *Ecology*, 85:2436–2445.

Morales, J., P. Moorcroft, J. Matthiopoulos, J. Frair, J. Kie, R. Powell, E. Merrill, and D. Haydon. 2010. Building the bridge between animal movement and population dynamics. *Philosophical Transactions of the Royal Society of London B: Biological Sciences*, 365:2289–2301.

Mueller, T. and W. Fagan. 2008. Search and navigation in dynamic environments—From individual behaviors to population distributions. *Oikos*, 117:654–664.

Murray, D. 2006. On improving telemetry-based survival estimation. *Journal of Wildlife Management*, 70:1530–1543.

Nathan, R., W. Getz, E. Revilla, M. Holyoak, R. Kadmon, D. Saltz, and P. Smouse. 2008. A movement ecology paradigm for unifying organismal movement research. *Proceedings of the National Academy of Sciences*, 105:19052–19059.

Nielson, R., B. Manly, L. McDonald, H. Sawyer, and T. McDonald. 2009. Estimating habitat selection when GPS fix success is less than 100%. *Ecology*, 90:2956–2962.

Nielson, R. M. and H. Sawyer. 2013. Estimating resource selection with count data. *Ecology and Evolution*, 3:2233–2240.

Northrup, J., M. Hooten, C. Anderson, and G. Wittemyer. 2013. Practical guidance on characterizing availability in resource selection functions under a use-availability design. *Ecology*, 94:1456–1464.

Nussbaum, M. 1978. *Aristotle's De Motu Animalium: Text with Translation, Commentary, and Interpretive Essays*. Princeton University Press, Princeton, New Jersey, USA.

Okubo, A., D. Grünbaum, and L. Edelstein-Keshet. 2001. The dynamics of animal grouping. In A. Okubo and S.A. Levin, editors, *Diffusion and Ecological Problems: Modern Perspectives*, pages 197–237. Springer, New York, New York, USA.

Otis, D. and G. White. 1999. Autocorrelation of location estimates and the analysis of radiotracking data. *Journal of Wildlife Management*, 63:1039–1044.

Ovaskainen, O. 2004. Habitat-specific movement parameters estimated using mark-recapture data and a diffusion model. *Ecology*, 85:242–257.

Ovaskainen, O. and S. Cornell. 2003. Biased movement at a boundary and conditional occupancy times for diffusion processes. *Journal of Applied Probability*, 40: 557–580.

Ovaskainen, O., D. Finkelshtein, O. Kutoviy, S. Cornell, B. Bolker, and Y. Kondratiev. 2014. A general mathematical framework for the analysis of spatiotemporal point processes. *Theoretical Ecology*, 7:101–113.

Ovaskainen, O., H. Rekola, E. Meyke, and E. Arjas. 2008. Bayesian methods for analyzing movements in heterogeneous landscapes from mark-recapture data. *Ecology*, 89: 542–554.

Paciorek, C. 2010. The importance of scale for spatial-confounding bias and precision of spatial regression estimators. *Statistical Science*, 25:107–125.

Paciorek, C. and M. Schervish. 2006. Spatial modelling using a new class of nonstationary covariance functions. *Environmetrics*, 17:483–506.

Parker, K., P. Barboza, and M. Gillingham. 2009. Nutrition integrates environmental responses of ungulates. *Functional Ecology*, 23:57–69.

Patil, G. and C. Rao. 1976. On size-biased sampling and related form-invariant weighted distributions. *Indian Journal of Statistics*, 38:48–61.

Patil, G. and C. Rao. 1977. The weighted distributions: A survey of their applications. In Krishnaiah, P., editor, *Applications of Statistics*, pages 383–405. North Holland Publishing Company, Amsterdam, The Netherlands.

Patil, G. and C. Rao. 1978. Weighted distributions and size-biased sampling with applications to wildlife populations and human families. *Biometrics*, 34:179–189.

Patterson, H. and R. Thompson. 1971. Recovery of inter-block information when block sizes are unequal. *Biometrika*, 58:545–554.

Patterson, T., M. Basson, M. Bravington, and J. Gunn. 2009. Classifying movement behaviour in relation to environmental conditions using hidden Markov models. *Journal of Animal Ecology*, 78:1113–1123.

Patterson, T., L. Thomas, C. Wilcox, O. Ovaskainen, and J. Matthiopoulos. 2008. State-space models of individual animal movement. *Trends in Ecology and Evolution*, 23:87–94.

Patterson, T. A., B. J. McConnell, M. A. Fedak, M. V. Bravington, and M.A. Hindell. 2010. Using GPS data to evaluate the accuracy of state–space methods for correction of Argos satellite telemetry error. *Ecology*, 91:273–285.

Penteriani, V. and M. Delgado. 2009. Thoughts on natal dispersal. *Journal of Raptor Research*, 43:90–98.

Pérez-Escudero, A., J. Vicente-Page, R. Hinz, S. Arganda, and G. de Polavieja. 2014. idTracker: Tracking individuals in a group by automatic identification of unmarked animals. *Nature Methods*, 11:743–748.

Plummer, M. 2003. JAGS: A program for analysis of Bayesian graphical models using Gibbs sampling. In *Proceedings of the 3rd International Workshop on Distributed Statistical Computing*, volume 124, page 125. Technische Universit at Wien Wien, Austria.

Potts, J., G. Bastille-Rousseau, D. Murray, J. Schaefer, and M. Lewis. 2014a. Predicting local and non-local effects of resources on animal space use using a mechanistic step selection model. *Methods in Ecology and Evolution*, 5:253–262.

Potts, J. R., K. Mokross, and M. A. Lewis. 2014b. A unifying framework for quantifying the nature of animal interactions. *Journal of The Royal Society Interface*, 11:20140333.

Powell, J. and N. Zimmermann. 2004. Multiscale analysis of active seed dispersal contributes to resolving Reid's paradox. *Ecology*, 85:490–506.

Powell, R. 1994. Effects of scale on habitat selection and foraging behavior of fishers in winter. *Journal of Mammalogy*, 75:349–356.

Powell, R. 2000. Animal home ranges and territories and home range estimators. *Research Techniques in Animal Ecology: Controversies and Consequences*, 442.

Powell, R. and M. Mitchell. 2012. What is a home range? *Journal of Mammalogy*, 93:948–958.

Pozdnyakov, V., T. Meyer, Y.-B. Wang, and J. Yan. 2014. On modeling animal movements using Brownian motion with measurement error. *Ecology*, 95:247–253.

Prange, S., T. Jordan, C. Hunter, and S. Gehrt. 2006. New radiocollars for the detection of proximity among individuals. *Wildlife Society Bulletin*, 34:1333–1344.

Preisler, H., A. Ager, B. Johnson, and J. Kie. 2004. Modeling animal movements using stochastic differential equations. *Environmetrics*, 15:643–657.

Pyke, G. 2015. Understanding movements of organisms: It's time to abandon the lévy foraging hypothesis. *Methods in Ecology and Evolution*, 6:1–16.

R Core Team. 2013. *R: A Language and Environment for Statistical Computing*. R Foundation for Statistical Computing, Vienna, Austria.

Rahman, M., J. Sakamoto, and T. Fukui. 2003. Conditional versus unconditional logistic regression in the medical literature. *Journal of Clinical Epidemiology*, 56:101–102.

Ramos-Fernández, G. and J. Morales. 2014. Unraveling fission-fusion dynamics: How subgroup properties and dyadic interactions influence individual decisions. *Behavioral Ecology and Sociobiology*, 68:1225–1235.

Ratikainena, I., J. Gill, T. Gunnarsson, W. Sutherland, and H. Kokko. 2008. When density dependence is not instantaneous: Theoretical developments and management implications. *Ecology Letters*, 11:184–198.

Rhodes, J., C. McAlpine, D. Lunney, and H. Possingham. 2005. A spatially explicit habitat selection model incorporating home range behavior. *Ecology*, 86:1199–1205.

Ricketts, T. 2001. The matrix matters: Effective isolation in fragmented landscapes. *The American Naturalist*, 158:87–99.

Ripley, B. 1976. The second-order analysis of stationary point processes. *Journal of Applied Probability*, 13:587–602.

Risken, H. 1989. *The Fokker–Planck Equation: Methods of Solution and Applications*. Springer, New York, New York, USA.

Rivest, L.-P., T. Duchesne, A. Nicosia, and D. Fortin. 2015. A general angular regression model for the analysis of data on animal movement in ecology. *Journal of the Royal Statistical Society: Series C (Applied Statistics)*, 65:445–463.

Ronce, O. 2007. How does it feel to be like a rolling stone? Ten questions about dispersal evolution. *Annual Review of Ecology, Evolution, and Systematics*, 38:231–253.

Rooney, S., A. Wolfe, and T. Hayden. 1998. Autocorrelated data in telemetry studies: Time to independence and the problem of behavioural effects. *Mammal Review*, 28:89–98.

Royle, J., R. Chandler, R. Sollmann, and B. Gardner. 2013. *Spatial Capture-Recapture*. Academic Press, Amsterdam, The Netherlands.

Royle, J. and R. Dorazio. 2008. *Hierarchical Modeling and Inference in Ecology: The Analysis of Data from Populations, Metapopulations and Communities*. Academic Press, London, United Kingdom.

Rubin, D. 1987. *Multiple Imputation for Nonresponse in Surveys*. Wiley, New York, New York, USA.

Rubin, D. 1996. Multiple imputation after 18+ years. *Journal of the American Statistical Association*, 91:473–489.

Rue, H. and L. Held. 2005. *Gaussian Markov Random Fields: Theory and Applications*. Chapman & Hall/CRC, Boca Raton, Florida, USA.

Rue, H., S. Martino, and N. Chopin. 2009. Approximate Bayesian inference for latent Gaussian models by using integrated nested Laplace approximations. *Journal of the Royal Statistical Society: Series B (Statistical Methodology)*, 71(2):319–392.

Russell, D., S. Brasseur, D. Thompson, G. Hastie, V. Janik, G. Aarts, B. McClintock, J. Matthiopoulos, S. Moss, and B. McConnell. 2014. Marine mammals trace anthropogenic structures at sea. *Current Biology*, 24:R638–R639.

Russell, D., B. McClintock, J. Matthiopoulos, P. Thompson, D. Thompson, P. Hammond, and B. McConnell. 2015. Intrinsic and extrinsic drivers of activity budgets in sympatric grey and harbour seals. *Oikos*, 124:1462–1472.

Russell, J. C., E. M. Hanks, and M. Haran. 2016a. Dynamic models of animal movement with spatial point process interactions. *Journal of Agricultural, Biological, and Environmental Statistics*, 21:22–40.

Russell, J. C., E. M. Hanks, M. Haran, and D. P. Hughes. 2016b. A spatially-varying stochastic differential equation model for animal movement. *Annals of Applied Statistics*, In Press.

Rutz, C. and G. Hays. 2009. New frontiers in biologging science. *Biology Letters*, 5:289–292.

Schabenberger, O. and C. Gotway. 2005. *Statistical Methods for Spatial Data Analysis*. Chapman & Hall/CRC, Boca Raton, Florida, USA.

Scharf, H., M. Hooten, B. Fosdick, D. Johnson, J. London, and J. Durban. 2015. Dynamic social networks based on movement. *Annals of Applied Statistics*, In Press.

Schick, R., S. Kraus, R. Rolland, A. Knowlton, P. Hamilton, H. Pettis, R. Kenney, and J. Clark. 2013. Using hierarchical Bayes to understand movement, health, and survival in the endangered North Atlantic right whale. *PloS One*, 8:e64166.

Schick, R. S., S. R. Loarie, F. Colchero, B. D. Best, A. Boustany, D. A. Conde, P. N. Halpin, L. N. Joppa, C. M. McClellan, and J. S. Clark. 2008. Understanding movement data and movement processes: Current and emerging directions. *Ecology Letters*, 11:1338–1350.

Schlägel, U. and M. Lewis. 2016. A framework for analyzing the robustness of movement models to variable step discretization. *Journal of Mathematical Biology*, 73:815–845.

Schoenberg, F., D. Brillinger, and P. Guttorp. 2002. Point processes, spatial-temporal. In El-Shaarawi, A. and W. Piegorsch, editors, *Encyclopedia of Environmetrics*, volume 3, pages 1573–1577. John Wiley & Sons, Ltd, New York, New York, USA.

Schoener, T. 1981. An empirically based estimate of home range. *Theoretical Population Biology*, 20:281–325.

Schtickzelle, N. and M. Baguette. 2003. Behavioural responses to habitat patch boundaries restrict dispersal and generate emigration–patch area relationships in fragmented landscapes. *Journal of Animal Ecology*, 72:533–545.

Schultz, C. and E. Crone. 2001. Edge-mediated dispersal behavior in a prairie butterfly. *Ecology*, 82:1879–1892.

Scott, D. 1992. *Multivariate Density Estimation: Theory, Practice, and Visualization*. Wiley, New York, New York, USA.

Shepard, E., R. Wilson, F. Quintana, A. Laich, N. Liebsch, D. Albareda, and C. Newman. 2008. Identification of animal movement patterns using tri-axial accelerometry. *Endangered Species Research*, 10:2.1.

Shepard, E., R. Wilson, W. Rees, E. Grundy, S. Lambertucci, and S. Vosper. 2013. Energy landscapes shape animal movement ecology. *The American Naturalist*, 182:298–312.

Shigesada, N. and K. Kawasaki. 1997. *Biological Invasions: Theory and Practice*. Oxford University Press, UK.

Shumway, R. and D. Stoffer. 2006. *Time Series and Its Applications*. Springer, New York, New York, USA.

Signer, J., N. Balkenhol, M. Ditmer, and J. Fieberg. 2015. Does estimator choice influence our ability to detect changes in home-range size? *Animal Biotelemetry*, 3(1):1–9.

Silva, M., I. Jonsen, D. Russell, R. Prieto, D. Thompson, and M. Baumgartner. 2014. Assessing performance of Bayesian state-space models fit to Argos satellite telemetry locations processed with Kalman filtering. *PLoS One*, 9:e92277.

Silverman, B. 1986. *Density Estimation*. Chapman & Hall, London, UK.

Skalski, G. and J. Gilliam. 2000. Modeling diffusive spread in a heterogeneous population: A movement study with stream fish. *Ecology*, 81:1685–1700.

Skalski, G. and J. Gilliam. 2003. A diffusion-based theory of organism dispersal in heterogeneous populations. *The American Naturalist*, 161:441–458.

Skellam, J. 1951. Random dispersal in theoretical populations. *Biometrika*, 38:196–218.

Smouse, P., S. Focardi, P. Moorcroft, J. Kie, J. Forester, and J. Morales. 2010. Stochastic modelling of animal movement. *Philosophical Transactions of the Royal Society of London B: Biological Sciences*, 365:2201–2211.

Stamps, J., V. Krishnan, and M. Reid. 2005. Search costs and habitat selection by dispersers. *Ecology*, 86:510–518.

Stamps, J., B. Luttbeg, and V. Krishnan. 2009. Effects of survival on the attractiveness of cues to natal dispersers. *The American Naturalist*, 173:41–46.

Strandburg-Peshkin, A., D. Farine, I. Couzin, and M. Crofoot. 2015. Shared decision-making drives collective movement in wild baboons. *Science*, 348:1358–1361.

Swihart, R. and N. Slade. 1985. Testing for independence of observations in animal movements. *Ecology*, 66:1176–1184.

Swihart, R. and N. Slade. 1997. On testing for independence of animal movements. *Journal of Agricultural, Biological and Environmental Statistics*, 2:48–63.

Taylor, J. R. 2005. *Classical Mechanics*. University Science Books.

Thomas, C., A. Cameron, R. Green, M. Bakkenes, L. Beaumont, Y. Collinghamne, B. Erasmus et al. 2004. Extinction risk from climate change. *Nature*, 427:145–148.

Thomas, C. and W. Kunin. 1999. The spatial structure of populations. *Journal of Animal Ecology*, 68:647–657.

Tracey, J., J. Zhu, and K. Crooks. 2005. A set of nonlinear regression models for animal movement in response to a single landscape feature. *Journal of Agricultural, Biological and Environmental Statistics*, 10:1–18.

Trakhtenbrot, A., R. Nathan, G. Perry, and D. Richardson. 2005. The importance of long-distance dispersal in biodiversity conservation. *Diversity and Distributions*, 11:173–181.

Turchin, P. 1998. *Quantitative Analysis of Animal Movement*. Sinauer Associates, Inc. Publishers, Sunderland, Massachusetts, USA.

Turchin, P. 2003. *Complex Population Dynamics: A Theoretical/Empirical Synthesis*. Princeton University Press, Princeton, New Jersey, USA.

Venables, W. and B. Ripley. 2002. *Modern Applied Statistics with S*. Springer, New York, New York, USA.

Ver Hoef, J. 2012. Who invented the delta method? *The American Statistician*, 66:124–127.

Ver Hoef, J. and P. Boveng. 2007. Quasi-Poisson vs. negative binomial regression: How should we model overdispersed count data? *Ecology*, 88:2766–2772.

Ver Hoef, J. and E. Peterson. 2010. A moving average approach for spatial statistical models of stream networks. *Journal of the American Statistical Association*, 105:6–18.

Ver Hoef, J., E. Peterson, M. Hooten, E. Hanks, and M.-J. Fortin. Spatial autoregressive models for ecological inference. *Ecological Monographs*. In Review.

Ver Hoef, J. M., E. M. Hanks, and M. B. Hooten. On the relationship between conditional (CAR) and simultaneous (SAR) autoregressive models. *Stat*. In Review.

Waller, L. and C. Gotway. 2004. *Applied Spatial Statistics for Public Health Data*, volume 368. John Wiley & Sons Ltd., Hoboken, New Jersey, USA.

Warton, D. and L. Shepherd. 2010. Poisson point process models solve the "pseudo-absence problem" for presence-only data in ecology. *Annals of Applied Statistics*, 4:1383–1402.

White, G. and R. Bennetts. 1996. Analysis of frequency count data using the negative binomial distribution. *Ecology*, 77:2549–2557.

White, G. and R. Garrott. 1990. *Analysis of Wildlife Radio-Tracking Data*. Academic Press, San Diego, California, USA.

Wiens, J. 1997. Metapopulation dynamics and landscape ecology. *Metapopulation Biology: Ecology, Genetics, and Evolution*, pages 43–62. Academic Press, San Diego, California, USA.

Wikle, C. 2002. Spatial modeling of count data: A case study in modelling breeding bird survey data on large spatial domains. In Lawson, A. and D. Denison, editors, *Spatial Cluster Modeling*, pages 199–209. Chapman & Hall/CRC, Boca Raton, Florida, USA.

Wikle, C. 2003. Hierarchical Bayesian models for predicting the spread of ecological processes. *Ecology*, 84:1382–1394.

Wikle, C. 2010a. Low-rank representations for spatial processes. In Gelfand, A., P. Diggle, M. Fuentes, and P. Guttorp, editors, *Handbook of Spatial Statistics*, pages 107–118. Chapman & Hall/CRC, Boca Raton, Florida, USA.

Wikle, C. 2010b. Hierarchical modeling with spatial data. In Gelfand, A., P. Diggle, M. Fuentes, and P. Guttorp, editors, *Handbook of Spatial Statistics*, pages 89–106. Chapman & Hall/CRC, Boca Raton, Florida, USA.

Wikle, C. and M. Hooten. 2010. A general science-based framework for nonlinear spatio-temporal dynamical models. *Test*, 19:417–451.

Williams, T., L. Wolfe, T. Davis, T. Kendall, B. Richter, Y. Wang, C. Bryce, G. Elkaim, and C. Wilmers. 2014. Instantaneous energetics of puma kills reveal advantage of felid sneak attacks. *Science*, 346:81–85.

Wilson, R., M. Hooten, B. Strobel, and J. Shivik. 2010. Accounting for individuals, uncertainty, and multiscale clustering in core area estimation. *The Journal of Wildlife Management*, 74:1343–1352.

Wilson, R., N. Liebsch, I. Davies, F. Quintana, H. Weimerskirch, S. Storch, K. Lucke et al. 2007. All at sea with animal tracks; Methodological and analytical solutions for the resolution of movement. *Deep Sea Research Part II: Topical Studies in Oceanography*, 54:193–210.

Wilson, R., E. Shepard, and N. Liebsch. 2008. Prying into the intimate details of animal lives: Use of a daily diary on animals. *Endangered Species Research*, 4:123–137.

Winship, A., S. Jorgensen, S. Shaffer, I. Jonsen, P. Robinson, D. Costa, and B. Block. 2012. State-space framework for estimating measurement error from double-tagging telemetry experiments. *Methods in Ecology and Evolution*, 3:291–302.

Wood, S. 2011. Fast stable restricted maximum likelihood and marginal likelihood estimation of semiparametric generalized linear models. *Journal of the Royal Statistical Society (B)*, 73:3–36.

Wood, S. N. 2003. Thin plate regression splines. *Journal of the Royal Statistical Society: Series B (Statistical Methodology)*, 65:95–114.

Worton, B. 1987. A review of models of home range for animal movement. *Ecological Modelling*, 38:277–298.

Worton, B. 1989. Kernel methods for estimating the utilization distribution in home-range studies. *Ecology*, 70:164–168.

Zucchini, W., I. L. MacDonald, and R. Langrock. 2016. *Hidden Markov Models for Time Series: An Introduction Using R*, Second Edition. CRC Press. Boca Raton, Florida, USA.

Author Index

C

D

E

F

I

J

K

L

M

S

T

U

V

Subject Index

D

E

F

G

H

Q

R

S

Printed in the United States
by Baker & Taylor Publisher Services